Smart Energy Systems

Series Editors

Yuanzheng Li, School of AI and Automation, Huazhong University of Science and Technology, Wuhan, Hubei, China

Zhigang Zeng, School of AI and Automation, Huazhong University of Science and Technology, Wuhan, Hubei, China

Jinyu Wen, School of Electrical and Electronics Engineering, Huazhong University of Science and Technology, Wuhan, Hubei, China

The book series Smart Energy Systems comprehensively explores the application of advanced theories and cutting-edge technologies in the evolving energy landscape. The series emphasizes the strategic practical application in smart grid management, renewable energy integration, and multi-source energy coordination. The series explores topics such as data analysis, system optimization, machine learning, and AI-driven control methods, highlighting their crucial role in ensuring efficient and sustainable energy system operations. Additionally, the series broadens its focus to encompass hybrid energy solutions and distributed energy management, offering a comprehensive perspective on how smart, integrated energy systems can transform future power networks.

The book series Smart Energy Systems seeks to provide researchers, engineers, and professionals with in-depth insights into the implementation of integrated energy frameworks and renewable energy technologies. Each volume focuses on distinct aspects such as advancements in smart grid technology, real-time energy optimization, and the integration of renewable into traditional energy systems. Collectively, the series acts as an indispensable resource for academia, energy industry professionals, and policymakers, providing technical depth to bridge the gap between research and practical implementation in pursuing a sustainable energy future.

Qiang Yang · Yanchong Zheng · Yuanyi Chen ·
Siyang Sun

Integration of Massive Electric Vehicles to Smart Grid

Advances, Applications and Perspectives

Springer

Qiang Yang
Zhejiang University
Hangzhou, Zhejiang, China

Yuanyi Chen
Zhejiang University
Hangzhou, China

Yanchong Zheng
Zhejiang University
Hangzhou, China

Siyang Sun
China Southern Power Grid
Guangzhou, China

ISSN 3059-4359 ISSN 3059-4367 (electronic)
Smart Energy Systems
ISBN 978-981-92-0556-1 ISBN 978-981-92-0557-8 (eBook)
https://doi.org/10.1007/978-981-92-0557-8

This Springer imprint is published by the registered company Springer Nature Singapore Pte Ltd.
The registered company address is: 152 Beach Road, #21-01/04 Gateway East, Singapore 189721, Singapore

Foreword

The global energy landscape is undergoing a profound transformation, driven by the twin imperatives of decarbonization and sustainable development. At the heart of this revolution lies the rapid electrification of the transportation sector. Electric vehicles (EVs), once a niche technology, have emerged as a mainstream force, reshaping not only how we move but also how we produce, distribute, and consume energy. As we stand in 2025, the penetration of EVs in global car sales has surpassed a quarter, with projections indicating an even more accelerated trajectory in the years to come. This paradigm shift, while essential for a low-carbon future, presents an unprecedented set of challenges and opportunities for the very foundation of our modern society: the electric power grid.

The integration of massive fleets of electric vehicles into the existing power infrastructure is far from a simple extension of demand. It represents a complex, dynamic, and deeply coupled interaction between two previously distinct systems—the power grid and the transportation network. The charging behavior of millions of EVs, characterized by significant spatial and temporal randomness, can introduce severe stresses on the grid, including voltage instability, frequency deviations, harmonic distortions, and the overloading of critical infrastructure like transformers and transmission lines. Left unmanaged, this "uncoordinated charging" threatens the secure and economical operation of the entire power system.

However, within this challenge lies an extraordinary opportunity. If intelligently managed, EVs are not merely loads but vast, distributed, and flexible energy resources. Through concepts like Vehicle-to-Grid (V2G) and smart charging, the batteries of parked EVs can act as a giant virtual power plant, absorbing excess renewable energy during periods of low demand and feeding power back to the grid during peak hours. They can provide crucial ancillary services such as frequency regulation and spinning reserve, enhancing grid resilience and facilitating the integration of intermittent renewable sources like wind and solar. Realizing this potential, however, requires a fundamental rethinking of our planning, operational, and market frameworks.

It is within this complex and exciting context that this book, *Integration of Massive Electric Vehicles to Smart Grid: Advances, Applications and Perspectives*, makes its seminal contribution. This work is not merely a collection of research findings; it is a comprehensive and forward-looking exploration of the entire ecosystem of vehicle-grid integration. The authors have masterfully structured the book to guide the reader from the fundamental building blocks of EV integration to the most advanced concepts of coupled network operation and electricity market participation.

The book begins by establishing a solid foundation. It provides a detailed overview of the current state of EV development, market trends, and the various charging modes, setting the stage for a thorough analysis of the impacts—both positive and negative—of EV charging on the power grid. This introductory section is crucial for understanding the scale of the challenge and the necessity of the solutions proposed later. The comprehensive literature review that follows demonstrates the authors' deep engagement with the existing body of knowledge, meticulously mapping the research landscape across charging demand prediction, infrastructure planning, scheduling strategies, and market mechanisms. This critical appraisal not only contextualizes the book's contributions but also clearly identifies the pressing research challenges that the subsequent chapters aim to address.

The core of the book is structured around three interconnected pillars, each representing a critical frontier in the field. The first pillar focuses on the foundational issues of planning and scheduling. It delves into sophisticated methods for characterizing and forecasting the complex charging demand of large EV populations using advanced analytical techniques such as Monte Carlo simulation, Markov models, and hidden Markov models. Building on these predictions, the book proposes novel optimization strategies for the collaborative planning of EV charging infrastructure and distributed energy resources within active distribution networks. Furthermore, it explores optimal scheduling strategies for EV charging and discharging, employing game-theoretic approaches to manage interactions between individual users and system operators, thereby unlocking the potential of V2G for providing grid services like peak shaving.

The second pillar elevates the discussion to the level of electricity markets. Recognizing that individual EVs are too small to participate directly, the book masterfully develops the concept of the EV aggregator. It provides a rigorous analysis of how these aggregators can optimally participate in both demand response programs and competitive electricity spot markets. Key contributions here include the formulation of optimal bidding strategies for day-ahead and real-time markets, the use of Conditional Value-at-Risk (CVaR) for risk-averse decision-making under uncertainty, and the design of multi-level game-theoretic models to capture the strategic interactions between aggregators, grid operators, and EV users. This section is particularly valuable for its practical insights into how market mechanisms can be designed to harness the flexibility of EVs while ensuring economic efficiency and fairness.

The third and final pillar pushes the boundaries of current research by tackling the grand challenge of systemic integration—the coupled power and transportation network. The authors break new ground by incorporating the realistic concept of "bounded rationality" into their models, acknowledging that users do not always make perfectly optimal decisions. This leads to the development of sophisticated dynamic network equilibrium models that capture the intricate feedback loops between traffic flow, charging demand, and power flow. The book then explores how fast-charging stations, as the physical nexus of this coupling, can use dynamic pricing strategies to guide user behavior and enhance system-wide performance. This analysis is further extended to the emerging concept of integrated energy service stations, which co-locate charging with hydrogen and natural gas refueling, demonstrating how multi-energy coordination can unlock even greater efficiencies and support the goal of a truly carbon-neutral transportation system.

In conclusion, *Integration of Massive Electric Vehicles to Smart Grid* is an indispensable resource for researchers, engineers, planners, and policymakers navigating the complexities of our electrified future. It masterfully synthesizes state-of-the-art research with forward-looking perspectives, providing not only a deep understanding of the challenges but also a robust toolkit of analytical and optimization methods to forge a path toward a resilient, efficient, and sustainable energy-transportation nexus. This book is a testament to the power of interdisciplinary thinking and will undoubtedly serve as a cornerstone reference for years to come, guiding the critical work of integrating millions of electric vehicles into the smart grids of tomorrow.

Hangzhou, China Qiang Yang
Hangzhou, China Yanchong Zheng
Hangzhou, China Yuanyi Chen
Guangzhou, China Siyang Sun

Preface

The inspiration for this book stems from a convergence of observations, questions, and a profound sense of urgency. Over the past decade, as a researcher in the field of power systems, I have witnessed the electrification of transportation evolve from a promising concept into a global, unstoppable reality. The sight of electric vehicles (EVs) on our roads is no longer a novelty but an everyday occurrence. The data speaks for itself: from a marginal player just a few years ago, EVs now command a significant and rapidly growing share of the global automobile market. This transformation is a critical victory in the fight against climate change and fossil fuel dependence, yet it presents our profession with its most complex and fascinating challenge to date: how to seamlessly and intelligently integrate tens of millions of these moving batteries into the electrical grid that powers our civilization.

The initial questions that motivated this work were seemingly simple: What happens when a neighborhood of a hundred homes all plug in their EVs to charge after returning from work? What happens to a city's distribution transformers when a popular fast-charging station sees a surge of demand during a holiday weekend? The answers, as our early simulations showed, were concerning—voltage drops, overloaded lines, and increased system losses painted a clear picture of the risks of unmanaged, or "uncoordinated," charging. It became immediately apparent that the grid, designed for a predictable and passive demand, could not simply absorb this new, highly dynamic load without consequence.

However, as I delved deeper, the narrative shifted from one of pure challenge to one of immense opportunity. It became clear that an EV is not just a load; it is a kilowatt-hour battery on wheels, a distributed energy storage unit that is parked 95% of the time. This duality—being both a consumer and a potential supplier of energy—is what makes the vehicle-grid integration problem so uniquely compelling. The central question evolved from "How do we protect the grid from EVs?" to "How can we use EVs to make the grid better, cleaner, and more resilient?" This book is the result of a multi-year journey to explore the full spectrum of answers to this evolved question.

The journey of writing this book has been a collaborative endeavor, reflecting the interdisciplinary nature of the topic itself. It required not only a deep dive into power system engineering, with its optimal power flows and market designs, but also an exploration into transportation network modeling, behavioral economics, and game theory. Understanding how an EV user chooses a route and a charging station is just as critical as understanding the resulting power flow on a distribution feeder. The concept of "bounded rationality," where we acknowledge that users act on imperfect information and with varying degrees of sensitivity to costs, became a cornerstone of our analysis, bridging the gap between theoretical models and real-world behavior.

This book is structured to guide the reader through this complex landscape in a logical and progressive manner. We begin by establishing a solid foundation, examining the current state of EV development, their charging characteristics, and their multifaceted impacts on power system operation. We then move to the foundational tools required for any integration strategy: the accurate prediction of charging demand and the optimal planning of charging infrastructure. From there, we ascend to the level of the electricity market, introducing the crucial role of the EV aggregator. Here, we explore how these entities can navigate the complexities of day-ahead and real-time markets, manage risks, and design incentive mechanisms to unlock the flexibility of EV fleets.

The final and most forward-looking part of the book tackles the grand challenge of systemic integration. We model the deep coupling between power and transportation networks, revealing the intricate feedback loops that govern their combined behavior. This framework allows us to explore the optimization of fast-charging stations, not just as points of sale for electricity, but as active hubs that can use dynamic pricing to guide both traffic and power flow. Finally, we extend our vision to the integrated energy service station of the future, a multi-energy hub capable of providing electricity, hydrogen, and natural gas, demonstrating how multi-energy coordination can pave the way for a truly sustainable transportation ecosystem.

This book is intended for a wide audience. For graduate students and young researchers, it offers a comprehensive overview of the state of the art and identifies numerous fertile areas for future investigation. For practicing engineers and planners, it provides a robust set of analytical tools and optimization methods to tackle the practical challenges of grid planning and operation in the age of e-mobility. For policymakers and energy market designers, it offers insights into the mechanisms and frameworks needed to harness the full potential of EVs as a flexible grid resource.

I am deeply grateful to my colleagues, collaborators, and students whose insights, questions, and dedication have shaped this work. Their contributions, whether through rigorous data analysis, spirited debates on modeling assumptions, or the simple act of asking "why," are woven into the fabric of this book. I also extend my appreciation to the numerous researchers whose foundational work is cited throughout these pages; building a new understanding is only possible on the shoulders of giants.

It is my sincere hope that this book serves not only as a record of our current understanding but also as a catalyst for future innovation. The journey toward a fully integrated, intelligent, and sustainable energy-transportation future is just beginning. The pages that follow are an invitation to join this critical and exciting endeavor.

Hangzhou, China Qiang Yang
Hangzhou, China Yanchong Zheng
Hangzhou, China Yuanyi Chen
Guangzhou, China Siyang Sun

Acknowledgments The journey of conceiving, researching, and writing a book on a topic as dynamic and interdisciplinary as the integration of electric vehicles with smart grids is a monumental task that cannot be accomplished in isolation. This work is the culmination of years of research, countless discussions, and the unwavering support of numerous individuals and institutions. It is with profound gratitude and humility that I take this opportunity to acknowledge those who have contributed to this endeavor.

First and foremost, I wish to express my deepest gratitude to my collaborators and fellow researchers, whose insights and expertise have shaped the very foundation of this book. The exploration of charging demand characterization in Chap. 4, with its development of Markov and hidden Markov models, would not have been possible without the rigorous data analysis and modeling contributions of my colleagues. Their dedication to uncovering the inherent patterns in vast datasets of travel behavior laid the groundwork for our predictive capabilities. Similarly, the work on distributed generation modeling and the K-means clustering approach in Chap. 5 was a collective effort, refining our understanding of uncertainty and enabling the robust planning frameworks presented.

The exploration of electricity markets and the strategic behavior of aggregators in Chaps. 6, 7, and 8 represents a significant portion of this book. I am indebted to the team of researchers who delved deep into the complexities of the Guangdong and PJM markets. Their tireless efforts in formulating the Stackelberg game models, proving the existence and uniqueness of equilibria using potential game theory, and developing the risk-averse bidding strategies based on Conditional Value-at-Risk (CVaR) were nothing short of inspirational. The late nights spent debugging optimization code and debating the nuances of real-time penalty structures are a testament to their commitment.

The conceptual leap to modeling the coupled power and transportation network in Chap. 10 was one of the most intellectually stimulating phases of this project. I am profoundly grateful to the colleagues who embraced the concept of "bounded rationality" and helped translate it into the sophisticated mathematical frameworks of multi-period static and dynamic user equilibrium. The development of the improved point queue model for fast-charging stations, with its ability to capture the intricate dance of entry, queuing, charging, and departure, was a particular point of pride and a direct result of their innovative thinking. This work naturally led to the dynamic pricing strategies explored in Chap. 11, and I thank the team for their tenacity in solving the complex tri-level Stackelberg game using the Gauss-Seidel iterative method.

The extension of our research to fast-charging station aggregators in Chap. 12 and the visionary concept of integrated energy service stations in Chap. 13 pushed the boundaries of this book into new and exciting territory. I am thankful for the researchers who tackled the challenges of large-scale system simplification through path merging and flow separation, and who bravely ventured into the coordination of electricity, hydrogen, and natural gas. Their work on modeling electrolyzers, carbon capture, and synthetic natural gas, and on applying CVaR for risk-averse multi-energy dispatch, demonstrates a remarkable ability to think holistically about a sustainable energy future.

Beyond the direct contributors to the chapters, I owe a debt of gratitude to the broader academic community. The works cited throughout this book represent the collective wisdom of countless scholars who have pioneered ideas in power systems, transportation engineering, economics, and operations research. Their foundational research provided the building blocks upon which this book is constructed. I am also grateful to the institutions that provided the data essential for our case studies, including the China Passenger Car Association, the State Grid Corporation, and the operators of the PJM and Guangdong electricity markets. Their commitment to transparency and data sharing is invaluable for applied research.

On a personal and professional level, I must acknowledge the support of my home institution and department. The provision of computing resources, a stimulating research environment, and the administrative support to manage a project of this scale have been indispensable. I am grateful to my colleagues for their encouragement and for creating a culture where ambitious, interdisciplinary research can flourish. The numerous seminars and informal discussions we shared helped refine countless ideas presented in these pages.

To the team at the publisher, I extend my sincere appreciation for their guidance, patience, and professionalism throughout the publication process. Their belief in the value of this project and their meticulous attention to detail have been instrumental in bringing this book to fruition.

Finally, the authors would like to thank for the funding support of the National Key Research and Development Program of China (2023YFB3107603), the National Natural Science Foundation of China (52177119), and the Special Support Plan for Zhejiang Province High-level Talents (2022R52012).

While this book bears the author names on the cover, its true authorship is shared with all those mentioned and countless others who offered advice, data, or a simple word of encouragement. Any errors or omissions that remain are, of course, my own. It is my sincere hope that this work serves to advance the field and inspire the next generation of solutions for a cleaner, more efficient, and deeply integrated energy and transportation system.

Competing Interests The authors have no competing interests to declare that are relevant to the content of this manuscript.

Contents

About the Authors

Professor Qiang Yang received a Ph.D. degree in Electronic Engineering and Computer Science from Queen Mary, University of London, London, U.K., in 2007, and worked in the Department of Electrical and Electronic Engineering at Imperial College London, U.K., from 2007 to 2010. He visited the University of British Columbia and the University of Victoria, Canada, as a visiting scholar in 2015 and 2016. He is currently a full Professor at the College of Electrical Engineering, Zhejiang University, China, and has published more than 400 technical papers, filed more than 100 national patents, co-authored 2 books, and edited 2 books and several book chapters. His research interests over the years include smart energy systems, large-scale complex network modeling, control and optimization, learning-based optimization and control. He is an IEEE PCCC Distinguished Engineer, a Member of the National Academy of Artificial Intelligence (MNAAI), a lifetime Fellow of the British Computer Society (FBCS), a Fellow of the Institution of Engineering and Technology (FIET) and International Association of Advanced Materials (FIAAM), and a Distinguished Member of China Computer Federation (CCF).

Dr. Yanchong Zheng received the B.S. degree and M.S. degree in Statistics and Applied Mathematics from the Wuhan University of Technology, Wuhan, China, in 2012 and 2015, respectively. He received the Ph.D. degree in Electrical Engineering from Zhejiang University, Hangzhou, China, in 2025. From 2015 to 2021, he was an Assistant Research Fellow at the Department of Electrical and Electronic Engineering, Southern University of Science and Technology, Shenzhen, China. He is currently an Assistant Professor at the Shenzhen University of Information Technology, Shenzhen, China. His research interests include coordinated charging of electric vehicles in smart grids, vehicle-to-grid, low-carbon transportation, energy economic, and sustainability.

Dr. Yuanyi Chen received the B.S. degree in Electrical Engineering from Beijing Jiaotong University in 2020 and the Ph.D. degree in Electrical Engineering from Zhejiang University in 2025. He was a visiting Ph.D. student with the School of Electrical and Electronic Engineering (EEE), Nanyang Technological University in 2024, Singapore. He is currently a Lecturer (Associated Researcher) with the NARI School of Electrical and Automation Engineering, Nanjing Normal University. His research interest includes the dynamic modeling and coordinated optimization of coupled power and transportation networks.

Dr. Siyang Sun received the B.S. degree and the Ph.D. degree both in Electrical Engineering from Zhejiang University, Hangzhou, China, in 2014 and 2019, respectively. Now he is with China Southern Power Grid as a manager. His research interests include electric vehicle-grid integration and the emerging industries of power grid corporations.

Chapter 1
Introduction

Abstract This chapter provides a comprehensive introduction to the current landscape and future trajectory of electric vehicles (EVs), emphasizing their rapid global proliferation and the accompanying technical, infrastructural, and systemic challenges. By 2025, global EV sales are projected to exceed 20 million units, capturing over 25% of the automotive market, with China emerging as the dominant force, contributing nearly 70% of global growth and achieving a domestic penetration rate of 48%. This expansion is underpinned by significant advancements in battery technology, including energy densities reaching 300 Wh/kg, ultra-fast charging capabilities, and extended driving ranges, alongside substantial improvements in charging infrastructure and user satisfaction. The chapter categorizes EVs into battery electric, plug-in hybrid, and fuel cell variants, while also detailing standardized charging modes and their technical specifications. A critical focus is placed on the potential adverse impacts of large-scale uncoordinated EV charging on power grids, such as voltage instability, harmonic distortion, transformer overloads, and transmission congestion. In response, the concept of vehicle-grid integration is introduced, highlighting orderly charging and bidirectional vehicle-to-grid (V2G) interactions as key strategies for mitigating grid stress and enhancing system resilience. Furthermore, the chapter explores the emerging paradigm of the electricity-transportation coupled network, wherein EV traffic flows and power flows converge at fast-charging stations, creating complex interdependencies between transportation and distribution systems. It underscores the role of fast-charging station aggregators in managing these interactions through price signals and power regulation, particularly under conditions of user bounded rationality and high penetration of renewable energy. Ultimately, this chapter sets the stage for investigating optimal scheduling and regulation strategies within multi-energy, multi-actor urban environments.

Q. Yang et al., *Integration of Massive Electric Vehicles to Smart Grid*, Smart Energy Systems, https://doi.org/10.1007/978-981-92-0557-8_1

1.1 Overview of the Development of Electric Vehicles

Electric vehicles (EVs) are automobiles that use electricity as their power source and are driven by electric motors. Some people also refer to them as new energy vehicles to highlight their difference from internal combustion engine vehicles.

In 2025, the global electric vehicle market continued its rapid growth trend of the past few years and is undergoing a profound restructuring. According to data from the China Passenger Car Association, from January to August 2025, the total global car sales reached 61.98 million units, of which new energy vehicle sales reached 13.82 million units, accounting for 28.2% of the total global car sales. This figure is 1.9 percentage points higher than that of the whole year of 2024. The penetration rate of new energy vehicles in the world is generally showing a rapid upward trend, climbing from 13% in 2022, 16% in 2023, and 19.9% in 2024 to 24.4% in the third quarter of 2025. This means that one out of every four new cars sold globally is a new energy vehicle. The Global Electric Vehicle Outlook 2025 report released by the International Energy Agency (IEA) further points out that global electric vehicle sales are expected to exceed 20 million units in 2025, accounting for more than 25% of global car sales. This prediction is consistent with the data from the China Passenger Car Association, highlighting the irreversible trend of global automotive electrification.

In the booming global electric vehicle market, the uneven development among regions is particularly evident. In August 2025, the electric penetration rates of various countries showed significant differences: China's new energy penetration rate reached 48.2%, Germany's was 26.4%, Norway's was as high as 80.2%, the UK's was 31.9%, while the US's was only 11.1%, and Japan's was as low as 1.8%. This uneven development pattern mainly stems from differences in policy support, infrastructure construction level, and consumer acceptance among countries. Among them, the Chinese market is particularly eye-catching. From January to August 2025, China's new energy passenger vehicles accounted for 67.6% of the global market share, and it remained at a high level of 68.7% from July to August. Especially in the plug-in hybrid market, China accounted for 75.7% of the global share, showing an absolute leading advantage. China is not only the world's largest electric vehicle market but also the core engine of global electric vehicle growth. From January to August 2025, China contributed 71% to the global increase in new energy vehicles, fully demonstrating the pivotal position of the Chinese market in the global electric vehicle industry.

China, as the world's largest electric vehicle market, has made remarkable progress. From January to August 2025, China's new energy vehicle sales exceeded 9.6 million units, a year-on-year increase of 36.7%. New car sales accounted for 45.5% of total new car sales. In August alone, China produced 1.33 million new energy vehicles, a year-on-year increase of 23%, and the penetration rate further climbed to 48%. These achievements mean that electric vehicles in China have shifted from being policy-driven to market-driven, becoming one of the choices of mainstream consumers. Looking back at the development process, China's new energy vehicle production has jumped from around 1.4 million units in 2020 to around 13 million units in 2024, with the market size more than tripling in four years. This growth rate is also rare in the history of the global automotive industry.

The booming development of China's electric vehicle market is attributed to multiple factors. In terms of policy support, the Chinese government has continued to implement policies that promote automobile consumption, such as the trade-in program, injecting sustained momentum into the market. Meanwhile, local governments also continue to support the market through measures such as subsidies for charging infrastructure and relaxation of traffic restrictions. In terms of technological progress, China's electric vehicle industry chain has achieved remarkable achievements: the energy density of domestic batteries has increased to 300 Wh/kg, charging time has been shortened to 15 min (400 kW fast charging), and the cruising range has generally exceeded 700 km. According to data from the China Passenger Car Association, the average cruising range of pure electric passenger vehicles in China is expected to reach 528 km in 2025, an increase of 103 km compared to 425 km in 2020. Especially in terms of technological routes, various routes such as pure electric, plug-in hybrid, and extended-range electric coexist and develop in the market, meeting the needs of different consumers. In 2025, the average cruising range of plug-in hybrid models will reach 137 km, and the average cruising range of extended-range models will be as high as 205 km, showing the market vitality of parallel development of multiple technological routes. Table 1.1 presents the major technical indicators of the electric vehicle market in China from 2020 to 2025.

In terms of infrastructure construction, China has established the most complete and resilient new energy vehicle industry chain and supply chain system in the world. By 2025, infrastructure such as charging piles and battery swapping stations will be more perfect. According to data from the State Grid, the number of new charging piles reached 1.2 million from January to June in 2025, and the ratio of vehicles to charging piles was optimized from 3.5:1 to 2.8:1, significantly alleviating users' anxiety about battery life. It is particularly worth mentioning that the "parking and charging integration" policy promoted by the Chinese government and the partnership between original equipment manufacturers and grid operators have greatly improved users' energy replenishment experience. According to survey data, up to 94% of Chinese users are very or quite satisfied with the overall energy replenishment experience

Table 1.1 Technical indicators of China's electric vehicle market (2020–2025)

Parameters	Year 2020	Year 2025	Improvement
Average cruising range of pure electric vehicles (km)	425	528	103
Battery energy density	~180–200 Wh/kg	~300 Wh/kg	~50%
Fast charging time (30–80%)	~30–40 min	~15 min	Reduction of ~60%
Penetration rate of new energy vehicles	5.4%	48.2%	~8 times
Equipping rate of the assisted driving function	16.2%	62.1%	~3.8 times

(including household and public energy replenishment), which is much higher than in other markets and has become an important driving force for the popularization of electric vehicles.

The global electric vehicle market is showing a clear regional differentiation in its development process, with differences in policy environment, market foundation, and consumer preferences across countries, resulting in diversified development paths. As an important pole in the global electric vehicle development, the European market saw new energy passenger vehicle sales reach 2.26 million units from January to August 2025, an increase of 470,000 units or 26% compared to the same period last year. In August alone, European new energy passenger vehicle sales reached 224,000 units, a year-on-year increase of 29%. The electric penetration rate in major European countries remains at a high level: Germany reached 26.4%, the United Kingdom 31.9%, and Norway as high as 80.2%. However, the European electric vehicle market is also facing many challenges: consumers are facing high interest rates, slowing economic growth, and the gradual elimination of financial subsidies for new energy vehicles. In addition, the lack of charging infrastructure remains an important factor restricting its widespread development, leading to relatively weak market demand. Despite this, traditional European automakers such as Volkswagen, BMW, and Mercedes-Benz are accelerating their electric transformation. The sales of electric models such as the Volkswagen ID series increased by 30%, showing the European automotive industry's positive response to the wave of electrification.

The development trend of the US market is in sharp contrast to that of China and Europe. From January to August 2025, the sales volume of new energy vehicles in the US was 1.04 million units, with a growth rate of only 9%, which was the lowest in recent years. In terms of penetration rate, the penetration rate of new energy vehicles in the US in August was only 11.1%, far lower than that of China and major European countries. The sluggish growth of the US market is mainly due to two factors: on the one hand, the US government has cancelled the subsidy policy for new energy vehicles, which has a direct impact on consumers' willingness to purchase; on the other hand, early adopters and environmentalists in Europe and the US have already purchased electric vehicles, and subsequent consumers' acceptance of electric vehicles has progressed slowly. It is worth noting that due to the imminent implementation of high tariffs and concerns about price increases caused by the imminent cancellation of new energy subsidies, the sales volume of new energy vehicles in the US rebounded in August, reaching 164,000 units, an increase of 22% year-on-year and 22% month-on-month. This "policy-driven" growth model reflects that the US electric vehicle market has not yet formed an endogenous and spontaneous growth momentum.

In addition to the European and American markets, emerging markets such as Asia and South America are becoming new growth centers for electric vehicle sales. According to the International Energy Agency report, the total sales of electric vehicles in these emerging markets will increase by more than 60% year-on-year in 2024. Especially in the Asian market, Japan's new energy vehicle penetration rate is only 1.8%, with huge room for development; while Southeast Asia, South America and other regions have become important destinations for China's electric vehicle exports. This pattern of global market differentiation indicates that electric vehicles, as an emerging industry, are deeply influenced by local policy environment, industrial foundation and market stage. In the future, this uneven development trend will continue.

The booming development of the electric vehicle industry is strongly supported by technological advancements. In recent years, battery technology, as the core of electric vehicles, has made significant progress. The energy density of Chinese electric vehicle batteries has continued to increase, from about 180–200 Wh/kg in 2020 to about 300 Wh/kg in 2025. It is worth noting that the improvement in the speed of energy density of pure electric vehicle batteries has slowed down in recent years, but with the promotion of technology by the market in 2025, more models with an energy density of about 130–140 Wh/kg will be launched. In terms of driving range, the average driving range of pure electric passenger vehicles has approached 500 km, and the driving range of high-quality models has generally exceeded 700 km. This improvement in driving range has enabled electric vehicles to gradually overcome the shackles of "range anxiety" and expand into more mainstream consumer markets. The driving range performance of different technological routes also varies: in 2025, more pure electric models with a driving range of 300–400 km will be launched, while models with a driving range of less than 300 km will be significantly reduced, and the number of products with a driving range of more than 600 km will continue to increase; the average driving range of plug-in hybrid models will reach 137 km, mainly in the range of 100–200 km; the average driving range of extended range models will be higher, reaching 205 km.

The level of charging infrastructure development directly impacts the user experience and adoption rate of electric vehicles. In 2024, China's public charging infrastructure grew by 31%, with an expansion rate roughly in sync with the growth in electric vehicle sales. Due to multiple factors such as the type of residential property structure, national charging policies, and the strong public charging network of original equipment manufacturers, only 40% of Chinese electric vehicle users complete their charging at home, compared to 50% and 55% in Europe and North America, respectively. The Chinese government maintains strong support for the industry, with almost one in every two charging stations in China being a DC fast charger, accounting for 46%. Therefore, Chinese users have a high level of satisfaction with the charging experience—up to 94% of Chinese users are very or quite satisfied with the overall charging experience, which is higher than all other surveyed regions. At the same time, China's relatively low charging costs are also one of its highlights, with only 20% of users believing that electricity prices are too high (the global average is 33%), while the vast majority of users (75%) believe that the prices are reasonable.

In terms of intelligent technology, the integration of electric vehicles with intelligent networking and autonomous driving technology is becoming increasingly deep. Data shows that the proportion of new passenger cars with combined assisted driving functions has increased from 16.2% in 2020 to 62.1% in the first half of 2025. Key core technologies such as artificial intelligence, operating systems, and intelligent chassis by wire have become the focus of car companies' competitive layouts. Especially in

2025, 800 V high-voltage platforms, solid-state batteries (energy density 350 Wh/kg), and autonomous driving L4 technology will become the focus of competition among enterprises. This deep integration of electrification and intelligence is reshaping the essential attributes of automotive products, transforming them from traditional transportation tools into mobile intelligent terminals, which not only creates new product value but also reconstructs the industrial competition landscape.

1.2 Classification and Charging Mode of Electric Vehicles

At the beginning of the development of new energy vehicles in China, to highlight the difference between new energy vehicles and traditional fuel vehicles, the term "new energy vehicles" was introduced. In the document "Regulations on the Administration of the Access of New Energy Vehicle Manufacturers and Products", new energy vehicles are defined as vehicles that adopt new power systems and are driven entirely or mainly by new energy sources. In fact, with the clarification of China's technology and industry direction, electric vehicles have become the mainstay of new energy vehicles in China. In addition, pure electric vehicles and plug-in hybrid electric vehicles are the key directions for the development of new energy vehicles.

An Electric Vehicle (EV) refers to a vehicle that uses electric energy as its power source and is driven entirely or partially by an electric motor. As shown in Fig. 1.1, electric vehicles can be classified into two major categories, namely Hybrid Electric Vehicle (HEV) and All-Electric Vehicle (AEV). The Hybrid Electric Vehicle has an important branch, namely the Plug-in Hybrid Electric Vehicle (PHEV). The special feature of this type of electric vehicle is that it can be charged by connecting to the power grid and can also be refueled to extend its driving range. All-Electric Vehicles can be divided into Battery Electric Vehicle (BEV) and Fuel Cell Electric Vehicle (FCEV). The following sections will introduce Battery Electric Vehicle, Plug-in Hybrid Electric Vehicle, and Fuel Cell Electric Vehicle, respectively.

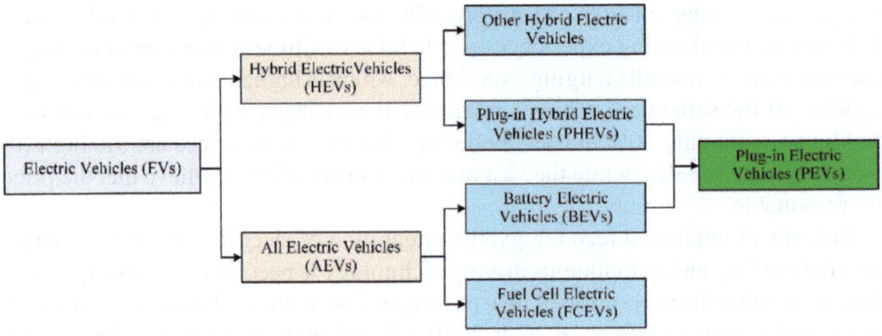

Fig. 1.1 Specific classifications of EVs

Battery electric vehicles (BEVs). BEVs rely solely on on-board traction batteries as their energy source, using electric motors to propel the wheels. Compared to traditional gasoline-powered vehicles, BEVs replace internal combustion engines with electric motors, fuel tanks with traction batteries, and traditional fossil fuels such as gasoline with electric energy. BEVs achieve "zero emissions" during operation, reducing local environmental pollution and significantly reducing dependence on fossil fuels. Additionally, the high efficiency of energy conversion in electric motors means less energy waste, making BEVs far more energy-efficient than traditional gasoline-powered vehicles. With continuous innovation in traction battery technology and gradual improvements in charging infrastructure, the range of BEVs continues to increase, and charging convenience has significantly improved, providing users with a more convenient, low-carbon, and environmentally friendly mode of transportation.

Plug-in hybrid electric vehicle (PHEV). A PHEV is a vehicle that is equipped with both an internal combustion engine and an electric motor, as well as a large-capacity battery pack and charging interface. Therefore, PHEVs have two energy replenishment methods: charging and refueling. In addition, PHEVs can operate in various modes, such as pure electric, hybrid, and fuel power. The driving range of PHEVs is usually higher than that of battery electric vehicles and fuel vehicles. However, their complex power systems result in higher manufacturing costs. PHEVs can not only achieve zero-emission driving in pure electric mode, but also switch to hybrid mode after the battery is depleted, relying on the internal combustion engine to extend the driving range. Therefore, the advantage of PHEVs lies not only in meeting users' demand for environmentally friendly travel, but also in considering the convenience of long-distance driving.

Fuel cell electric vehicles (FCEVs). FCEVs refer to a type of vehicle that uses hydrogen as fuel, converts chemical energy into electrical energy through electrochemical reactions, and relies on electric motors to drive the wheels. Their core power source is the fuel cell system. Compared with traditional fuel vehicles, FCEVs achieve true zero emissions, with the only by-product being water, which is environmentally friendly. At the same time, FCEVs exhibit excellent range performance, far exceeding the range of pure electric vehicles. In addition, the refueling process of FCEVs is fast, similar to the refueling time of fuel vehicles, greatly improving user convenience. In addition, the preparation, storage, and transportation of hydrogen also restrict the development of FCEVs. This leads to the large-scale commercial application of FCEVs being immature.

In summary, the three types of electric vehicles differ in terms of energy sources, drive systems, and energy supply infrastructure, as shown in Table 1.2. It can be seen that both battery electric vehicles and plug-in hybrid electric vehicles can be connected to the power grid for charging and energy replenishment. Therefore, these two types of electric vehicles are also collectively referred to as plug-in electric vehicles (PEV). The research direction of this book is the integration and interaction between large-scale electric vehicles and the power grid. Therefore, all references to electric vehicles in the following text refer to plug-in electric vehicles (PEV), also known as gridable electric vehicles (GEV).

When the battery state of charge of an electric vehicle drops to a critical threshold, it is necessary to charge the electric vehicle through an external power source to meet the user's travel needs. Considering the differentiated needs of users, such as commuting and long-distance travel, it is important to provide diversified charging modes. According to the latest Chinese national standard GB/T 20234.1-2023 "Connecting Devices for Conductive Charging of Electric Vehicles Part 1: General Requirements", there are four ways to connect an electric vehicle to the power grid to supply power to the vehicle: Mode 1, Mode 2, Mode 3, and Mode 4. The main parameters of these four modes are shown in Table 1.3.

Mode 1 connects to the AC power grid through a plug and socket that complies with GB2099.1 and GB1002 standards, without an In-Cable Control and Protection Device (IC-CPD), which may pose risks such as overcharging, leakage, and short circuit. Therefore, this mode is prohibited for electric vehicle charging.

Mode 2 is a method of charging electric vehicles by connecting them to the power grid using an on-board charger or a small charging device. Mode 2 adds on-cable control and protection devices to Mode 1, providing additional safety protection. Mode 2 uses low-voltage single-phase AC power to supply power to electric vehicles,

Table 1.2 Feature comparison among different types of EVs

Electric vehicle type	Energy source	Drive system	Energy supply infrastructure	Emissions
Battery electric vehicles (BEVs)	External rechargeable battery	Electric motor	Charging and swapping facilities	No emissions
Plug-in hybrid electric vehicle (PHEV)	Fuel and rechargeable battery	Internal combustion engine and electric motor	Gas stations and charging facilities	Limited emissions
Fuel cell electric vehicles (FCEVs)	Hydrogen fuel	Electric motor	Hydrogen refueling station	Water

Table 1.3 Technical specifications of charging standards in China

Charging standard		AC/DC	Maximum voltage (V)	Maximum current (A)	Charging type
GB/T 20234.1-2023	Mode 1	AC	250 (1–Ø)	8	–
	Mode 2	AC	250 (1–Ø)	8/32	AC slow charging
	Mode 3	AC	250 (1–Ø)	32	AC slow/fast charging
			440 (3–Ø)	63	
	Mode 4	DC	1500	800	DC fast charging

with a charging voltage generally between 220 and 250 V, a charging current between 8 and 32 A, and a charging power generally between 3 and 7 kW. Due to the small charging current, Mode 2 takes a very long time to charge, generally requiring several hours or even longer to fully charge the battery. The advantage of this mode is that charging is very convenient and the cost of charging equipment is relatively low. In addition, this mode has very little impact on the battery and the charging process is smooth, making it suitable for charging at night or during long periods of parking. Mode 2 is commonly used in residential areas, road parking spaces, and other places.

Mode 3 is a charging mode that uses dedicated AC power supply equipment to interconnect electric vehicles with the power grid. Mode 3 uses three-phase AC power to supply power to electric vehicles, with a charging current of up to 63 A and a charging power generally ranging from 10 to 25 kW. Compared to Mode 2, Mode 3 has a faster charging speed and shorter charging time, generally taking only 1–2 h to fully charge the electric vehicle battery. However, its equipment installation cost is relatively high, resulting in higher charging fees. Mode 3 is commonly found in shopping centers, public parking lots, highway service areas, and other places.

Mode 4 is the only DC charging mode that connects the grid directly to the electric vehicle (EV) using a dedicated DC charging cabinet. Unlike other charging modes, it bypasses the on-board charger (OBC) and delivers power directly to the EV battery, eliminating the need for AC–DC conversion. This mode offers very high charging voltage and current levels, with the voltage reaching up to 1500 V and the current up to 800 A. The charging power typically ranges from tens to hundreds of kilowatts, with a maximum of 800 kW. As a result, Mode 4 can quickly recharge an EV's battery, typically achieving an 80% state of charge (SOC) in under half an hour. However, this mode has significant drawbacks. It demands high safety standards for charging equipment and incurs high installation costs. Additionally, frequent high-power DC charging can degrade the battery's capacity, reducing its lifespan. Mode 4 is commonly used in highway service areas, fast-charging stations, and large parking lots.

1.3 Impact of EV Charging on the Power Grid

Disordered charging of electric vehicles refers to the immediate start of charging after the electric vehicle is connected to a charging pile, and the termination of charging when the battery is fully charged or disconnected. In this situation, the output power remains constant throughout the charging period and is not regulated by the power grid. Therefore, the charging load curve of electric vehicles will be determined by the charging behavior of electric vehicle users and the operating power of charging devices. At present, whether it is home charging, workplace charging, or public charging, most of them adopt a disorderly charging mode. The charging behavior of electric vehicle users exhibits strong randomness and is not constrained by the operation and regulation mechanism of the power grid. If a large number of electric vehicles are in a state of disorderly charging, it is highly likely to lead to a significant increase in the load peak of the power grid, and can affect the distribution and operation trend of power grid flow, thereby threatening the stable operation of the power grid.

When the penetration rate of electric vehicles reaches a certain proportion, the random charging load of a large number of electric vehicles will add a huge load to the existing power system, which in turn will affect power production and transmission from two aspects: (1) overall power demand increases; and (2) the demand pattern of power load is changed. The changes in load demand caused by large-scale electric vehicle disorderly charging will bring challenges to the planning and scheduling of transmission and distribution systems. For the transmission system, a large number of electric vehicle charging demands are prone to cause transmission congestion, which makes the transmission lines overwhelmed; at the same time, it will also cause many adverse effects on the distribution system, including power quality issues (e.g., unstable voltage, frequency fluctuations, harmonic distortion), increased peak load, increased grid loss, transformer overload, and line overload. This article summarizes the potential impacts of large-scale electric vehicle disorderly charging on the power system from multiple dimensions, as shown in Table 1.4.

1.4 Collaborative Interaction Between EVs and the Power Grid

While vigorously promoting the electric vehicle industry, China actively advocates for the orderly grid-connected charging of electric vehicles to mitigate the impact of high penetration rates on power grid operations. In the Energy Conservation and New Energy Vehicle Industry Development Plan (2012–2020) released in 2012, electric vehicles were proposed as an adjustable resource for research and application of vehicle-grid integration. In the New Energy Vehicle Industry Development Plan (2021–2035) released in 2020, it was proposed to build a vehicle-grid interaction demonstration system, reduce the electricity cost of electric vehicles through demand-side response, and enhance the resilience of the power system. In the Implementation Opinions on Strengthening the Integration and Interaction between New Energy Vehicles and the Power Grid released in 2023, the development goal of vehicle-grid interaction was proposed, which is to establish a vehicle-grid interaction system and market mechanism by 2030, realize the transformation of electric vehicle clusters into mobile energy storage units, and form a flexible adjustment capability of tens of millions of kilowatts for the power grid.

The collaborative interaction mode between electric vehicles and the power grid mainly adopts orderly charging and bidirectional charging and discharging as the main technical paths, which can support the power grid to achieve peak shaving, spinning reserve, frequency modulation and voltage regulation and other application scenarios. Orderly charging, also known as intelligent charging, refers to dynamically

Table 1.4 Potential impacts of uncoordinated EV charging on power grids

Power grid	Impact	Description
Electricity transmission system	System Transmission congestion	The increasing number of electric vehicles will significantly increase the load demand during peak hours, leading to congestion in the transmission system. A study based on the German power grid shows that the expanded transmission network in Germany is still insufficient to meet the power demand caused by the disorderly charging of the expected number of electric vehicles [1]
	Overload of transmission lines	The charging load of a large number of electric vehicles will significantly increase the power demand of the power grid, which may cause overload of the transmission lines. An evaluation of the Perth transmission system shows that electric vehicle charging will increase the load on transmission lines, but it is expected that its capacity will be sufficient to meet the future charging needs of electric vehicles without exceeding the prescribed limit [2]
Electricity distribution system	Voltage instability	The charging load of a large number of electric vehicles can significantly reduce the stability of the steady-state voltage of the distribution network. Research on the IEEE 34-bus test feeder system shows that when the penetration rate of plug-in electric vehicles reaches 30%, the voltage deviation can reach 10%, which is close to the allowable range of voltage deviation [3]
	Frequency deviation	The disorderly charging of large-scale electric vehicles will further exacerbate the imbalance of the power system, leading to significant deviations in grid frequency. Theoretically, large-scale electric vehicle charging operations will cause the grid frequency to exhibit a negative deviation; conversely, electric vehicle discharging operations will cause the grid frequency to exhibit a positive deviation [4]
	Harmonic distortion	Nonlinear devices such as rectifiers and high-frequency switching power supplies in electric vehicle charging systems generate harmonics during the AC–DC conversion process, which may lead to harmonic distortion. According to tests conducted by the International Council on Large Electric Systems (CIGRE) on power distribution systems, in the worst-case scenario, when the penetration rate of electric vehicles reaches 45%, the Total Harmonic Distortion (THD) may approach the limit value [5]

(continued)

Table 1.4 (continued)

Power grid	Impact	Description
	Increase peak load	The charging load of electric vehicles may coincide with the peak demand for electricity, further increasing the peak load of the system and potentially causing severe congestion in local power grids. A study of the power grid in Shenzhen showed that the disorderly charging of approximately 1 million electric vehicles could increase the existing peak load by 11.08% [6]
	Increase grid loss	If electric vehicle charging is not controlled, large-scale electric vehicle charging may further exacerbate system imbalances, leading to higher grid losses. Research on the IEEE 18-bus network shows that compared to other intelligent charging methods, the disorderly charging mode of electric vehicles can cause higher power losses [7]
	Transformer overload	Due to the fact that the load of electric vehicles was not considered in the initial planning of distribution transformers, the large-scale integration of electric vehicles into the power grid may significantly increase the probability of transformer overload. A case study has shown that if the charging of electric vehicles is not regulated, 4 out of 7 transformers studied will be overloaded, while through orderly charging regulation, the overload situation of 3 out of 4 transformers can be alleviated [8]
	Overload of the distribution line	Due to the high uncertainty in charging time of electric vehicles, a large amount of charging load may cause "thermal overload" of power distribution equipment (such as cables and transformers), posing a risk of grid failure. Research has shown that when the penetration rate of electric vehicles reaches 30%, uncoordinated charging of electric vehicles can significantly increase the peak temperature of cables [9]

regulating the charging power of electric vehicles based on user travel needs and real-time operating status of the power grid, with the help of advanced communication and energy management systems, to promote the balance between supply and demand of the power grid. Bidirectional charging and discharging (Vehicle-to-Grid, V2G), also known as intelligent charging and discharging technology, is an emerging energy management technology that allows electric vehicles to feed power back to the grid during peak hours, improving the peak load demand of the grid, thereby enhancing the stability and economy of the grid. The following introduces these two vehicle-grid collaborative interaction modes.

The core of the orderly charging mode lies in guiding electric vehicle (EV) users to charge during off-peak periods, absorbing excess power during low-demand periods, and thereby alleviating the pressure on the power grid during peak demand periods. This mode does not have the function of EV discharging to the grid, and therefore does not cause additional wear and tear on EV batteries. With the continuous development of smart grid technology, bidirectional communication between EVs and the grid has become possible, providing a solid technical foundation for the regulation of EV charging load. In addition, most EVs currently on the market are equipped with chargers and charging infrastructure that have the controllable ability for unidirectional transmission of electric energy. Therefore, the feasibility of large-scale application of the orderly charging mode for EVs in the short term is relatively high.

The core of the bidirectional charging and discharging mode (V2G) is to allow the electric energy stored in the electric vehicle battery to be fed back to the power grid while ensuring the daily travel needs of electric vehicle users. This mode not only requires information interconnection between electric vehicles and charging facilities, but also requires the charging infrastructure to have the ability to convert energy in both directions. Under V2G mode, electric vehicles can provide diversified auxiliary services such as spinning reserve and frequency modulation for the power grid, thereby enhancing the stability of the power grid.

Compared to the uncoordinated charging mode, the aforementioned two modes not only mitigate the impact of electric vehicle charging on the power grid but also provide support to the grid, as illustrated in Table 1.5. Given the issues associated with battery depletion due to electric vehicle discharging and the high requirements for V2G technology application, this paper primarily focuses on the market-oriented operation of electric vehicle aggregators and the charging scheduling strategies for electric vehicles in the electricity market environment.

Table 1.5 Feature comparison of different power interaction modes

Feature	Mode		
	Stochastic charging	Ordered charging	Bidirectional charging and discharging (V2G)
Energy flow direction	Grid → EV	Grid → EV	Grid ↔ EV
Technical specifications	/	Bidirectional communication Charging control	Bidirectional communication Charging/discharging control
Support for the power grid	/	Load fluctuation stabilization Valley filling	Load fluctuation stabilization Peak shaving and valley filling Backup generation
Impact on batteries	/	/	Affect the battery lifetime
Application difficulty	/	Easy	Difficult

1.5 Electricity-Transportation Coupled Network

The large-scale application of electric vehicles (EVs) is reshaping the coupling relationship between urban energy systems and transportation systems. Due to the high dependence of the spatial and temporal distribution of charging loads on user travel chains, a significant bidirectional coupling effect has been formed between the power grid and the road network. Specifically, when EV traffic flows pass through fast charging stations (FCS), charging loads are generated, which in turn affect the spatial and temporal distribution of power flow in the distribution network. In addition, the dynamic changes in power flow also cause fluctuations in the locational marginal price (LMP) of distribution network nodes, affecting charging costs, which in turn affect EV users' travel and charging choices, resulting in changes in the spatial and temporal distribution of traffic flows. The coupled operation of the power distribution network (PDN) and the transportation network (TN) effectively forms a coupled power and transportation network (CPTN).

In the electric power-transportation coupled network, the aggregation of electric vehicle traffic flow at fast charging stations will exacerbate congestion in the surrounding road network, while unguided and disorderly charging behavior may also lead to risks such as overload of the power distribution system and voltage violations, affecting the safe and stable operation of the electric power network. It is noteworthy that users' charging and travel behavior are largely governed by their decision-making rationality, which determines their response characteristics to charging prices and travel costs, thereby affecting the operating state of the coupled system. Users with absolute rationality will only choose the optimal travel and charging scheme. However, in reality, due to incomplete information acquisition, insensitivity to subtle costs, and other reasons, users find it difficult to make globally optimal decisions and instead accept suboptimal solutions within a certain range. This decision-making rationality is called bounded rationality (BR). The travel choices of bounded rational users are more diverse and differ significantly from the behavior patterns of absolute rational users, making the operating mechanism of the actual electric power-transportation coupled network more complex. At the same time, a high proportion of new energy power generation will be integrated into China's new power system in the future, and its intermittent and fluctuating output will further exacerbate the operational uncertainty of the electric power-transportation coupled network. Therefore, under the combined effects of the widespread application of electric vehicles, users' bounded rationality, and high penetration of new energy, the collaborative optimization and scheduling problem of the electric power-transportation coupled network faces unprecedented challenges. It is urgent to develop an interdisciplinary research paradigm that integrates behavioral science, systems engineering, and operations research, and select appropriate entry points for coupled network regulation.

The healthy development of the electric vehicle industry cannot be separated from the support of a high-quality charging infrastructure system. According to the "Guiding Opinions of the General Office of the State Council on Further Building a High-quality Charging Infrastructure System" in 2023, a high-quality charging network with wide coverage, reasonable structure, and perfect functions should be basically established by 2030, with a focus on building a fast charging network connecting major cities in the region. The popularization of DC fast charging technology has significantly shortened charging time, effectively solved the charging anxiety of electric vehicle users, and become the mainstream charging option in urban areas. In the fast charging mode of "charge and go", traffic flow and power flow will directly converge in the fast charging station, which becomes the core hub connecting the power and transportation systems, and also the key means to regulate power flow and traffic flow to achieve coupled network synergy.

In terms of specific regulation mechanisms, fast charging stations have dual functionalities of price guidance and power regulation: (1) price guidance: users can be guided to charge during off-peak hours through price signals, alleviating the load pressure on the power grid, reducing queuing and congestion within the station, and improving the operational efficiency of the coupled system; and (2) power regulation: the shape of the charging power curve can be directly regulated to achieve orderly grid access of charging loads, while promoting the local consumption of distributed renewable energy. Through unified aggregation management of fast charging station clusters, fast charging station aggregators (FCSA) can participate in energy trading and demand response as power market entities, which is expected to further optimize electricity costs and achieve a win–win situation for vehicle-grid coordination.

In summary, fast charging stations play a central role in connecting the power grid and road network, and harmonizing traffic flow and power flow in the electric-transportation coupled network. Their optimization and regulation strategies in the "price-power" dimension will directly affect the collaborative operation efficiency and safety of the electric-transportation coupled network. Therefore, considering the coupled relationship between electric power and transportation and the limited rationality of users, this paper takes the regulation strategy of fast charging stations as the core research object, and explores its optimal regulation method under the scenario of multi-source uncertainty and multi-scale interaction coupling, with a view to providing theoretical support and technical path for building an efficient, resilient, and low-carbon smart transportation energy system.

References

1. Staudt P, Schmidt M, Gärttner J et al (2018) A decentralized approach towards resolving transmission grid congestion in Germany using vehicle-to-grid technology. Appl Energy 230:1435–1446
2. McCarthy D, Wolfs P (2010) The HV system impacts of large-scale electric vehicle deployments in a metropolitan area. In: 2010 20th Australasian universities power engineering conference, pp 1–6
3. Clement-Nyns K, Haesen E, Driesen J (2009) The impact of charging plug-in hybrid electric vehicles on a residential distribution grid. IEEE Trans Power Syst 25:371–380
4. Lopes JAP, Soares FJ, Almeida PMR (2010) Integration of electric vehicles in the electric power system. Proc IEEE 99:168–183

5. Ul-Haq A, Perwaiz A, Azhar M et al (2018) Harmonic distortion in distribution system due to single-phase electric vehicle charging. In: 2018 2nd international conference on green energy and applications (ICGEA), pp 205–209
6. Zheng Y, Shao Z, Zhang Y et al (2020) A systematic methodology for mid-and-long term electric vehicle charging load forecasting: the case study of Shenzhen, China. Sustain Cities Soc 56:102084
7. Sortomme E, Hindi MM, MacPherson SDJ et al (2010) Coordinated charging of plug-in hybrid electric vehicles to minimize distribution system losses. IEEE Trans Smart Grid 2:198–205
8. van den Berg MA, Lampropoulos I, AlSkaif TA (2021) Impact of electric vehicles charging demand on distribution transformers in an office area and determination of flexibility potential. Sustain Energy Grids Netw 26:100452
9. Marshall JS, Hines PD, Zhang JD et al (2013) Modeling the impact of electric vehicle charging on heat transfer around underground cables. Electr Power Syst Res 97:76–83

Chapter 2
Literature Review

Abstract This literature review systematically examines the state-of-the-art research on electric vehicle (EV) integration with power systems, focusing on charging demand characterization, infrastructure planning, charging/discharging scheduling, and vehicle-to-grid (V2G) interactions under demand response mechanisms and electricity market environments. The review highlights that accurate prediction of EV charging load, considering both temporal and spatial uncertainties, is fundamental for grid stability and facility planning. Studies have evolved from using survey data and Monte Carlo simulations to advanced models incorporating Markov chains and traffic flow theory. In infrastructure planning, research emphasizes the dual attributes of EV charging stations as transport and electrical facilities, advocating for coordinated planning with distributed energy resources while considering both distribution and transportation network constraints. For charging scheduling, the literature distinguishes between aggregator-based coordination and price-driven strategies, with growing attention to V2G technologies that enable bidirectional power flow. The review extensively covers demand response frameworks, comparing price-based and incentive-based mechanisms for regulating EV charging behavior. In electricity market contexts, EV aggregators face challenges in day-ahead bidding and real-time charging regulation under uncertainties from both EV behavior and market prices. Recent advances include game-theoretic approaches for aggregator-user interactions and risk-averse bidding strategies. The review also explores emerging research on power-transportation coupling networks, incorporating user rationality analysis and dynamic traffic flow models. Furthermore, it examines charging pricing mechanisms, from fixed and nodal marginal pricing to optimization-based strategies within Stackelberg game frameworks. Finally, the review addresses multi-energy systems integrating EV charging with hydrogen and natural gas refueling in integrated energy service stations, highlighting the complexity of heterogeneous traffic flows and multi-energy coordination. Overall, the literature reveals critical research gaps in coupled network modeling, bounded user rationality, computational efficiency for real-time scheduling, and comprehensive theoretical frameworks for coordinated planning of EV facilities and distributed energy resources.

Q. Yang et al., *Integration of Massive Electric Vehicles to Smart Grid*, Smart Energy Systems, https://doi.org/10.1007/978-981-92-0557-8_2

2.1 Electric Vehicle Charging Demand Characterization

Accurate prediction of electric vehicle charging load is the foundation for scheduling electric vehicle charging behavior and planning electric vehicle charging facilities. Due to the significant individual differences and randomness of electric vehicle charging behavior, the charging load of large-scale electric vehicles has considerable uncertainty in both time and space dimensions, which can negatively affect the safe and stable operation of the power grid. Only by accurately predicting the charging load of large-scale electric vehicles can we address these issues in a targeted manner. In this regard, experts and scholars at home and abroad have conducted extensive research on the prediction of electric vehicle charging demand and have made some progress.

In the past years, most research work has utilized existing survey data on car travel time and electric vehicle battery parameters, combined with some reasonable assumptions about electric vehicle travel habits, to analyze the behavior of electric vehicle charging and predict the charging load of electric vehicles. The data used is often the time of travel start and end, daily mileage, etc. It is generally assumed that the end of travel time is the start of charging time, and the initial charge and charging duration are both related to daily mileage. For example, the authors in [1] systematically analyzed the key points of modeling electric vehicle charging load earlier, proposed that existing car travel survey data and reasonable assumptions can be used to predict electric vehicle charging load, and assumed that there is a linear relationship between electric vehicle daily mileage and charging duration, thus obtaining the charging load curve of electric vehicles. The study in [2] adopted the Monte Carlo simulation method to randomly generate the start time of electric vehicle charging and battery charge, and assumed that electric vehicles do not charge every night, and the battery charge when leaving may not be fully charged. Whether to charge depends on the battery charge at the end of the day's journey and the estimated mileage for the next day. Literature [3] obtained the joint distribution function between random variables such as electric vehicle travel time and daily mileage based on Copula function theory, and then used the Monte Carlo simulation method to randomly generate electric vehicle parameters, thus obtaining the charging load of electric vehicles.

At the same time, the charging load of electric vehicles not only has considerable uncertainty on the time scale but also exhibits significant randomness on the spatial scale. Currently, there are also some efforts to model and predict the spatial characteristics of electric vehicle charging load. Based on traffic flow theory and queuing theory, the work in [4] proposed a method for predicting electric vehicle charging load considering spatial and temporal variations for fast charging stations on main roads. Similarly, based on traffic flow theory and existing survey data, the authors in [5] used the Monte Carlo simulation method to predict the spatiotemporal characteristics of electric vehicle charging load in cities, and proved that reasonable scheduling strategies can effectively achieve peak shaving and valley filling.

In addition, a few studies used a Markov model to describe the changes in travel and charging behavior of electric vehicles, and then obtained the charging load of electric vehicles. The work in [6] used a Markov chain to describe the dynamic changes in travel behavior of electric vehicles based on existing car travel survey data, and then obtained the changes in charging load of electric vehicles, and analyzed the impact of electric vehicle charging load on the power grid. Literature [7] used a semi-Markov model to predict the travel behavior of electric vehicles, including the parking time at the current location and the next destination, and then obtained the charging behavior and charging load of electric vehicles. The authors in [8] used a Markov decision process to analyze the charging cost at different times, so that electric vehicles can save as much charging cost as possible, and used a reinforcement learning algorithm to solve the model, obtaining the optimal charging load curve of electric vehicles.

2.2 Planning of Electric Vehicle Charging Infrastructures

Electric vehicle charging stations are important infrastructure for ensuring the large-scale popularization of electric vehicles, directly affecting the convenience of electric vehicle charging and thus the enthusiasm of people to purchase electric vehicles. Therefore, reasonable planning of them is very important. A large amount of research work has been carried out both domestically and internationally on the planning of electric vehicle charging facilities, and some preliminary results have been achieved.

For public vehicles such as taxis and buses, their travel behavior is relatively fixed, so there are few research results on the planning of charging stations for public vehicles. For private electric vehicles, due to significant differences and randomness in users' driving habits and travel purposes, determining their charging load becomes complex and difficult. Therefore, the layout planning of electric vehicle charging facilities often needs to consider a large number of random factors. Generally speaking, current research on electric vehicle charging facility planning mainly focuses on two aspects: the location selection and capacity determination of electric vehicle charging stations. In terms of location selection, the research idea is mainly to model the spatial distribution characteristics of electric vehicles in the studied area through some geographic analysis models or traffic flow models, to obtain the optimal location of charging facilities. For example, the work in [9] considered the balance between the captured traffic flow of charging stations, node electricity prices, and user travel habits in the case of coupling between transportation networks and distribution networks, and established an optimal planning model for electric vehicle charging stations to maximize social benefits.

In terms of the capacity determination of electric vehicle charging facilities, only a small number of studies have utilized queuing theory to model electric vehicle charging stations. For example, the work in [10] established an electric vehicle charging station planning model based on queuing theory to maximize system revenue, and considered various types of electric vehicles and battery loss costs.

The authors in [11] used queuing theory to describe the dynamic change process of electric vehicles in charging stations, and based on this, analyzed the uncertainty of charging station load, thereby calculating the probabilistic power flow of the distribution network. The study in [12] took the service cost and charging cost of electric vehicle charging stations as optimization objectives, while considering the endurance mileage of electric vehicles and the arrival rate of arriving at charging stations, and obtained the optimal charging pile configuration of electric vehicle charging stations through queuing theory. The work in [13] established a model based on queuing theory to maximize the revenue of electric vehicle charging stations, and considered the charging cost, waiting and punishment cost of rejecting electric vehicles, thereby determining the optimal configuration of charging piles and waiting positions.

The above literature only considers the location or capacity of electric vehicle charging facilities. Currently, many research works have combined the location and capacity of electric vehicle charging facilities to determine the optimal location and capacity of electric vehicle charging facilities. For example, the authors in [14] firstly used a traffic flow capture model to obtain the charging demand of electric vehicles, and then considers the constraints of the transportation network and distribution network simultaneously, modeling the planning problem of electric vehicle charging stations as a mixed integer linear programming problem, and using the corresponding optimization algorithm to obtain the optimal location and capacity of electric vehicle charging stations. The work in [15] considered various charging facilities for electric vehicles, established an optimization model to minimize the charging cost of electric vehicle charging facilities using Voronoi diagram theory, and used a particle swarm optimization algorithm to solve it. The study in [16] established a two-level planning model, where the first level aims to minimize the cost of electric vehicle charging stations, thus determining the configuration of electric vehicle charging stations, and the second level considers the impact of using some scheduling strategies on the planning problem. The work in [17] considered the constraints of urban transportation networks and distribution network capacity, established an optimization model to minimize the cost and power overflow of electric vehicle charging stations, and used a genetic algorithm to obtain the optimal location and capacity of electric vehicle charging stations that meet the stable operation of the distribution network. The authors in [18] first adopted geographic information theory to obtain the loss rate of electric vehicle charging stations, and then established a nonlinear integer programming problem that includes factors such as charging station construction cost, electric vehicle loss rate, and distribution network loss, thus obtaining the optimal location and capacity of electric vehicle charging stations. The authors in [19] constructed a multi-objective optimization model to maximize the traffic flow captured by charging stations and minimizing the sum of construction cost and distribution network loss, uses a user equilibrium traffic assignment model to analyze the traffic flow captured by charging stations, and calculates the charging waiting time during peak traffic periods based on queuing theory, thus configuring the optimal number of equipment for each charging station.

In addition, with the rapid development of distributed renewable energy sources, many studies have investigated the planning of distributed energy sources in distribution networks, mainly focusing on the location and capacity determination of distributed energy sources in distribution networks. In recent years, more and more literature has also focused on the collaborative planning of electric vehicle charging facilities and distributed energy sources. For example, literature [20] proposed a multi-objective optimization model that considers minimizing greenhouse gas emissions and total system costs, and uses a non-dominated sorting genetic algorithm to solve it, thereby determining the optimal location and capacity of distributed energy sources and electric vehicle charging stations. Literature [21] established a multi-objective optimization model for the location and capacity determination of electric vehicle charging stations and distributed energy sources, with the goals of reducing network losses, improving node voltage stability, and reducing costs, and uses a heuristic differential evolution algorithm to solve it. The authors in [22] considered the economic benefits of electric vehicle charging stations and the constraints of distribution networks, and proposed a bi-level planning method that first determines alternative solutions based on economic factors of electric vehicle charging stations, and then determines the final planning results based on the constraints of distribution networks. The work in [23] transformed the location and capacity determination of electric vehicle charging stations, distributed energy sources, and fixed energy storage into a second-order cone optimization problem, considering the dynamic characteristics of various optimization objects. Literature [24] proposed a multi-level optimization model that includes various costs such as construction, maintenance, and network losses, and considers the uncertainties of distributed energy sources and electric vehicle charging loads, thereby determining their optimal configurations. The planning model of electric vehicle charging stations and distributed energy sources in [25] considered economic factors, distribution network constraints, and system CO_2 emissions, and finally uses genetic algorithms to solve it.

However, the existing body of literature still presents several limitations and areas requiring improvement in the study of collaborative planning for electric vehicle (EV) charging facilities and distributed power sources. First, the majority of studies only account for constraints from either the distribution network or the transportation network, but not both. Since EV charging facilities function both as public transport infrastructure and as electrical loads, their dual attributes must be considered during planning. This necessitates an integrated approach that addresses the coupling relationship between the distribution and transportation networks, ensuring that planning outcomes comply with the constraints of both systems. Second, electric vehicles support two primary charging modes, i.e. conventional and fast charging, each requiring different types of infrastructure. To adequately address the diverse charging demands of EV users, both types of charging facilities should be planned concurrently. Yet, most existing studies focus on only one charging mode. Third, with the growing penetration of distributed power sources, opportunities arise to supply EVs with clean energy, thereby improving the utilization efficiency of renewable resources. This synergy makes the coordinated planning of EV charging facilities and distributed power sources increasingly important. Despite this importance, current literature lacks a systematic and comprehensive theoretical and modeling framework for such collaboration. Finally, when the siting and sizing of EV charging stations and distributed power sources are considered together with the various constraints and uncertainties arising from the distribution-transportation network coupling, the planning problem becomes a highly complex and challenging optimization task [26]. Both the computational complexity and the time required to solve such a model increase exponentially.

2.3 Charging and Discharging Scheduling for Electric Vehicles

Due to the potential negative impact of uncoordinated charging of large-scale electric vehicles on the safe and stable operation of the power grid, the scheduling strategy for electric vehicle charging behavior has received significant attention from researchers both domestically and internationally in recent years, and a large number of research results have emerged.

Regarding the methods of electric vehicle (EV) charging scheduling, existing research primarily focuses on two approaches: aggregators coordinating EV charging behavior and grid operators directly using electricity pricing incentives to guide EVs to charge in an orderly manner. The former approach is adopted mainly because allowing individual EVs to directly participate in grid scheduling operations would significantly increase the system's communication burden. Moreover, the charging load of a single EV is difficult to reach a magnitude that can alter the load characteristics of the grid. Therefore, aggregators are used to uniformly schedule EV clusters. Specifically, reference [27] introduced the concept of an EV aggregator, which acts as an intermediary system between individual EVs and grid operators. This aggregator schedules EV charging behavior in a unified manner, satisfying EV charging needs while also participating in grid operation scheduling. The authors in [28] compared the feasibility and economic efficiency of EV charging scheduling with and without aggregators. It finds that while the cost for individual EVs increases with an aggregator, the reliability and feasibility of scheduling are greatly enhanced.

The latter mainly focuses on the economic aspect, guiding the charging behavior of electric vehicles through dynamic electricity pricing, thereby achieving the effects of reducing the charging cost of electric vehicle users, improving the load characteristics of the distribution network, reducing the operating cost of the power system, and coordinating and complementing with distributed power sources. This approach considers both economic and technical issues. For example, literature [29] studied the random distribution characteristics of parameters such as initial battery capacity and charging start time based on the charging power curve and driving characteristics of electric vehicles, and proposed a sequential scheduling model for electric vehicle charging, thereby achieving the effect of reducing system load fluctuations. Literature [30] established an optimal charging strategy for electric vehicles from three perspectives: minimizing distribution network losses, maximizing load factor, and minimizing load variance, to minimize the impact of large-scale electric vehicle charging on the distribution network. The work in [31] proposed a two-stage electric vehicle charging scheduling method to address the fluctuation problem of large-scale renewable energy grid integration. Firstly, a day-ahead plan for unit output is formulated based on the prediction of electric vehicle charging demand, and then the charging plan of electric vehicles is adjusted in real time through a dynamic algorithm, thereby improving the robustness of the distribution network under uncertain renewable energy output.

In recent years, more and more research work has not only considered the charging behavior of electric vehicles, but also studied how to further control electric vehicles to feed power to the grid in a suitable time period, namely V2G. V2G regards electric vehicle batteries as both controllable energy storage units and distributed power sources. During low load periods, they store excess power from the grid as energy storage units, while during high load periods, they discharge power to the grid as distributed power sources. A reasonable V2G scheduling strategy can not only solve the negative effects of large-scale electric vehicle disorderly charging on the grid, but also improve the load curve fluctuation and power quality of the grid, providing various auxiliary services such as frequency modulation, peak shaving, and spinning reserve for the grid. This significantly reduces user charging costs and grid operation costs, and also improves the utilization efficiency of distributed renewable energy sources. For example, the study in [32] used the Monte Carlo method to simulate the travel conditions of electric vehicles, proving that the charging and discharging behavior of electric vehicles significantly affects the economic benefits obtained from providing frequency modulation auxiliary services. The work in [33] considered the probability of electric vehicles providing reserve capacity and proposed a spinning reserve day-ahead optimization model to determine the reserve plan for generating units and electric vehicles to minimize system operation costs, and proved that the system reserve service costs have decreased. Literature [34] proposed a distributed electric vehicle charging and discharging day-ahead management strategy using distributed iterative algorithms and droop control, enabling electric vehicle clusters to provide certain auxiliary services to the grid while meeting charging demands. The work in [35] proposed a scheduling method using electric vehicles to consume the output of distributed photovoltaic power sources, and used a dynamic adaptive algorithm to solve the problem of output fluctuations in distributed photovoltaic power sources.

In addition, some research work has introduced game theory to solve the scheduling problem of electric vehicle charging and discharging, mainly focusing on how to achieve a game equilibrium between electric vehicles or between electric vehicles and grid operators. For example, the work in [36] proposed a non-cooperative game-based electric vehicle charging and discharging scheduling strategy, which minimizes the charging cost of each electric vehicle while satisfying the relevant constraints of safe and stable operation of the power grid, and uses an accelerated gradient algorithm to solve for the Nash equilibrium. The work in [37] established a cooperative game model between electric vehicles, enabling the electric vehicle cluster to participate in V2G auxiliary services in a unified manner and pursue common revenue maximization, thus achieving peak shaving and valley filling effects. The authors in [38] proposed a Stackelberg game model where the grid operator acts as the leader to set electricity prices, and electric vehicle users act as followers to formulate charging and discharging strategies, enabling both parties to maximize revenue and minimize costs as much as possible.

2.4 Vehicle-to-Grid Interaction Under Demand Response Mechanism

Based on existing research, the implementation architecture of vehicle-to-grid interaction mainly falls into two categories: demand response mechanism-based vehicle-to-grid interaction and power market environment-based vehicle-to-grid interaction. The former utilizes electricity price signals or incentive mechanisms to guide electric vehicle users to adjust their charging behavior, achieving grid supply and demand regulation while reducing electricity costs. The latter introduces electric vehicle aggregators to integrate distributed resources of electric vehicles, and conducts electricity arbitrage and grid service supply through two-stage operations: day-ahead market bidding and real-time charging scheduling. Both architectures need to address the dual uncertainties of randomness in electric vehicle charging behavior and market volatility in order to develop optimal charging regulation strategies, thereby reducing the electricity costs of electric vehicles or enhancing the economic benefits of electric vehicle aggregators. The demand response mechanism aims to transfer charging loads during peak periods to alleviate grid pressure, while the power spot market mechanism aims to plan electric vehicle charging loads one day in advance, achieving a balance between power supply and electricity consumption on the operation day through real-time charging regulation. Research teams at home and abroad have conducted relevant research on this issue. This section summarizes the research context and briefly elaborates as follows.

As suggested in [39], demand response (DR) can be divided into two categories: price-based demand response (PDR) and incentive-based demand response (IDR). PDR guides power users to adjust their electricity consumption behavior through electricity price signals, mainly including time-of-use (ToU), real-time pricing (RTP), and critical peak pricing (CPP). IDR encourages power users to change their electricity consumption behaviors to support reliable system operation through economic incentives, mainly including interruptible load (IL) and direct load control (DLC). In the scenario of electric vehicle (EV) interaction with the power grid, the PDR mechanism can influence the charging behavior of EV users through electricity price elasticity, while the IDR mechanism supports grid operators to actively schedule charging loads. The two mechanisms work together to provide a practical and feasible operational framework for large-scale EVs to charge in an orderly manner. This article summarizes the application scenarios and research progress of vehicle-grid interaction from the perspectives of PDR and IDR.

Vehicle-to-infrastructure interaction under the PDR mechanism. Establishing a scientific charging pricing system is crucial for guiding electric vehicles to charge in an orderly manner. In a system that uses the PDR mechanism to achieve vehicle-grid integration, grid operators need to broadcast electricity price signals to charging station operators or electric vehicle users to shift the charging load of electric vehicles from peak hours to off-peak hours, as shown in Fig. 2.1. This solution only requires the deployment of smart meters to achieve time-of-use metering, which is easy to operate and has low implementation costs. However, under this architecture, grid operators cannot actively regulate electric vehicle charging. In addition, there are differences and uncertainties in the response of electric vehicle users to electricity prices [40]. Therefore, implementing electricity price policies to shift electric vehicle load also carries certain risks. Studies have shown that differentiated electricity pricing policies may induce a large number of electric vehicle users to charge during low-price periods, forming a rebound effect, which may trigger new load spikes and increase the risk of transformer overload [41]. Therefore, when formulating charging price strategies, it is necessary to comprehensively evaluate multiple factors such as electric vehicle users' charging preferences, electricity price sensitivity, and user response [42].

The researchers have conducted extensive research on using electricity pricing mechanisms to guide the charging behavior of electric vehicles. For example, literature [43] proposes a real-time charging pricing strategy for electric taxi charging scheduling to match the charging load with the established load curve. Simulation results show that the proposed real-time pricing mechanism can effectively regulate the overall load of electric taxis to avoid new load spikes and promote the balance between supply and demand in the power grid. In addition, research data from the Smart Electric Power Alliance in the United States shows that 72% of electric vehicle users are willing to charge during low electricity pricing periods to reduce charging

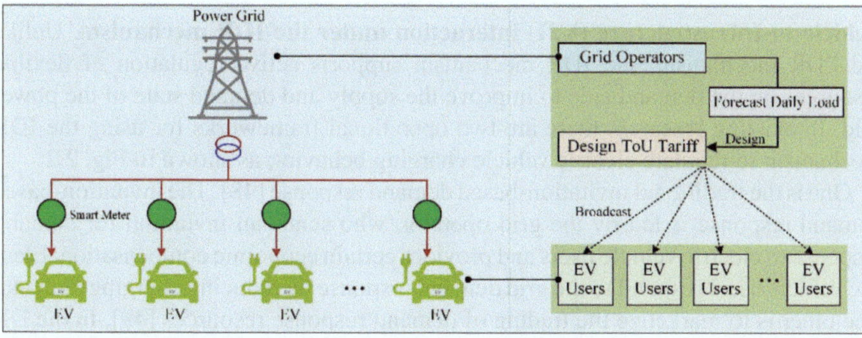

Fig. 2.1 Operating framework for coordinated interaction between EVs and power grids in PDR programs

costs [44]. Furthermore, studies have shown that increasing the difference between peak and valley electricity prices can further incentivize electric vehicle users to charge during low electricity usage periods, thereby enhancing the ability of electric vehicles to transfer load [45].

Under the differentiated electricity pricing policy, regulating the charging of electric vehicles has achieved notable results in improving the state of the power grid. For instance, the study in [46] proposed an optimal charging scheduling approach for electric vehicles using the time-of-use (TOU) pricing mechanism. This approach further enhances voltage conditions by switching capacitors and adjusting transformers, and has been validated using the IEEE 31-bus network. The results demonstrate that the proposed method effectively reduces the charging cost of electric vehicles and minimizes power losses and voltage deviations on the grid side. The work in [47] simulated the charging behavior of electric vehicles using probability distribution functions. Based on this, it analyzes the impacts of uncoordinated charging and TOU pricing charging modes on the power grid. Simulation results indicate that uncoordinated charging significantly increases system peak loads and affects system reliability, whereas the coordinated charging mode based on TOU pricing helps smooth the system load curve and enhances the reliability of the power system.

Based on the analysis of the above literature, it can be seen that the PDR mechanism mainly affects the charging behavior of electric vehicles through electricity price signals, thereby achieving the effect of peak shaving and load leveling. However, the PDR mechanism relies too much on the sensitivity of electric vehicle users to electricity price signals. In addition, the price elasticity differences between different user groups are significant, resulting in a high degree of dispersion in response rates. Furthermore, some users find it difficult to actively adjust their charging behavior due to rigid charging needs or information asymmetry, weakening the effect of peak shaving and load leveling. On the other hand, the PDR mechanism is mainly based on one-way regulation by grid operators, lacking the ability to control charging loads in a closed loop. It cannot dynamically adjust charging power or interrupt high-load charging requests according to real-time grid status, limiting the ability of the grid to actively intervene in load management.

Vehicle-to-Infrastructure (V2I) interaction under the IDR mechanism. Unlike the PDR mechanism, the IDR mechanism supports active regulation of flexible resources on the demand side to improve the supply and demand state of the power grid. In existing research, there are two operational frameworks for using the IDR mechanism to regulate electric vehicle charging behavior, as shown in Fig. 2.2.

One is the traditional invitation-based demand response [48]. The invitation-based demand response is led by the grid operator, who sends an invitation for demand response to electric vehicle users and provides certain economic compensation. Electric vehicle users respond to the grid demand response by reducing charging demand. The other is to marketize the trading of demand response resources [49]. In the DR trading market, DR buyers submit their DR capacity demand, and DR sellers submit their DR supply plan. Through a certain market-clearing mechanism, the maximum

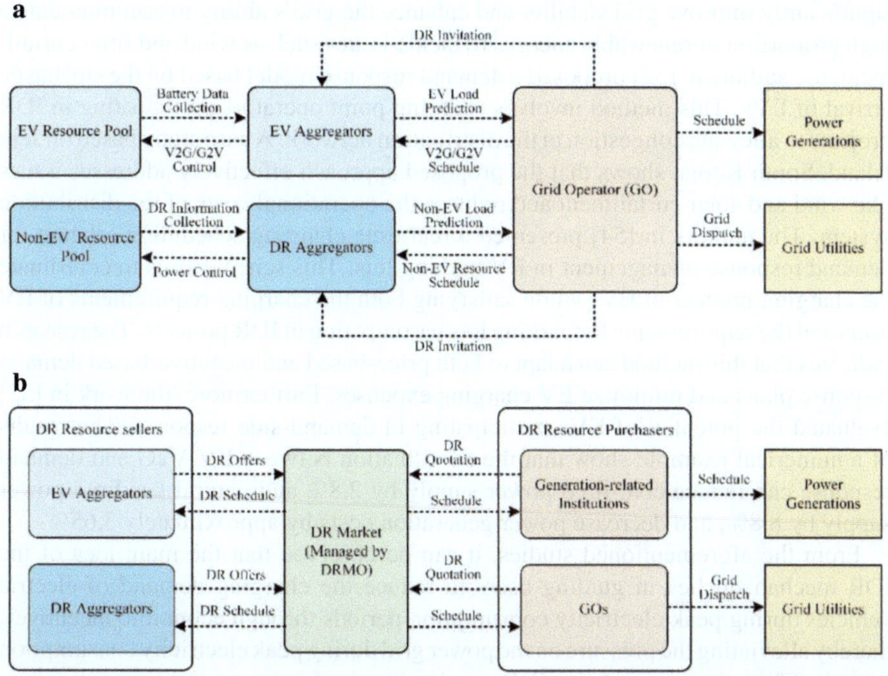

Fig. 2.2 Coordinated interaction between EVs and power grids in IDR programs: **a** invitation-based DR structure; **b** market-operated DR structure

matching of DR demand and supply is achieved to optimize DR resource allocation. Electric vehicle aggregators need to schedule electric vehicles according to the clearing results to complete the established plan. In addition, to meet the minimum standards for market participation, electric vehicle aggregators need to establish a pool of schedulable resources for electric vehicles to form a large-scale DR trading capacity [50].

Numerous studies indicated that electric vehicles (EVs) participating in IDR projects can lead to a win–win situation for both EV users and the power grid. For instance, the work in [51] introduced a stochastic programming method to optimize the operational strategies of EV charging stations involved in IDR and PDR projects. The findings reveal that EV charging stations can enhance their economic returns by selectively engaging in demand response projects. Additionally, the operational strategies of EV charging stations have an impact on the power grid's earnings in both the energy and reserve markets. The work in [52] also focused on optimizing the charging scheduling strategies for EVs participating in IDR and PDR projects. However, this approach differs in that it considers uncertainties arising from renewable energy generation, load patterns, parking patterns, and transmission line reliability from the perspective of an independent system operator. The research demonstrates that optimizing EV charging and discharging behaviors can significantly improve grid stability and enhance the grid's ability to accommodate a high proportion of renewable energy. To tackle issues such as wind and solar curtailment, the authors in [53] proposed a demand response model based on the stochastic arrival of EVs. This method involves charging point operators participating in IDR projects to alleviate congestion in the distribution network. A case study based on Jeju Island, South Korea, shows that the proposed approach effectively addresses issues like wind and solar curtailment and reduces the operational costs of the distribution system. The authors in [54] presented a real-time charging scheduling scheme for demand response management in EV parking lots. This scheme aims to coordinate the charging process of EVs while satisfying both the charging requirements of EV users and the requirements for parking lots participating in IDR projects. The research indicates that this method can adapt to both price-based and incentive-based demand response plans and minimize EV charging expenses. Furthermore, the work in [55] evaluated the potential of EVs participating in demand-side response. The results of a numerical example show that the coordination between EV V2G and demand response can reduce coal-fired power supply by 2.8% and natural gas-fired power supply by 8.8%, and decrease power generation costs by approximately 3.65%.

From the aforementioned studies, it can be observed that the main idea of the IDR mechanism lies in guiding users to reduce the charging demand of electric vehicles during peak electricity consumption periods through economic incentives, thereby alleviating the pressure on the power grid during peak electricity consumption periods. The advantage of the IDR mechanism is that power grid operators can actively intervene in the charging and discharging behavior of electric vehicles. In addition, under the IDR mechanism, electric vehicles can coordinate with new energy output to enhance the local consumption capacity of renewable energy. Furthermore, electric vehicles can form large-scale distributed energy storage resources through aggregation, participate in IDR projects, reduce the cost of peak shaving for the power grid, and improve the operational efficiency of the power system.

2.5 Vehicle-to-Grid Interaction in the Electricity Market

The electricity spot market typically consists of a day-ahead market and a real-time market. In the context of the electricity market, the market participation process of electric vehicle aggregators can be divided into two stages: day-ahead bidding and real-time charging regulation. During the day-ahead bidding stage, electric vehicle aggregators need to predict the charging demand of the electric vehicle cluster for the next day in order to formulate the optimal energy declaration strategy. During the real-time charging regulation stage, electric vehicle aggregators need to consider the charging demand of grid-connected electric vehicles and real-time electricity price signals to formulate the optimal energy management strategy. It is worth noting that when the deviation between the actual electricity consumption and the day-ahead bidding electricity exceeds a certain threshold, the independent system operator (market operator) will impose penalties on the electric vehicle aggregator [56].

In order to avoid unnecessary penalty costs, electric vehicle aggregators need to schedule the charging load of electric vehicles to track the day-ahead bidding curve, thereby achieving orderly grid-connected charging of electric vehicles.

The operational framework for implementing vehicle-to-grid interaction in a competitive electricity market environment is shown in Fig. 2.3. It can be seen that electric vehicle aggregators, as a special type of load aggregator, are key to electric vehicles participating in energy trading or ancillary services. In the day-ahead market, electric vehicle aggregators face uncertainties from both electric vehicles and the electricity market, including the mobility characteristics of electric vehicles, the charging demand of electric vehicles, day-ahead clearing prices, and real-time clearing prices. These uncertainties may lead to deviations between the day-ahead bidding electricity volume of electric vehicle aggregators and the actual electricity consumption of electric vehicles, resulting in high penalty costs [57]. In the real-time market, electric vehicle aggregators do not need to bid again and can optimize the scheduling of electric vehicle charging loads to achieve optimal market participation. Therefore, developing optimal day-ahead bidding and real-time charging regulation strategies is key to reducing the operating costs of electric vehicle aggregators. In addition, electric vehicle aggregators must integrate sufficient electric vehicle charging transaction capacity to meet the threshold for market participation [58]. Therefore, how to develop incentive strategies to obtain flexibility in electric vehicle charging is also a problem that electric vehicle aggregators need to consider. The following provides an overview of the market-oriented operational strategies of electric vehicle aggregators from three aspects: day-ahead bidding strategies, real-time charging regulation strategies for electric vehicles, and interaction strategies with electric vehicle users.

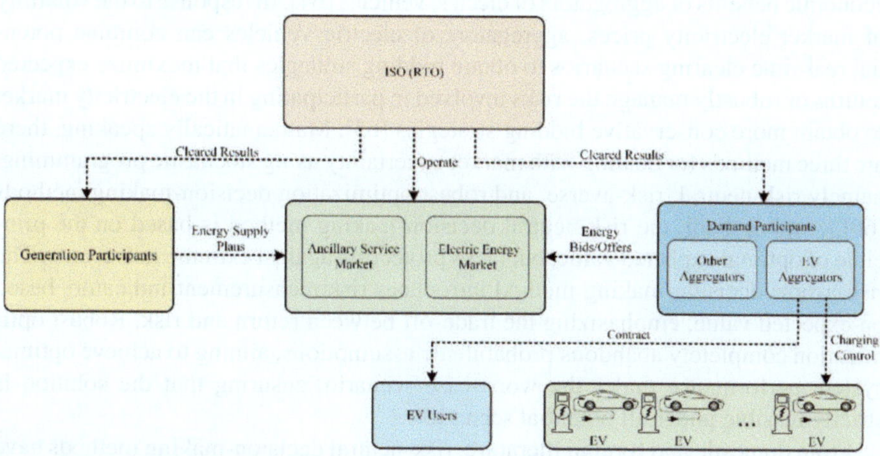

Fig. 2.3 Operating framework for coordinated interaction between EVs and power grids in competitive electricity markets

Day-ahead bidding strategy. According to the trading rules of the electricity market, electric vehicle aggregators need to bid for the charging demand of the next day in the day-ahead market. The day-ahead bidding process accounts for the majority of the operating costs of electric vehicle aggregators' market participation. However, the randomness of electric vehicle charging and the volatility of market electricity prices pose significant challenges to electric vehicle aggregators' day-ahead bidding [59]. As a means of transportation, electric vehicles exhibit strong randomness in both travel and charging behavior. Therefore, it is difficult to accurately obtain electric vehicle charging demand (including charging time and charging capacity) one day in advance [60]. In response to the uncertainty of electric vehicle charging, existing research work mainly adopts probabilistic modeling or scenario simulation methods. For example, literature [61] used collected electric vehicle charging data to probabilistically model electric vehicle users' charging behavior, and in the optimization model, uses probability distribution functions to describe the uncertainty of electric vehicle-related variables, thereby obtaining the optimal day-ahead bidding strategy for electric vehicle aggregators. Literature [62] accurately modelled the driving distance of electric vehicles to construct several charging scenarios for electric vehicles, providing a reference for electric vehicle aggregators' day-ahead bidding strategy. It is worth noting that when using simulation scenarios to model the uncertainty of electric vehicles, if the scenarios are not generated properly, the charging needs of electric vehicle users may not be met [63].

In addition to the random charging behavior of electric vehicles, aggregators of electric vehicles also need to consider the volatility of market electricity prices. During the day-ahead bidding process, aggregators of electric vehicles not only need to consider the day-ahead clearing price, but also need to consider the clearing results of the real-time market. This is because the real-time clearing price may also affect the economic benefits of aggregators of electric vehicles [64]. In response to the volatility of market electricity prices, aggregators of electric vehicles can combine potential real-time clearing scenarios to obtain bidding strategies that maximize expected returns or robustly manage the risks involved in participating in the electricity market to obtain more conservative bidding strategies [65]. Mathematically speaking, there are three methods for dealing with market uncertainty using stochastic programming, namely risk-neutral, risk-averse, and robust optimization decision-making methods [66]. Among them, the risk-neutral decision-making method is based on the principle of optimal expected value, but lacks protection against extreme risk events. The risk-averse decision-making method introduces risk measurement indicators based on expected value, emphasizing the trade-off between return and risk. Robust optimization completely abandons probabilistic assumptions, aiming to achieve optimal system performance under the worst-case scenario, ensuring that the solution is strictly feasible under all potential scenarios.

From domestic and foreign literature, risk-neutral decision-making methods have been widely applied in the day-ahead bidding problem of electric vehicle aggregators. For example, literature [67] uses a scenario-based method to maximize the expected profit of electric vehicle aggregators and bids for the charging demand of electric vehicle clusters under two contract models. The numerical results show that the proposed bidding strategy can not only meet the charging demand of electric vehicles, but also reduce the deviation between actual electricity consumption and day-ahead bidding electricity. The authors in [68] presented a two-stage linear stochastic programming model to formulate an optimal bidding strategy for electric vehicle aggregators to participate in the day-ahead energy and frequency modulation markets. The model systematically considers the uncertainties of the market and the uncertainties of the electric vehicle cluster. Empirical analysis shows that the proposed strategy can not only reduce the deviation between day-ahead bidding electricity and actual electricity consumption, but also ensure the revenue and reserve capacity supply of electric vehicle aggregators. Taking the LAAFB V2G demonstration project in the United States as the research object, literature [69] proposed a day-ahead optimization scheduling model for planning the charging scheduling and auxiliary service bidding strategy of electric vehicle clusters. The model aims to minimize the operating costs of electric vehicle aggregators and maximize the revenue obtained from participating in the CAISO frequency modulation market. This study shows that many potential factors, such as frequency modulation price, overall load level, and battery state of charge, significantly affect the revenue of electric vehicle load aggregators.

Existing research on using risk-averse decision-making methods for day-ahead bidding is very limited, mainly concentrated in foreign literature. For example, the work in [70] proposed a day-ahead bidding method based on conditional value at risk (CVaR)-constrained. This method uses conditional value at risk to constrain the risk costs caused by market electricity price volatility, and on this basis, constructs a mathematical model with the goal of minimizing the operating costs of electric vehicle aggregators. A case study based on the PJM market shows that the day-ahead bidding strategy of electric vehicle aggregators is influenced by factors such as day-ahead bidding prices, confidence levels, and penalty prices. The work in [71] proposed a mixed integer stochastic linear programming model aimed at maximizing the profits of electric vehicle aggregators while controlling their risks at an acceptable level. The research results show that using V2G technology can not only increase the profits of electric vehicle aggregators but also reduce carbon emissions from the power grid, achieving a win–win situation for both the power grid and electric vehicles. The authors in [72] proposed a decision-making tool based on risk constraints, aimed at supporting electric vehicle aggregators in developing optimal bidding strategies in day-ahead and frequency modulation markets. This decision-making tool comprehensively considers the real-time clearing electricity price, frequency modulation services, and uncertainties of electric vehicles, and adopts conditional value at risk to constrain potential risks. The results show that the proposed bidding strategy can not only significantly reduce the operational risks of electric vehicle aggregators in uncertain market environments, but also increase their earnings in the electricity energy and frequency modulation markets.

Compared with risk-neutral and risk-averse decision-making methods, the bidding method for electric vehicle aggregators based on robust optimization is more conservative. For example, the work in [73] established a robust optimization scheduling method considering electricity price uncertainty for the charging scheduling of electric vehicle aggregators. The research shows that compared with deterministic methods, robust optimization technology performs well in dealing with electricity price uncertainty, effectively reducing the loss risk in the worst case while improving the profits of electric vehicle aggregators in the remaining cases. The work in [74] proposed a collaborative bidding model for electric vehicle aggregators considering distribution network constraints, aiming to solve the collaborative optimization problem when multiple electric vehicle aggregators participate in the electricity market. The model uses a hybrid stochastic robust optimization method to model uncertainties such as day-ahead electricity prices, ancillary service prices, and real-time electricity prices. The model uses a distributed algorithm based on a standardized Nash equilibrium to solve the game between multiple electric vehicle aggregators. The study found that network congestion and the selection of robust control parameters have a significant impact on the bidding strategies and operating costs of electric vehicle aggregators. The authors in [63] proposed a decision-making method based on stochastic robust optimization to optimize the bidding strategies of electric vehicle aggregators in the day-ahead electricity market. This method comprehensively considers the uncertainties of day-ahead electricity prices and electric vehicle charging demand. Through case studies, it verifies the effectiveness of the proposed method in optimizing the charging costs of electric vehicle aggregators and improving the robustness of market participation, revealing that network congestion and robust parameters have a significant impact on the charging scheduling strategies of electric vehicle aggregators.

Through analysis of the above literature, it is found that day-ahead bidding methods under different risk preferences exhibit significant differences. The risk-neutral decision-making method takes expected return maximization as the objective function, but ignores the tail risk under extreme events, which may lead to huge losses for electric vehicle aggregators in certain extreme situations [75]. The robust optimization method can achieve optimal system performance in the worst-case scenario, but it ignores the overall distribution information of uncertain factors. This characteristic leads to overly conservative bidding strategies for electric vehicle aggregators, which may miss opportunities for high returns in actual bidding. The risk-averse decision-making method integrates the probability distribution of uncertain factors and constructs conditional value at risk by introducing confidence level parameters, thereby controlling the bidding risk of electric vehicle aggregators within a certain threshold [76]. Therefore, the risk-averse day-ahead bidding strategy does not appear to be overly conservative or overly aggressive.

Real-time charging regulation strategy for electric vehicles. In real-time operation, electric vehicle aggregators must respond quickly after receiving grid dispatch instructions. Due to the short response instruction cycle, electric vehicle aggregators need to quickly develop charging and discharging strategies based on grid signals and the state of charge of electric vehicles, and issue them to terminal execution units through a distributed control architecture. Therefore, the computational complexity of the proposed optimization scheduling algorithm is crucial for real-time energy management of electric vehicles [77]. In the framework of the electricity market, the time scale of electric vehicle charging scheduling is generally on the order of 15 min [78]. Therefore, electric vehicle aggregators must obtain optimized scheduling strategies for electric vehicle clusters within a minute-level time window. For auxiliary services such as frequency modulation and peak shaving, the response time of the system needs to reach the second level. The variables that electric vehicle aggregators need to optimize are closely related to the scale of electric vehicles, which places more stringent requirements on the computational complexity of charging optimization scheduling algorithms.

In the electricity market, aggregators of electric vehicles can maximize their economic benefits by regulating charging loads in real-time. For instance, the study in [79] introduced a quasi-real-time charging management approach for electric vehicles. This method aims to minimize discrepancies between the market's purchased electricity and the actual electricity usage of electric vehicles, thereby reducing market-related penalties. The results of a numerical example demonstrate that the proposed method meets the timeliness requirements of applications and shows significant effectiveness in optimizing electric vehicle charging and improving grid voltage. The work in [80] presented a mixed-integer linear programming model aimed at maximizing the profits of aggregators of electric vehicles, optimizing real-time charging strategies for these vehicles. However, solving this model can be challenging for large-scale electric vehicle charging planning problems. Therefore, this study proposes a heuristic algorithm based on linear programming rounding techniques to quickly obtain an approximately optimal solution for the model. The simulation results demonstrated that the proposed method yields higher profits for aggregators of electric vehicles compared to unregulated charging behaviors.

In the auxiliary service market, electric vehicle aggregators can increase their revenue by actively regulating the charging and discharging behavior of electric vehicles in response to grid auxiliary signals. For example, literature [81] proposed a three-tier real-time charging framework architecture. The first tier is the market participation of electric vehicle aggregators, the second tier is the five-minute-level charging power regulation, and the third tier is the response to the adjustment signals of system operators. The research results show that using dynamic adjustment signals can significantly improve the reliability of electric vehicle aggregators in market participation. The authors in [82] designed a real-time controller to regulate the charging and discharging of electric vehicles to provide auxiliary services for the grid. The controller uses a model predictive control algorithm to predict future states to optimize the charging and discharging behavior of electric vehicles, thereby improving the tracking accuracy and regulation capacity of AGC signals. Compared with other controllers of different complexities, the proposed real-time controller can significantly improve the accuracy of frequency modulation signals and reduce battery cycling.

Through analyzing the above literature, it is found that in the complex and dynamic environment of the electricity market, the computational complexity of the optimized charging scheduling algorithm constructed is crucial for real-time charging scheduling of electric vehicles. In real-time operation, electric vehicle aggregators need to develop optimal charging strategies for each electric vehicle in a very short time based on real-time electricity price signals and electric vehicle demand. The computational complexity of the algorithm directly determines its calculation speed and accuracy. If the computational complexity is too high, the algorithm will consume a lot of time and computing resources when dealing with large amounts of data and complex constraints, resulting in the inability to quickly provide charging regulation strategies in real-time operation, which will affect the market operation of electric vehicle aggregators. Therefore, reducing the computational complexity of real-time charging scheduling algorithms and improving their computational efficiency and accuracy are important challenges and key issues that need to be urgently addressed in the current electricity market environment for formulating real-time charging strategies.

Electric vehicle aggregator and electric vehicle interaction strategy. Currently, most work focuses on day-ahead bidding and real-time charging scheduling strategies for electric vehicle aggregators. Based on different optimization objectives, application scenarios, and market operating mechanisms, corresponding market operation strategies for electric vehicle aggregators have been designed. However, these research efforts do not fully consider the willingness of electric vehicle users to provide flexible services, but only regard electric vehicles as an adjustable and optimizable grid resource. As the party interacting between electric vehicles and the grid, the willingness of electric vehicle users to participate in electric energy trading or ancillary services is an important social attribute issue. In the framework of the electricity market, how electric vehicle aggregators balance the interests between electric vehicle users and themselves is one of the key issues in implementing vehicle-grid interaction. Therefore, it is necessary to explore methods to capture the charging flexibility of electric vehicles from the perspective of electric vehicle aggregators to meet the threshold for market participation.

Existing research mainly explores the interaction mechanism between electric vehicle aggregators and users based on a two-tier game architecture. The main results are summarized as follows. The work in [83] proposed a robust master–slave game model based on non-cooperative optimization and cooperative optimization to study the problem of electric vehicle charging scheduling. In this game model, a single electric vehicle aggregator is the leader, and multiple electric vehicle users are followers. A distributed algorithm is used to prove the existence and uniqueness of robust Nash equilibria under two optimization modes. Simulation experiments verify the effectiveness of the proposed robust optimization model in uncertain environments and compare the performance of non-cooperative games and cooperative games under different conditions. The work in [84] established a master–slave game model between multiple individual electric vehicles and their aggregators. In this model, the electric vehicle aggregator is the leader, responsible for setting the charging service price and providing the price of frequency modulation services to incentivize electric vehicle users; the electric vehicle users are followers, aiming to balance the charging cost and the benefits obtained from frequency modulation services. Case studies show that the proposed day-ahead charging scheduling and frequency modulation strategies can bring considerable economic benefits to electric vehicle aggregators and electric vehicle users under the uncertainty of frequency modulation signals. For the problem of interest coordination among grid operators, electric vehicle aggregators, and users, existing work adopts a multi-level optimization coordination method to study it. For example, the work in [85] proposed a three-tier vehicle-grid interaction framework consisting of the grid, electric vehicle aggregators, and electric vehicles. In this framework, a non-cooperative master–slave game model is established to obtain the optimal pricing of electric vehicle aggregators and the optimal charging and discharging strategies of electric vehicles, and a contract-based incentive mechanism is proposed to promote electric vehicles to provide their charging flexibility. Case study results show that the contract-based incentive mechanism can effectively guide electric vehicle users to participate in electricity trading while improving the operational efficiency of the grid.

Based on the analysis of the above literature, it can be seen that how to balance the interest relationship between electric vehicle aggregators and users is the key to enhancing the willingness of electric vehicle users to participate in electricity trading. However, there are differences in interest demands between electric vehicle aggregators and users. Specifically, electric vehicle aggregators pursue maximization of economic benefits, while electric vehicle users focus on the trade-off between economic compensation, battery wear and tear, and travel flexibility. This difference in interest inevitably leads to contradictions between the two parties in the interaction process. Therefore, the pursuit of the tradeoff between two parties is important to enhancing the willingness of electric vehicle users to participate in electricity trading.

2.6 Modeling of the Electric Power and Transportation Coupling Network and User Rationality Analysis

In recent years, with the acceleration of the electrification of transportation and the rapid increase in the penetration rate of electric vehicles, the research paradigm of power and transportation systems is evolving from traditional independent analysis to deep coupling. Among them, fast charging stations play an important role as key regulatory hubs. Scholars at home and abroad have conducted extensive interdisciplinary research on related issues of power-transportation coupled networks and have achieved fruitful results in theoretical models, algorithmic methods, and engineering applications. This paper focuses on the optimal scheduling of fast charging stations under the background of power-transportation coupling and user bounded rationality, covering key issues, e.g., coupled network modeling, user decision-making rationality characterization, dynamic pricing strategies, cluster scheduling and market participation, as well as multi-energy complementary collaboration with hydrogen and natural gas energy replenishment facilities.

The modeling of the power-transportation coupled network depicts the spatiotemporal distribution characteristics and interaction mechanisms of power flow and traffic flow, where the model of fast charging stations determines the interaction and coupling relationship between power flow and traffic flow [86]. Therefore, the modeling of the power-transportation coupled network is an important prerequisite for analyzing and predicting the spatiotemporal distribution of charging load and traffic flow, providing key model support for the optimal scheduling of fast charging stations and the collaborative planning and operation of coupled systems [87]. In the transportation network, although gasoline vehicles (GV) and electric vehicles without charging requirements do not generate power load, they compete and restrict with electric vehicles that require charging in terms of road resource occupation. Therefore, the traffic flow model in the power-transportation coupled network usually considers three basic vehicle types: electric vehicles with charging requirements, electric vehicles without charging requirements, and gasoline vehicles. On the other hand, users' travel and charging behavior patterns are closely related to their decision-making rationality. Against this background, scholars at home and abroad have conducted a large number of studies on the modeling of the power-transportation coupled network, and introduced relevant considerations of users' rational behavior into the modeling. The current modeling methods of the power-transportation coupled network can be mainly divided into three categories: single-time-period static modeling, multi-time-period static modeling, and dynamic modeling.

In the single-period static modeling method, the power flow distribution problem in the distribution network is usually modeled as a single-period optimal power flow model, including an alternating current optimal power flow (ACOPF) model [88] and a direct current optimal power flow (DCOPF) model [14]. The AC optimal power flow model has higher accuracy, while the DC optimal power flow model has higher computational efficiency. Both models can calculate the marginal electricity price of the node where the fast charging station is located and use it as the charging price, given the charging load. In the transportation network, the calculation of static traffic flow distribution is defined as a static traffic assignment (STA) problem [89], which ignores the dynamic propagation process of traffic flow moving through the network over time, but reflects the average steady-state distribution of traffic flow within a certain period, often assuming that the charging load is proportional to traffic flow. The dynamic propagation of traffic flow refers to the spatial displacement and time delay experienced by vehicles from the starting point to the end point, and the static traffic flow model cannot depict this dynamic characteristic and is only used to represent the traffic distribution under a steady state. In the single-period static model, users' travel decisions only include the selection of travel paths and fast charging stations. According to different allocation rules, the static traffic flow distribution model includes a single-period static user equilibrium (SUE) model [90] and a single-period static social optimum (SSO) model [91]. In the user equilibrium model, users aim to minimize their own travel costs and form a non-cooperative game with each other. When all users cannot further reduce costs by unilaterally changing their travel and charging schemes, traffic flow reaches an equilibrium state. In the social optimum model, it is assumed that users cooperate to minimize the total travel cost of the entire network, corresponding to the ideal traffic operation state. Due to the fact that users in reality usually do not have a complete willingness to cooperate, the user equilibrium model is more widely applied. Literature [92] pointed out that there is a mutual coupling relationship between the distribution network optimal power flow model and the transportation network static user equilibrium model, and their interaction ultimately leads to the entire system reaching a static network equilibrium (SNE) state, which can be solved by a fixed-point iteration algorithm. Therefore, the optimization problem in the power-transportation coupled network usually manifests as a mathematical program with equilibrium constraints (MPEC) problem with a bi-level structure, belonging to the NP-hard problem.

Single-period static modeling methods are difficult to effectively depict the dynamic evolution characteristics of power-transportation coupled networks in the short term, especially the spatiotemporal dynamic changes of key factors such as power flow, electricity price, traffic flow, and charging load. Therefore, such methods are mostly used for long-term planning problems of coupled networks. The authors in [93] proposed a mixed integer convex programming method to solve the expansion planning problem of power-transportation coupled networks, coordinately optimizing the configuration of key facilities such as engines, feeders, roads, and charging piles to enhance the system's adaptability to future travel and charging demand growth. The work in [94] relaxed the road capacity constraints in the transportation network and proposed a coordinated planning method for power-transportation coupled networks, simultaneously determining the optimal layout schemes for feeders, substations, lanes, and charging stations. When modeling the congestion characteristics of transportation network segments, both studies adopted the Nesterov & de Palma (NdP) model [95] to depict the functional relationship between travel time and traffic flow in conventional road segments and fast charging station segments, enabling the static user equilibrium problem to be transformed into a Linear Programming (LP) problem for solution. In the NdP model, road traffic flow must not exceed road capacity: when road traffic flow is less than road capacity, travel time is constant, representing a free-flow state; when road traffic flow equals road capacity, travel time is the free-flow travel time plus the dual variable of the road capacity constraint, representing a congested state. In addition, the BPR model [96] is also widely used for modeling the congestion characteristics of conventional road segments and fast charging station segments. It does not limit road traffic flow to be lower than road capacity, but assumes that road travel time is a power function of road traffic flow. Due to the absence of solving dual variables, the BPR model is simpler to calculate and more widely applied.

In addition, the work in [97] further explores the coordinated optimization problem of congestion pricing strategies for road segments and power generation scheduling strategies for distribution networks based on static network equilibrium, extending the optimal power flow model to the entire power-transportation coupled network and proposing the concept of optimal power-transportation flow. The work in [9] uses a DC optimal power flow model to depict the power network and obtain node marginal electricity prices, based on which an optimal location method for fast charging stations in the power-transportation coupled network is proposed. The above studies adopt the same modeling approach (NdP or BPR model) for conventional road segments and fast charging station segments, but the two types of models were originally designed mainly for estimating the travel time of conventional road segments, and may have certain deviations when directly applied to fast charging station segments. In addition, it is generally assumed in the research that charging load is proportional to traffic flow in fast charging station segments, and the proportionality coefficient is constant, which can easily cause errors in repeated calculations of charging load when traffic flow traverses multiple fast charging station segments. To solve the aforementioned problems, the work in [98] introduces the Davidson model [99] to independently model the congestion characteristics of fast charging station segments (conventional road segments still use the BPR model), and studies the optimal planning problem of a smart multi-energy system considering the coupling of active distribution networks, transportation networks, and natural gas pipeline networks. The Davidson model adopts a fractional function form, and when traffic flow approaches the service capacity limit, travel time will rapidly approach infinity, reflecting the high sensitivity of fast charging station segments to congestion. In addition, the work in [100] introduced the M/M/c/K queuing theory model [101] to separately depict the travel time function of fast charging segments, studies the deployment strategies of charging subsidies and transportation subsidies, and achieves the minimization of operation costs of power networks and transportation networks and promotes the reduction of carbon emissions. The M/M/c/K model assumes that each fast charging station is equipped with c charging piles, and allows a maximum of K electric vehicle users (including those who are charging and waiting in line) to enter at any given time period, fully reflecting the physical limitations of fast charging stations in terms of spatial capacity. Compared with the Davidson model, the M/M/c/K model exhibits a more gradual increase in travel time characteristics.

In the multi-period static modeling approach, the single-period optimal power flow model and the single-period static user equilibrium model are extended to multi-period forms to capture the time-varying characteristics of traffic flow and charging load across different periods. This method operates independently in each operational period, offering parallel computing capabilities and high computational efficiency. However, it also fails to consider the dynamic propagation characteristics of traffic flow.

Compared with single-period static modeling methods, multi-period static modeling methods can reveal the evolutionary laws of traffic flow and power flow in the temporal dimension, and are suitable for analyzing more complex dynamic time-varying scenarios. The work in [102] extended the concept of optimal power traffic flow to a multi-period static power traffic coupling network model, and proposes an optimal road congestion pricing and distribution network generation scheduling strategy based on mixed integer second-order cone programming (SOCP), to achieve optimal operation of the multi-period static power traffic coupling system. The work in [103] studied the collaborative pricing mechanism of retail electricity prices at fast charging stations and road congestion pricing, using price signals to guide the reasonable distribution of traffic flow in time and space, achieving joint minimization of distribution network loss and traffic network congestion, and evaluating the economic value of information sharing between distribution network operators (DNOs) and traffic network schedulers. The authors in [104] constructed a three-layer planning model for optimal expansion and operation management of key facilities in the power traffic coupling network, significantly enhancing the robustness and resilience of the coupled system to random natural disasters.

Considering the dynamic propagation characteristics of traffic flow, the multi-period static traffic network model will evolve into a more complex dynamic traffic network model, where the calculation of dynamic traffic flow is also known as the Dynamic Traffic Assignment (DTA) problem [105]. It can be used to track the real-time movement, congestion, intersection diversion, and charging stops of traffic flow from the starting point to the end point, reflecting a more realistic traffic operation state. Users' travel decisions not only include the selection of paths and fast charging stations, but also the planning of departure time, so this model can more flexibly depict users' behavioral characteristics. The distribution models of dynamic traffic flow include the Dynamic User Equilibrium (DUE) model [106] and the Dynamic Social Optimum (DSO) model [107], which have similar meanings to the static user equilibrium model and the static social optimum model, and are not repeated here. In the dynamic traffic network model considering users' departure time selection, there is no longer an explicit functional relationship between the travel time of road segments and the traffic flow of road segments, as in the static model. The road segment model is widely used to depict the physical propagation characteristics of traffic flow on the road (such as movement and queuing). Therefore, it is necessary to use the Dynamic Network Loading (DNL) method, with the traffic flow at the input nodes of the traffic network (path departure flow) as the boundary condition, to solve the differential equations (or discrete difference equations) corresponding to the dynamic propagation process of traffic flow, and then obtain the traffic flow evolution process and road segment travel time of the whole network at each time period. Common road segment models include the Vickrey model [108], the cellular transmission model [109], and the bidirectional wave model [110]. This paper defines the joint modeling method based on the multi-period optimal power flow model and dynamic traffic flow model as the dynamic modeling method of the electric power-transportation coupled network. Compared with the multi-period static modeling method, this method has higher accuracy, but the computational complexity and solution time also increase significantly.

Due to the difficulty in transforming dynamic traffic network models that consider user departure time choices into mathematical analytical models with closed feasible regions, traditional optimization methods cannot be directly applied to solve them, and research applications in power-transportation coupled networks are still limited. Literature [111] pointed out that there is also a stable interactive feedback mechanism between the multi-period optimal power flow model of the distribution network and the dynamic user equilibrium model of the transportation network, which is defined as the Dynamic Network Equilibrium (DNE) state and can be solved through fixed-point iteration methods. In addition, this work proposed a point queue model based on a difference equation system, modeling conventional road segments and fast charging road segments separately. This model can depict the queuing phenomenon caused by congestion and calculate the real-time queue length, the traffic flow entering the queue at the tail, and the traffic flow leaving the queue at the head. Its dynamic behavior is highly consistent with real-world queuing processes. The studies in [112, 113] studied the joint solution problem of the multi-period optimal power flow model and the dynamic user equilibrium model, but considering the complexity of modeling the dynamic propagation of traffic flow, these studies often use function approximation to simplify the calculation of road segment travel time, without explicitly modeling the propagation process of traffic flow in road segments. The study in [114] proposed a charging facility location model based on spatiotemporal captured traffic flow, which estimates the time-varying distribution of road network traffic flow using a dynamic user equilibrium model to determine the optimal layout scheme for slow charging, fast charging, and wireless charging facilities, but does not consider the bidirectional coupling effects between traffic flow and power flow.

Through analyzing the above literature, it is found that existing research provides important theoretical support and methodological references for modeling the coupled network of electric power and transportation at different scales. However, these studies generally assume that users have complete information perception capabilities when making travel decisions, are absolutely sensitive to travel costs, and can accurately predict the spatiotemporal evolution process of key factors such as congestion, queues, and charging prices in the network in advance. Therefore, users can accurately assess the travel costs corresponding to different travel schemes (including departure time, route, fast charging stations along the way), and choose the scheme with the lowest cost, thus achieving user equilibrium under non-cooperative game conditions. Under such assumptions, users are regarded as absolutely rational behavioral agents, that is, they can accurately perceive travel costs and always tend to choose the globally optimal scheme. However, in practical applications, the information obtained by users (such as through navigation software or historical experience) is often incomplete and may have certain deviations, making it difficult for them to accurately judge the true costs of various travel schemes.

When the actual travel costs of multiple options are similar, users often cannot distinguish which one has a lower cost. In addition, users usually have a tolerance for subtle cost differences. That is, they are not sensitive to small changes in cost, so they subjectively accept a certain degree of suboptimal options rather than always choosing the option with the lowest cost. Therefore, users' travel behavior in reality is more in line with the bounded rationality hypothesis. Current traffic flow modeling methods for bounded rational users mainly include three categories, i.e. cumulative prospect theory, cognitive theory and user tolerance method.

Cumulative Prospect Theory constructs a behavior model of user bounded rationality by introducing reference points (such as expected arrival time), an S-shaped value function, and an inverted S-shaped decision weight function. Under this framework, the proportion of users choosing a certain travel plan is positively correlated with the perceived utility of the plan, and this utility is influenced by users' subjective risk preferences. The work in [115] further integrated Prospect Theory into a dynamic user equilibrium model and proposed a bi-level programming model for charging station location selection. Although Prospect Theory can more accurately reflect user behavior, its introduction significantly increases model complexity. Additionally, due to the involvement of a large number of subjective parameters, it is difficult to obtain effective data in practical engineering, limiting its application and promotion.

Cognitive theory divides users into different rational levels based on their information processing capabilities: low-level users make travel decisions first, while high-level users make decisions based on observing the behavior of the former. The work in [116] adopted this theory to characterize the cognitive differences between pure electric vehicle users and hydrogen hybrid electric vehicle users when choosing charging stations and hydrogen refueling stations, and designed corresponding dynamic pricing strategies accordingly. However, such models are based on individual vehicle modeling rather than traffic flow, and therefore have high computational complexity when facing large-scale electric vehicle access scenarios.

The user tolerance method introduces a tolerance parameter based on the traditional user equilibrium model to characterize the user's acceptance of suboptimal solutions. Specifically, tolerance is defined as the maximum perceived cost difference between the suboptimal solution and the optimal solution that the user can accept. After introducing tolerance, the (single-period/multi-period) static user equilibrium model and dynamic user equilibrium model can be extended to the Boundedly Rational Static User Equilibrium (BRSUE) model [117] and Boundedly Rational Dynamic User Equilibrium (BRDUE) model [118], respectively. This method does not significantly increase the computational complexity of the model, while being more closely aligned with actual user behavior patterns. However, these bounded rational user equilibrium models are mostly used in traditional transportation research fields and have not fully considered the charging behavior of electric vehicles during travel, making it difficult to directly apply them to the modeling of electric power-transportation coupled networks. Further research and expansion are needed.

2.7 Charging Pricing Mechanism for EV Charging Stations

A reasonable retail charging pricing strategy is not only the key to ensuring the profitability of fast charging station operators, but also the price signal it transmits indirectly guides the spatial and temporal distribution of traffic flow and power flow in the power-transportation coupled network. This guidance can regulate users' charging behavior, achieve orderly charging, alleviate congestion problems in the distribution network, and avoid road congestion caused by traffic flow aggregation. Therefore, the pricing strategy of fast charging stations has become one of the important means to regulate the coordinated operation of the power-transportation coupled network [119, 120]. According to different traffic network modeling methods, retail pricing methods can be divided into two categories: static pricing and dynamic pricing. In static pricing strategies, the retail charging price is a constant value that does not change with time. While dynamic pricing strategies allow the electricity price to be flexibly adjusted according to different time periods, thereby enhancing the ability to regulate user behavior. On the other hand, according to different ways of determining electricity prices, retail charging pricing strategies can be further divided into three categories: fixed pricing method, node marginal electricity pricing method (also known as cost electricity pricing method), and optimized pricing method.

In the fixed pricing method, the retail charging electricity price is a constant directly set externally and remains unchanged during different charging periods. This method has the advantages of simple implementation and convenient operation. For example, in the two-stage robust optimization model proposed in literature [121] for collaborative expansion planning of active distribution networks and transportation networks, it is assumed that fast charging stations adopt a fixed electricity price, and multiple uncertain factors such as conventional load, new energy output, and transportation demand are considered. However, the price relied on by this method is artificially set and lacks the support of an optimization model, making it difficult to ensure maximum operational profitability. At the same time, because the electricity price remains consistent across all periods, the fixed pricing strategy cannot effectively regulate power flow and traffic flow through price signals.

The node marginal pricing method sets the retail electricity price based on the node marginal electricity price (or other electricity cost prices) of fast charging stations in the power network, usually multiplied by a pricing coefficient. Since the node marginal price can dynamically reflect the marginal cost of electricity used by fast charging stations, this pricing method has strong cost reflection ability: during peak load periods, the node marginal price increases, and the corresponding retail charging price also increases, thus suppressing peak-time charging demand and guiding some users to shift to off-peak periods, achieving peak shaving and valley filling of charging load. At the same time, the profit margin of charging station operators during peak periods is relatively higher, which helps to improve the overall income level. The authors in [122] proposed a pricing mechanism for coordinated management of electric vehicle charging and road congestion, which adjusts traffic flow from static user equilibrium to static social optimal state through price adjustment to minimize the operating cost of the coupled system, where the charging price is derived from the node marginal price in the DC optimal power flow model. The work in [123, 124] studied the short-term operation and pricing issues of wireless charging facilities in the power-transportation coupled network, also using the node marginal price based on the DC optimal power flow model as the charging price. The study in [125] adopted the same pricing method in the context of a pricing game among multiple charging station operators. In addition, the studies in [126, 127] further explored the coordinated pricing mechanism of charging price and road congestion fee, where the charging price is set based on the node marginal price in the AC optimal power flow model. Although the node marginal pricing method can better reflect the operating state of the distribution network and has a certain peak-valley regulation function, its core focus is on the operating efficiency of the distribution system rather than maximizing the profits of charging station operators. This method usually restricts the flexibility and autonomy of charging station operators in setting prices, making it difficult to fully utilize their pricing strategies in market competition. Therefore, it still has certain limitations in regions or application scenarios where the charging marketization mechanism is not well developed.

Optimization pricing method is a strategy to dynamically generate retail charging electricity prices by solving the mathematical optimization model for maximizing the profits of fast charging stations. This type of method is usually based on the Stackelberg game framework between fast charging station operators and electric vehicle users. The lower-level user decision-making behavior can be characterized by a single-period or multi-period static user equilibrium model, and further transformed into nonlinear complementary constraints, thereby simplifying the original bi-level optimization problem into a directly solvable single-level optimization model. Compared with fixed pricing and node marginal pricing methods, the optimization pricing method considers the response mechanism of electric vehicle users to prices, and can flexibly formulate differentiated pricing strategies based on actual charging demand, thus achieving the goal of maximizing the profits of fast charging stations or minimizing the operating costs of the electric power and transportation coupling system. The authors in [128] proposes a comprehensive demand response model under the electric power and transportation coupling network, studies the response mechanism of user travel and charging demand to charging prices and tolls, and covers both fixed charging facilities and road dynamic wireless charging facilities. The constructed bi-level collaborative pricing model, after nonlinear complementary constraints transformation and relaxation, can be iteratively solved as a nonlinear programming problem, effectively achieving system congestion management. The authors in [129] constructed a joint economic-emission dispatch model for the electric power and transportation coupling network, evaluated the carbon emission levels of electric vehicles and fuel vehicles through macro-emission modeling, and optimized the generation of differentiated charging prices and tolls to guide traffic and charging behavior, achieving low-carbon economic operation of the system.

The work in [130] also studied the collaborative optimization pricing problem and modeled it as a bi-level optimization problem, which was solved by using the manifold suboptimal algorithm and pattern search method, achieving dual optimization of distribution network loss and traffic time consumption. The work in [131] focused on the pricing and energy sharing problem of fast charging station clusters, using a relaxed iterative algorithm to solve the pricing model with complementary constraints. The literature [132] explored how to promote the collaborative operation of the electric power and transportation coupling system through the pricing strategy of fast charging stations, constructed a bi-level optimization model including uncertainties in wind power output and user travel demand, and proposed a solution framework based on deep reinforcement learning for efficiently approximating the optimal solution of the stochastic game model. The experiment verified the effect of reducing congestion and improving new energy consumption. The authors in [133] further constructed a three-level game pricing model, corresponding to the AC optimal power flow scheduling of multiple distribution networks, the profit maximization retail electricity pricing strategy of fast charging station operators, and the static travel equilibrium model of electric vehicle users. Among them, the node marginal electricity price was regarded as the wholesale electricity price for charging station electricity consumption, and the entire pricing game model was solved by iterative methods, verifying its advantages in improving profits, guiding load transfer across networks, and optimizing traffic efficiency. The work in [134] introduced an optimization iteration and simulated annealing algorithms to solve the optimal charging strategy.

In addition, the work in [135] proposed an incentive-compatible pricing and guidance mechanism for differentiated charging services to minimize the operating costs of the coupled network. The authors in [136] explored the optimal pricing model of charging stations under monopoly and duopoly market conditions, considering two charging service options: fast charging and slow charging. The authors in [137] introduced a location-based incentive mechanism in the optimized pricing model, using price signals to guide users to participate in demand response, improving the consumption rate of new energy generation. The work in [138] proposed two types of optimized pricing strategies: centralized and distributed, achieving peak shaving and valley filling of charging load through electricity price signals. Although the above research has achieved significant results, most optimization pricing methods are still based on static power-transportation coupling models with single or multiple time periods. For dynamic modeling methods that can better reflect the dynamic evolution characteristics of coupled networks, the internal dynamic user equilibrium model often introduces a large number of binary variables when converted into nonlinear complementary constraints, making the relaxed optimization problem difficult to solve effectively, limiting the application of such models in actual fast charging station pricing strategies. In addition, the introduction of nonlinear complementary constraints also makes the overall model computationally inefficient, and solutions for large-scale instances often rely on numerical iterative methods or data-driven machine learning methods for approximate solutions.

The pricing behavior of fast charging station operators falls within the scope of market-oriented operations, and their pricing strategies directly affect the overall operational status of the power-transportation coupling network. To ensure the orderly development of the charging market, the reasonable rights and interests of electric vehicle users, and the safe and stable operation of the coupling system, charging pricing behavior is usually supervised and constrained by third-party market regulators. In existing research, charging pricing regulation mainly includes price upper and lower limit constraints and price mean value constraints, and the regulatory targets are usually fast charging station operators who adopt optimization pricing methods. Among the above-mentioned optimization pricing-related research, only a small number of studies introduce price regulation mechanisms. However, such regulatory mechanisms can only indirectly regulate charging behavior at the price level, making it difficult to fundamentally limit the actual charging expenditure of electric vehicle users. Fast charging station operators can still maximize profits by adjusting the retail electricity prices during peak or low-peak periods, transferring electricity price risks to users to a certain extent. Especially in the scenario of monopoly operation by a single operator, electric vehicle users face a more significant risk of paying high charging fees.

Through analyzing the aforementioned literature, it can be found that existing research on the pricing of fast charging stations is generally based on the assumption that electric vehicle users have a fully rational response to electricity prices, that is, users can fully perceive price information and make cost-optimal decisions. However, in practical applications, users often exhibit limited rationality, with behavioral characteristics such as incomplete information acquisition, delayed decision-making, or psychological preferences. This means that pricing strategies formulated based on the assumption of absolute rationality may not achieve the expected profit optimization goals in practical applications, and it is difficult to effectively guide the operation trend of the coupled network. In addition, the regulatory mechanisms currently adopted in charging pricing research mostly restrict operators' pricing behavior through indirect constraints, which have significant deficiencies in ensuring user charging fairness and payment acceptability. Therefore, there is an urgent need to construct a new market regulatory mechanism with more comprehensive regulatory effects and the ability to directly regulate users' actual charging cost levels, to more effectively balance the contradiction between the profit goals of charging stations and the protection of user rights, and promote the coordinated and optimized operation of the electric power and transportation coupled network.

2.8 EV Charging Scheduling and Participation in the Electricity Market

In the power-transportation coupled network, the indirect guidance of a single price signal has become difficult to cope with operational risks such as voltage overruns and line congestion brought by high-penetration electric vehicles. Therefore, it is particularly important to combine direct power regulation strategies to achieve fine management of the spatial and temporal distribution of charging load. Among them, active control of charging power in charging station clusters is one of the core means to ensure the orderly integration of large-scale electric vehicles into the power grid. Charging station clusters participating in aggregation management usually have a large amount of flexible load that can be scheduled. These managers are usually defined as electric vehicle aggregators (EVAs) or charging load aggregators. On the premise of meeting the charging needs of electric vehicle users, electric vehicle aggregators have the qualifications and potential to participate in power market transactions. They can reduce the impact of charging load on the power grid through reasonable power bidding and charging scheduling strategies, thereby reducing wholesale electricity purchasing and consumption costs. At the same time, with the development of electric vehicle grid-connected technology (Vehicle-to-grid, V2G), electric vehicle aggregators can also utilize the mobile energy storage characteristics of electric vehicles to provide auxiliary services for the power grid and enhance the ability to consume new energy, achieving benign vehicle-grid interaction.

The electricity market can be divided into two types based on different levels of competition: the imperfect competition electricity market (also known as the retail electricity market) and the competitive electricity spot market. In the retail electricity market, users are mainly billed through time-of-use or real-time electricity pricing, but there is a lack of a mechanism for active user participation. In the competitive electricity spot market, users who meet the entry requirements first bid for electricity usage in various time periods the next day during the day-ahead electricity market stage. The electricity trading center calculates the clearing price (also known as the day-ahead electricity price) of the day-ahead market based on the bidding schemes of all users and power plants. The user's bidding cost will be settled based on this price for the first time. Then, in the real-time electricity market (balance market), users can adjust the actual electricity usage in various time periods according to the actual situation. The electricity trading center calculates the clearing electricity price (real-time electricity price) of the real-time market based on the real-time supply and demand balance. The difference between the user's real-time electricity usage and the day-ahead bidding electricity usage will be settled based on the real-time electricity price. If the difference is too large, the user will also need to pay additional fines. Due to the extensive participation and early declaration of electricity demand for the next day by users, including various load aggregators, the competitive electricity spot market has significant advantages in promoting energy structure transformation and improving energy utilization efficiency. In recent years, multiple countries and regions have begun to study and implement competitive electricity spot markets. China has launched a new round of electricity market reform since 2015 and has taken the lead in establishing a corresponding competitive electricity spot market in Guangdong Province. In 2023, the National Development and Reform Commission and the National Energy Administration jointly issued the "Basic Rules for Electricity Spot Market (Trial)", providing a guiding document for the operation and management of the electricity spot market. Against this backdrop, electric vehicle aggregators need to deal with operational uncertainties such as price differences between the two stages of the market and changes in charging demand when participating in competitive electricity spot market transactions, to ensure the orderly conduct of electricity transactions.

Based on the modeling of the power-transportation coupled network, the academic community has further explored the energy optimization and regulation methods of charging stations under the electric vehicle aggregator model, aiming to improve the economy and safety of the power-transportation coupled network through centralized charging scheduling and market transaction coordination. Specifically, the research on electric vehicle aggregators can be divided into slow charging scenarios (L1 and L2 level AC charging, with power below 19.2 kW and charging time of about 5 h) and fast charging scenarios (L3 level DC charging, with power ranging from 20 to 350 kW and charging time within 1 h) according to their scheduling objects [139].

Slow charging typically occurs in public parking lots or private residences at night. Although slow charging takes longer, in slow charging scenarios, the parking duration of electric vehicles is usually longer than the required charging time. Therefore, slow charging loads have high scheduling flexibility, which also leads to most research on electric vehicle aggregators focusing on slow charging scenarios. The work in [70] proposed a risk-averse electric vehicle aggregator bidding model considering electricity market price fluctuations, which minimizes the expected cost of purchasing electricity in the electricity market through charging scheduling. It considers the penalty cost for the deviation between the bidding electricity consumption in the day-ahead electricity market and the electricity consumption in the real-time electricity market, avoiding large power changes and arbitrage behavior of electric vehicle aggregators. In [140], a two-stage coordinated charging scheduling scheme is developed for electric vehicle aggregators participating in electricity market transactions, using stochastic programming methods to deal with the uncertainty of real-time electricity prices. Literature [141] studied the optimal bidding strategy of electric vehicle aggregators in the distribution network-level day-ahead electricity market, considering both the charging and discharging scheduling of electric vehicles, and proposes an incentive mechanism to encourage electric vehicle users to participate in aggregated scheduling. Literature [142] established a three-level robust optimization model to solve the profit maximization problem of electric vehicle aggregators in the day-ahead spinning reserve market, considering the uncertainty of electric vehicle users' access to the grid state. Literature [143] proposed a game-theoretic-based energy exchange method between electric vehicle aggregators and wind power operators in the day-ahead, real-time, and frequency modulation markets. Literature [144] studied the application of vehicle-grid interaction technology in stochastic security-constrained unit commitment problems, using aggregated electric vehicle fleets to promote the consumption of large-scale new energy generation, where the mobility of electric vehicles is modeled as a multi-period static user equilibrium model. Since slow charging mainly occurs in parking areas of residential or workplace locations, with relatively fixed charging time and location, and less correlation with traffic flow mobility, most research on electric vehicle aggregators in slow charging scenarios ignores the mobility of electric vehicles in the transportation network, and instead adopts probability-based methods (e.g., Gaussian distribution, Monte Carlo sampling or historical statistical data) to model the uncertainty of electric vehicles arriving and leaving charging stations.

Unlike slow charging at night, electric vehicle (EV) users who travel during the day usually choose faster public fast-charging stations. In the fast-charging mode of charging-and-go, the spatiotemporal distribution of charging demand is highly correlated with the mobility of EV traffic flow in the road network. The random movement characteristics of traffic flow make the uncertainty of charging demand more prominent, posing challenges to the charging scheduling of EV aggregators [145]. On the other hand, due to the more urgent demand for fast charging, EVs have a shorter dwell time, which reduces the flexibility of charging scheduling. Therefore, the application of EV aggregator charging scheduling and electricity trading strategies developed for slow-charging scenarios in fast-charging scenarios faces certain difficulties. To this end, some scholars have conducted research specifically on EV aggregators in fast-charging scenarios. Literature [146] proposed a collaborative charging scheduling and electricity purchasing strategy for multiple EV aggregators, allocating corresponding fast-charging stations based on the driving routes of EVs. It considers users' charging cancellation events and establishes corresponding penalty mechanisms, but does not consider the impact of uncertainty. Literature [147] constructed a comprehensive day-ahead scheduling framework that includes EVs, fast-charging stations, and electricity suppliers, considering the random travel trajectories and on-the-way charging behaviors of EV users. Literature [148] studied the joint bidding strategies of EV aggregators in the electricity energy market and spinning reserve ancillary market, where aggregators simultaneously take over EVs and on-site Energy Storage System (ESS) resources at fast-charging stations. To distinguish aggregators that only aggregate EVs, it defined EV aggregators that simultaneously aggregate on-site resources, e.g., on-site energy storage and new energy generation at fast-charging stations, as fast charging station aggregators.

Through analyzing the aforementioned studies, it is found that the charging scheduling and power market trading issues of electric vehicle aggregators in fast charging scenarios have not received sufficient attention. With the increase in the proportion of fast charging facilities, it is necessary to further carry out targeted research, especially on fast charging station aggregator models that integrate distributed renewable energy generation and energy storage resources. In addition, the access of large-scale electric vehicle clusters in fast charging scenarios poses higher requirements on the time granularity and computational performance of electric vehicle aggregator scheduling algorithms. However, most of the literature in the above research considers independent charging scheduling for individual electric vehicles, which has high accuracy but low computational efficiency, making it difficult to meet the scenario of large-scale electric vehicle access; while some research treats all electric vehicles as a virtual battery and regulates the charging and discharging behavior of all electric vehicles from an overall perspective, which introduces significant approximation errors. Finally, the above research on electric vehicle aggregators assumes that electric vehicle users are absolutely rational when calculating the spatial and temporal distribution of electric vehicle charging demand, but the impact of user bounded rationality on electric vehicle aggregators has not been deeply analyzed and discussed.

2.9 Multi-energy Dispatch and Electricity Trading in Integrated Energy Service Stations

In the construction of a green, low-carbon, and sustainable integrated transportation and energy system, hydrogen fuel cell vehicles (HFCVs) and natural gas vehicles (NGVs) also play an important role in reducing carbon emissions [149–151]. Hydrogen fuel cell vehicles have the advantage of zero carbon emissions, with combustion products only being water. They also have high energy density, long range, and short refueling time, making them have broad application prospects in long-distance transportation scenarios for heavy-duty vehicles [152]. Although natural gas vehicles produce some tailpipe emissions, their carbon emissions can be reduced by 27% compared to fuel vehicles, and the exhaust does not contain harmful substances such as benzene and lead. At the same time, they have the advantages of high economy and long engine life [153]. The coexistence of multiple energy-replenishment demand entities, e.g., electric vehicles, hydrogen fuel cell vehicles, and natural gas vehicles, has led to the evolution of the electric power transportation coupling network into an energy-transportation coupling network that includes multiple energy-replenishment demand entities and heterogeneous traffic flows. Among them, fast charging stations, hydrogen injection stations, and gas filling stations jointly constitute the coupling hub of the integrated energy network and transportation network, which is the key facility for regulating the coordinated operation of the coupling system.

Due to the relatively small number of hydrogen fuel cell vehicles and natural gas vehicles [154], many countries and regions have adopted the plan of combining fast charging stations, battery swapping stations, hydrogen refueling stations, gas filling stations, and gas stations to improve the efficiency of land resource utilization and the convenience of energy replenishment services [155, 156]. The combined stations are equipped with new energy power generation and energy storage facilities, further promoting carbon reduction in the transportation network, and constitute an integrated energy service station (IESS) that provides services such as charging, battery swapping, hydrogen refueling, gas filling, refueling, shopping, dining, leisure, etc., providing one-stop energy replenishment services for electric vehicles, hydrogen fuel cell vehicles, and natural gas vehicle users in a low-carbon transportation network.

Compared with traditional fast charging stations, the operation and optimization scheduling of integrated energy service stations are more complex. Their service targets have also expanded from single electric vehicle users to heterogeneous user groups, including electric vehicles, hydrogen fuel cell vehicles, and natural gas vehicles. Electric vehicles, hydrogen fuel cell vehicles, and natural gas vehicles have different energy replenishment characteristics, with hydrogen injection and gas filling significantly less time-consuming than charging. Under the same energy replenishment mode of charging and leaving immediately, the diverse energy replenishment needs of users are closely related to the mobility of traffic flows. The homogeneous traffic flow model is no longer suitable for modeling the mobility of vehicles with different energy replenishment characteristics in the traffic network, and heterogeneous traffic flow models are needed [157]. Correspondingly, the energy regulation and control objects of integrated energy service stations have shifted from single electric energy to electric-hydrogen-gas multi-energy coordination and complementarity, involving the coordination of various energy production, mutual conversion, storage, and sales processes. Electricity remains the main form of energy consumption in integrated energy service stations, while hydrogen and natural gas can be prepared through pipeline transportation or new energy power generation within the station (for example, hydrogen production from water electrolysis [158], and natural gas synthesis from hydrogen and carbon dioxide [159]).

At the same time, hydrogen and natural gas can be converted into electric energy through hydrogen fuel cells and gas turbines, involving various energy conversion processes such as Power-to-hydrogen (P2H), Hydrogen-to-natural-gas (H2G), Hydrogen-to-power (H2P), and Natural-gas-to-power (G2P). In public places with dense traffic flows, the access conditions for natural gas and hydrogen pipelines are relatively stringent and involve high construction costs. Therefore, using new energy power generation within the station to achieve on-site preparation of hydrogen and natural gas can play a certain role in pipeline substitution and promote the local consumption of new energy. Higher electricity demand makes integrated energy service station clusters also have the potential to aggregate and participate in electricity market transactions [160], but the Integrated Energy Service Station Aggregator (IESSA) needs to address the challenges brought by electric-hydrogen-gas multi-energy coordination scheduling and multi-dimensional operational uncertainty (e.g., EV users, electricity markets, new energy power generation) [161]. Therefore, it is necessary to extend the relevant models and algorithms for fast charging station aggregation energy optimization and regulation to the integrated energy service station scenario to ensure the economic and stable operation of integrated energy service station clusters and support the green, clean, and low-carbon construction of the traffic network [162]. In recent years, scholars at home and abroad have carried out a series of explorations and research on the energy optimization and regulation of integrated energy service stations. According to the types of integrated energy, relevant research mainly involves three types of multi-energy coordinated integrated energy service stations: electric-gas, electric-hydrogen, and electric-hydrogen-gas.

The literature [163] studied the energy optimization and regulation method of electric-gas multi-energy collaborative integrated energy service stations for electric vehicle and natural gas vehicle users, established a two-stage robust optimization model, and calculated the optimal planning and operation strategies for photovoltaic power generation, gas turbines, natural gas synthesis devices, battery energy storage, hydrogen storage tanks (Hydrogen Storage System, HSS), and other equipment to cope with uncertain user energy replenishment needs and photovoltaic output. In the above research, the integrated energy service station was simultaneously connected to the power grid and natural gas pipeline network, and the mobility and energy replenishment demand distribution of electric vehicles and natural gas vehicles in the road network were calculated based on traffic flow models, but the constraint of user equilibrium was not considered.

Existing literature has studied energy optimization and regulation methods for electric-hydrogen multi-energy collaborative integrated energy service stations for users of electric vehicles and hydrogen fuel cell vehicles. Literature [164] proposed a dynamic planning and energy regulation strategy for integrated energy service stations in highway scenarios, considering only the unidirectional energy conversion from electricity to hydrogen, and verifies the economic feasibility of using on-site new energy generation to prepare green hydrogen in the station. Literature [165] considered integrated energy service stations equipped with photovoltaic power generation, electrolyzers, hydrogen fuel cells, battery energy storage, hydrogen storage tanks, and other facilities, and constructs a bi-level game model between regional integrated energy systems and stations. In this model, the upper level is determined by the integrated energy system operator to determine the electricity-heat-gas-storage scheduling scheme and energy exchange price, while the lower level is optimized by the integrated energy service station operator to optimize the electricity-hydrogen-storage scheduling strategy, and uses distributed robust optimization methods to cope with the randomness of user demand for electric vehicles and hydrogen fuel cell vehicles. Literature [166] designed an off-grid integrated energy service station, and its optimal planning and energy regulation problems are modeled as stochastic programming problems, considering the uncertainty of user demand and photovoltaic output, to ensure the minimization of investment and operation costs of the integrated energy service station. Literature [167] studied the coordinated optimization problem of photovoltaic power generation units, charging units, and hydrogen injection units within the integrated energy service station, using model predictive control methods to reduce scheduling deviations caused by uncertainties in charging demand, hydrogen refueling demand, electricity prices, and photovoltaic output.

Also, the work in [168] proposed a day-ahead and real-time two-stage energy optimization and regulation method for grid-connected integrated energy service station clusters participating in electricity market transactions, also using model predictive control to cope with multi-dimensional operational uncertainties. However, the above studies did not consider the mobility modeling of electric vehicles and hydrogen fuel cell vehicles in the road network. In terms of absolute rationality, the travel decision-making and traffic flow distribution analysis of electric vehicle and hydrogen fuel cell vehicle users, literature [169] proposed a multi-source information-based energy demand prediction method in the optimal scheduling model of the integrated energy service station, predicting user preferences for selecting the integrated energy service station through a hybrid utility-regret decision model, assuming that users will choose the station with the lowest hybrid utility-regret value. The work in [170] used a single-period static user equilibrium model to characterize the mobility and distribution characteristics of heterogeneous traffic flows (including fuel vehicles, electric vehicles, and hydrogen fuel cell vehicles), providing support for the optimal operation of integrated energy service stations. The proposed energy optimization and regulation strategy enhances the coordinated operation of urban traffic networks and integrated energy service stations, significantly alleviates traffic congestion, and improves user and integrated energy service station operator satisfaction.

On this basis, further exploration was conducted on the energy optimization and regulation method for the electricity-hydrogen-gas multi-energy collaborative integrated energy service station that simultaneously serves electric vehicle, hydrogen fuel cell vehicle, and natural gas vehicle users. The work in [171] studied the model of an off-grid integrated energy service station powered by new energy and its risk scheduling strategy, which involved three types of energy conversion processes: electricity-to-hydrogen, hydrogen-to-natural gas, and hydrogen-to-electricity. Based on the optimization method of hybrid stochastic/information gap decision theory, scheduling strategies for risk aversion and risk pursuit were formulated for the uncertainty of energy replenishment demand of three types of heterogeneous users and the fluctuation of new energy output. However, it did not further explore the potential of integrated energy service stations to operate in parallel with the grid and participate in electricity market transactions. The authors in [172] proposed an economic scheduling method for integrated energy service stations, mainly involving the energy conversion process of electricity-to-hydrogen, while natural gas supply relied on pipeline transportation. In this study, the integrated energy service station was modeled as a microgrid, and energy exchange was conducted with the main grid based on real-time electricity prices. The energy replenishment demand prediction came from historical data clustering. Reference [173] established a microgrid model for electricity-hydrogen-gas integrated energy service stations considering local electricity/natural gas demand and vehicle energy replenishment demand, and proposed corresponding energy management methods that allowed bidirectional energy conversion between electricity, hydrogen, and natural gas subsystems. However, the carbon neutrality potential in the hydrogen-to-natural gas process has not been fully explored. Despite this, the above research did not consider the heterogeneous traffic flow mobility modeling problem of electric vehicle, hydrogen fuel cell vehicle, and natural gas vehicle users, and ignored the influence of uncertain factors such as user demand, electricity market prices, and new energy generation.

Based on the aforementioned literature, it is found that the operation of integrated energy service stations involves the mutual conversion of multiple energy forms and coordinated equipment scheduling, and its stability is closely related to the local new energy generation level. Due to the high energy loss in the energy conversion process, in some extreme scenarios (such as low user demand or insufficient new energy output), integrated energy service station operators/aggregators may face the risk of profit decline, especially when participating in two-stage electricity market transactions. Although existing research widely adopts stochastic programming methods to optimize expected returns, these methods are mainly based on maximizing average returns across all scenarios and fail to effectively address operational risks in extreme scenarios. In contrast, robust optimization methods can provide more robust scheduling strategies, but may lead to overly conservative scheduling schemes. In addition, user-differentiated energy replenishment needs are influenced by their travel behavior and decision-making rationality, but current research on the mobility and limited rational decision-making of electric vehicles, hydrogen fuel cell vehicles, and natural gas vehicles has not received sufficient attention. Existing research lacks a heterogeneous traffic flow model that can uniformly represent the travel and energy replenishment behaviors of three types of vehicle users.

References

1. Taylor J, Smith JW, Dugan R (2011) Distribution modeling requirements for integration of PV, PEV, and storage in a smart grid environment. In: 2011 IEEE power and energy society general meeting, pp 1–6
2. Pashajavid E, Shahnia F (2014) Sensitivity of electric vehicles demand profile to the batteries departure state-of-charge. In: 2014 Australasian universities power engineering conference (AUPEC), pp 1–6
3. Lojowska A, Kurowicka D, Papaefthymiou G et al (2012) Stochastic modeling of power demand due to EVs using copula. IEEE Trans Power Syst 27:1960–1968
4. Bae S, Kwasinski A (2012) Spatial and temporal model of electric vehicle charging demand. IEEE Trans Smart Grid 3:394–403
5. Mu Y, Wu J, Jenkins N et al (2014) A spatial–temporal model for grid impact analysis of plug-in electric vehicles. Appl Energy 114:456–465

6. Johnson BJ, Starke MR, Dimitrovski AD (2015) Examining the potential impact of plug-in electric vehicles on residential sector power demand. In: 2015 IEEE power and energy society general meeting, pp 1–5
7. Sundstroem O, Corradi O, Binding C (2012) Toward electric vehicle trip prediction for a charging service provider. In: 2012 IEEE international electric vehicle conference, pp 1–6
8. Chiş A, Lundén J, Koivunen V (2013) Scheduling of plug-in electric vehicle battery charging with price prediction. In: IEEE PES ISGT Europe 2013, pp 1–5
9. He F, Wu D, Yin Y et al (2013) Optimal deployment of public charging stations for plug-in hybrid electric vehicles. Transp Res Part B Methodol 47:87–101
10. Kong C, Bayram IS, Devetsikiotis M (2015) Revenue optimization frameworks for multi-class PEV charging stations. IEEE Access 3:2140–2150
11. Li G, Zhang XP (2012) Modeling of plug-in hybrid electric vehicle charging demand in probabilistic power flow calculations. IEEE Trans Smart Grid 3:492–499
12. Feng L, Ge S, Liu H et al (2012) The planning of charging stations on the urban trunk road. In: IEEE PES innovative smart grid technologies, pp 1–4
13. Ismail M, Bayram IS, Abdallah M et al (2015) Optimal planning of fast PEV charging facilities. In: 2015 First workshop on smart grid and renewable energy (SGRE), pp 1–6
14. Zhang H, Moura SJ, Hu Z et al (2018) PEV fast-charging station siting and sizing on coupled transportation and power networks. IEEE Trans Smart Grid 9:2595–2605
15. Zhang H, Hu Z, Xu Z et al (2016) An integrated planning framework for different types of PEV charging facilities in urban area. IEEE Trans Smart Grid 7:2273–2284
16. Zhang H, Hu Z, Xu Z et al (2017) Optimal planning of PEV charging station with single output multiple cables charging spots. IEEE Trans Smart Grid 8:2119–2128
17. Mohsenzadeh A, Pang C, Pazouki S et al (2015) Optimal siting and sizing of electric vehicle public charging stations considering smart distribution network reliability. In: 2015 North American power symposium (NAPS), pp 1–6
18. Sadeghi-Barzani P, Rajabi-Ghahnavieh A, Kazemi-Karegar H (2014) Optimal fast charging station placing and sizing. Appl Energy 125:289–299
19. Liu Z, Wen F, Ledwich G (2013) Optimal planning of electric-vehicle charging stations in distribution systems. IEEE Trans Power Deliv 28:102–110
20. Shaaban MF, El-Saadany EF (2014) Accommodating high penetrations of PEVs and renewable DG considering uncertainties in distribution systems. IEEE Trans Power Syst 29:259–270
21. Moradi MH, Abedini M, Tousi SMR et al (2015) Optimal siting and sizing of renewable energy sources and charging stations simultaneously based on differential evolution algorithm. Int J Electr Power Energy Syst 73:1015–1024
22. Amini MH, Moghaddam MP, Karabasoglu O (2017) Simultaneous allocation of electric vehicles' parking lots and distributed renewable resources in smart power distribution networks. Sustain Cities Soc 28:332–342
23. Erdinç O, Taşcıkaraoğlu A, Paterakis NG et al (2018) Comprehensive optimization model for sizing and siting of DG units, EV charging stations, and energy storage systems. IEEE Trans Smart Grid 9:3871–3882
24. de Quevedo PM, Muñoz-Delgado G, Contreras J (2017) Impact of electric vehicles on the expansion planning of distribution systems considering renewable energy, storage and charging stations. IEEE Trans Smart Grid
25. Pazouki S, Mohsenzadeh A, Ardalan S et al (2015) Simultaneous planning of PEV charging stations and DGs considering financial, technical, and environmental effects. Can J Elect Comput Eng 38:238–245
26. Zhang H, Moura SJ, Hu Z et al (2018) Joint PEV charging network and distributed PV generation planning based on accelerated generalized benders decomposition. IEEE Trans Transp Electrif 4:789–803
27. Kempton W, Tomic J, Letendre S et al (2001) Vehicle-to-grid power: battery, hybrid, and fuel cell vehicles as resources for distributed electric power in California. Institute of Transportation Studies, UC Davis

28. Quinn C, Zimmerle D, Bradley TH (2010) The effect of communication architecture on the availability, reliability, and economics of plug-in hybrid electric vehicle-to-grid ancillary services. J Power Sources 195:1500–1509
29. Zhang P, Qian K, Zhou C et al (2012) A methodology for optimization of power systems demand due to electric vehicle charging load. IEEE Trans Power Syst 27:1628–1636
30. Sortomme E, Hindi MM, MacPherson SDJ et al (2011) Coordinated charging of plug-in hybrid electric vehicles to minimize distribution system losses. IEEE Trans Smart Grid 2:198–205
31. Wang R, Wang P, Xiao G (2016) Two-stage mechanism for massive electric vehicle charging involving renewable energy. IEEE Trans Veh Technol 65:4159–4171
32. Dallinger D, Krampe D, Wietschel M (2011) Vehicle-to-grid regulation reserves based on a dynamic simulation of mobility behavior. IEEE Trans Smart Grid 2:302–313
33. Zhao J, Wan C, Xu Z et al (2017) Spinning reserve requirement optimization considering integration of plug-in electric vehicles. IEEE Trans Smart Grid 8:2009–2021
34. Karfopoulos EL, Panourgias KA, Hatziargyriou ND (2016) Distributed coordination of electric vehicles providing V2G regulation services. IEEE Trans Power Syst 31:2834–2846
35. Alam MJE, Muttaqi KM, Sutanto D (2016) Effective utilization of available PEV battery capacity for mitigation of solar PV impact and grid support with integrated V2G functionality. IEEE Trans Smart Grid 7:1562–1571
36. Li J, Li C, Xu Y et al (2018) Noncooperative game-based distributed charging control for plug-in electric vehicles in distribution networks. IEEE Trans Ind Inform 14:301–310
37. Yu R, Ding J, Zhong W et al (2014) PHEV charging and discharging cooperation in V2G networks: a coalition game approach. IEEE Internet Things J 1:578–589
38. Guo D, Zhou C Realistic modeling of vehicle-to-grid in an enterprise parking lot: a Stackelberg game approach. In: 2018 IEEE Texas power and energy conference (TPEC), pp 1–6
39. Pourramezan A, Samadi M (2022) A novel approach for incorporating incentive-based and price-based demand response programs in long-term generation investment planning. Int J Electr Power Energy Syst 142:108315
40. Ma Q, Li R, Li F (2020) Differential customer responsiveness to time-of-use tariffs. In: 2020 IEEE power and energy society general meeting (PESGM), pp 1–5
41. Safdarian A, Fotuhi-Firuzabad M, Lehtonen M (2014) A distributed algorithm for managing residential demand response in smart grids. IEEE Trans Ind Inform 10:2385–2393
42. Waraich RA, Galus MD, Dobler C et al (2013) Plug-in hybrid electric vehicles and smart grids: investigations based on a microsimulation. Transp Res Part C Emerg Technol 28:74–86
43. Yang J, Xu Y, Yang Z (2016) Regulating the collective charging load of electric taxi fleet via real-time pricing. IEEE Trans Power Syst 32:3694–3703
44. Smart Electric Power Alliance (2019) Residential electric vehicle rates that work: attributes that increase enrollment
45. Hledik R, Faruqui A, Warner C (2017) The national landscape of residential TOU rates. The Brattle Group
46. Suyono H, Rahman MT, Mokhlis H et al (2019) Optimal scheduling of plug-in electric vehicle charging including time-of-use tariff to minimize cost and system stress. Energies 12:1500
47. Almutairi A, Salama MMA (2018) Assessment and enhancement frameworks for system reliability performance using different PEV charging models. IEEE Trans Sustain Energy 9:1969–1984
48. Yu M, Jiang J, Ye X et al (2022) Demand response flexibility potential trading in smart grids: a multileader multifollower Stackelberg game approach. IEEE Trans Syst Man Cybern Syst 53:2664–2675
49. Nguyen DT, Negnevitsky M, de Groot M (2013) Market-based demand response scheduling in a deregulated environment. IEEE Trans Smart Grid 4:1948–1956
50. Gkatzikis L, Koutsopoulos I, Salonidis T (2013) The role of aggregators in smart grid demand response markets. IEEE J Sel Area Commun 31:1247–1257
51. Shafie-Khah M, Heydarian-Forushani E, Osório GJ et al (2015) Optimal behavior of electric vehicle parking lots as demand response aggregation agents. IEEE Trans Smart Grid 7:2654–2665

52. Nezamoddini N, Wang Y (2016) Risk management and participation planning of electric vehicles in smart grids for demand response. Energy 116:836–850
53. Park SW, Cho KS, Hoefter G et al (2022) Electric vehicle charging management using location-based incentives for reducing renewable energy curtailment considering the distribution system. Appl Energy 305:117680
54. Yao L, Lim WH, Tsai TS (2016) A real-time charging scheme for demand response in electric vehicle parking station. IEEE Trans Smart Grid 8:52–62
55. Sarabi S, Davigny A, Courtecuisse V et al (2016) Potential of vehicle-to-grid ancillary services considering the uncertainties in plug-in electric vehicle availability and service/localization limitations in distribution grids. Appl Energy 171:523–540
56. Sarker MR, Dvorkin Y, Ortega-Vazquez MA (2015) Optimal participation of an electric vehicle aggregator in day-ahead energy and reserve markets. IEEE Trans Power Syst 31:3506–3515
57. Fallah-Mehrjardi O, Yaghmaee MH, Leon-Garcia A (2020) Charge scheduling of electric vehicles in smart parking-lot under future demands uncertainty. IEEE Trans Smart Grid 11:4949–4959
58. Vayá MG, Andersson G (2014) Optimal bidding strategy of a plug-in electric vehicle aggregator in day-ahead electricity markets under uncertainty. IEEE Trans Power Syst 30:2375–2385
59. Wang Q, Huang C, Qiu P et al (2022) A stochastic bilevel bidding strategy for electric vehicle charging station in day-ahead electricity market considering time uncertainty. In: 2022 IEEE/IAS industrial and commercial power system Asia (I&CPS Asia), pp 1900–1905. IEEE
60. Hemmatpour MH, Koochi MHR, Dehghanian P et al (2022) Voltage and energy control in distribution systems in the presence of flexible loads considering coordinated charging of electric vehicles. Energy 239:121880
61. Zheng Y, Yu H, Shao Z et al (2020) Day-ahead bidding strategy for electric vehicle aggregator enabling multiple agent modes in uncertain electricity markets. Appl Energy 280:115977
62. Takahashi T, Tamura S (2020) Day-ahead planning for EV aggregators based on statistical analysis of road traffic data in Japan. In: 2020 International conference on smart grids and energy systems (SGES), pp 117–122. IEEE
63. Baringo L, Amaro RS (2017) A stochastic robust optimization approach for the bidding strategy of an electric vehicle aggregator. Electr Power Syst Res 146:362–370
64. Wu H, Shahidehpour M, Alabdulwahab A et al (2015) A game theoretic approach to risk-based optimal bidding strategies for electric vehicle aggregators in electricity markets with variable wind energy resources. IEEE Trans Sustain Energy 7:374–385
65. Donadee J, Ilić MD (2014) Stochastic optimization of grid to vehicle frequency regulation capacity bids. IEEE Trans Smart Grid 5:1061–1069
66. Zheng Y, Wang Y, Yang Q (2023) Bidding strategy design for electric vehicle aggregators in the day-ahead electricity market considering price volatility: a risk-averse approach. Energy 283:129138
67. Zheng Y, Wang Y, Yang Q (2023) Two-phase operation for coordinated charging of electric vehicles in a market environment: from electric vehicle aggregators' perspective. Renew Sustain Energy Rev 171:113006
68. Vagropoulos SI, Bakirtzis AG (2013) Optimal bidding strategy for electric vehicle aggregators in electricity markets. IEEE Trans Power Syst 28:4031–4041
69. DeForest N, MacDonald JS, Black DR (2018) Day ahead optimization of an electric vehicle fleet providing ancillary services in the Los Angeles Air Force Base vehicle-to-grid demonstration. Appl Energy 210:987–1001
70. Yang H, Zhang S, Qiu J et al (2017) CVaR-constrained optimal bidding of electric vehicle aggregators in day-ahead and real-time markets. IEEE Trans Ind Inform 13:2555–2565
71. Al-Awami AT, Sortomme E (2011) Coordinating vehicle-to-grid services with energy trading. IEEE Trans Smart Grid 3:453–462
72. Habibifar R, Lekvan AA, Ehsan M (2020) A risk-constrained decision support tool for EV aggregators participating in energy and frequency regulation markets. Electr Power Syst Res 185:106367

73. Cao Y, Huang L, Li Y et al (2020) Optimal scheduling of electric vehicles aggregator under market price uncertainty using robust optimization technique. Int J Electr Power Energy Syst 117:105628
74. Moghaddam SZ, Akbari T (2018) Network-constrained optimal bidding strategy of a plug-in electric vehicle aggregator: a stochastic/robust game theoretic approach. Energy 151:478–489
75. Rockafellar RT, Uryasev S (2000) Optimization of conditional value-at-risk. J Risk 2:21–42
76. Afzali P, Rashidinejad M, Abdollahi A et al (2020) Risk-constrained bidding strategy for demand response, green energy resources, and plug-in electric vehicle in a flexible smart grid. IEEE Syst J 15:338–345
77. Xu B, Dvorkin Y, Kirschen DS et al (2016) A comparison of policies on the participation of storage in US frequency regulation markets. In: 2016 IEEE power and energy society general meeting (PESGM), pp 1–5
78. Zaneti LAL, Arias NB, de Almeida MC et al (2022) Sustainable charging schedule of electric buses in a university campus: a rolling horizon approach. Renew Sustain Energy Rev 161:112276
79. Soares FJ, Almeida PMR, Lopes JAP (2014) Quasi-real-time management of electric vehicles charging. Electr Power Syst Res 108:293–303
80. Jin C, Tang J, Ghosh P (2013) Optimizing electric vehicle charging with energy storage in the electricity market. IEEE Trans Smart Grid 4:311–320
81. Vagropoulos SI, Kyriazidis DK, Bakirtzis AG (2015) Real-time charging management framework for electric vehicle aggregators in a market environment. IEEE Trans Smart Grid 7:948–957
82. Wenzel G, Negrete-Pincetic M, Olivares DE et al (2017) Real-time charging strategies for an electric vehicle aggregator to provide ancillary services. IEEE Trans Smart Grid 9:5141–5151
83. Yang H, Xie X, Vasilakos AV (2015) Noncooperative and cooperative optimization of electric vehicle charging under demand uncertainty: a robust Stackelberg game. IEEE Trans Veh Technol 65:1043–1058
84. Cui Y, Hu Z, Luo H (2020) Optimal day-ahead charging and frequency reserve scheduling of electric vehicles considering the regulation signal uncertainty. IEEE Trans Ind Appl 56:5824–5835
85. Jember AG, Xu W, Pan C et al (2020) Game and contract theory-based energy transaction management for the internet of electric vehicle. IEEE Access 8:203478–203487
86. Chen Y, Hu S, Zheng Y et al (2024) Coordinated expansion planning of coupled power and transportation networks considering dynamic network equilibrium. Appl Energy 360:1–17
87. Yuan Q, Ye Y, Tang Y et al (2022) A novel deep-learning based surrogate modeling of stochastic electric vehicle traffic user equilibrium in low-carbon electricity-transportation nexus. Appl Energy 315:1–10
88. Shao C, Qian T, Wang Y et al (2021) Coordinated planning of extreme fast charging stations and power distribution networks considering on-site storage. IEEE Trans Intell Transp Syst 22:493–504
89. Chudak FA (2007) Static traffic assignment problem. A comparison between Beckmann (1956) and Nesterov & de Palma (1998) models. In: 7th Swiss transport research conference
90. Wardrop JG (1952) Some theoretical aspects of road traffic research. Proc Inst Civ Eng 1:325–362
91. Moradipari A, Tucker N, Alizadeh M (2021) Mobility-aware electric vehicle fast charging load models with geographical price variations. IEEE Trans Transp Electrif 7:554–565
92. Wei W, Wu L, Wang J et al (2018) Network equilibrium of coupled transportation and power distribution systems. IEEE Trans Smart Grid 9:6764–6779
93. Wei W, Wu L, Wang J et al (2017) Expansion planning of urban electrified transportation networks: a mixed-integer convex programming approach. IEEE Trans Transp Electrif 3:210–224
94. Wang X, Shahidehpour M, Jiang C et al (2019) Coordinated planning strategy for electric vehicle charging stations and coupled traffic-electric networks. IEEE Trans Power Syst 34:268–279

95. Nesterov Y, de Palma A (2003) Stationary dynamic solutions in congested transportation networks: summary and perspectives. Netw Spat Econ 3:371–395
96. United States. Bureau of Public Roads (1964) Traffic assignment manual for application with a large, high speed computer. United States
97. Wei W, Mei S, Wu L et al (2017) Optimal traffic-power flow in urban electrified transportation networks. IEEE Trans Smart Grid 8:84–95
98. Xie S, Hu Z, Wang J et al (2020) The optimal planning of smart multi-energy systems incorporating transportation, natural gas and active distribution networks. Appl Energy 269:1–20
99. Davidson KB (1966) A flow-travel time relationship for use in transportation planning. In: 3rd Australian Road Research Board (ARRB) conference, pp 183–194
100. Lv S, Chen S, Wei Z (2023) Coordinating urban power-traffic networks: a subsidy-based Nash-Stackelberg-Nash game model. IEEE Trans Ind Inform 19:1778–1790
101. Liu Y, Wang Y, Li Y et al (2021) Multi-agent based optimal scheduling and trading for multi-microgrids integrated with urban transportation networks. IEEE Trans Power Syst 36:2197–2210
102. Lv S, Wei Z, Sun G et al (2020) Optimal power and semi-dynamic traffic flow in urban electrified transportation networks. IEEE Trans Smart Grid 11:1854–1865
103. Sheng Y, Guo Q, Chen F et al (2021) Coordinated pricing of coupled urban power-traffic networks: the value of information sharing. Appl Energy 301:1–16
104. Gan W, Shahidehpour M, Guo J et al (2022) A tri-level planning approach to resilient expansion and hardening of coupled power distribution and transportation systems. IEEE Trans Power Syst 37:1495–1507
105. Wang Y, Szeto WY, Han K et al (2018) Dynamic traffic assignment: a review of the methodological advances for environmentally sustainable road transportation applications. Transp Res Part B Methodol 111:370–394
106. Friesz TL, Bernstein D, Smith TE et al (1993) A variational inequality formulation of the dynamic network user equilibrium problem. Oper Res 41:179–191
107. Tran CQ, Keyvan-Ekbatani M, Ngoduy D et al (2022) Dynamic wireless charging lanes location model in urban networks considering route choices. Transp Res Part C Emerg Technol 139:1–15
108. Han K, Friesz TL, Yao T (2013) A partial differential equation formulation of Vickrey's bottleneck model, part I: methodology and theoretical analysis. Transp Res Part B Methodol 49:55–74
109. Ukkusuri SV, Han L, Doan K (2012) Dynamic user equilibrium with a path based cell transmission model for general traffic networks. Transp Res Part B Methodol 46:1657–1684
110. Han K, Benedetto P, Szeto WY (2016) Continuous-time link-based kinematic wave model: formulation, solution existence, and well-posedness. Transportmetrica B 4:187–222
111. Xie S, Xu Y, Zheng X (2022) On dynamic network equilibrium of a coupled power and transportation network. IEEE Trans Smart Grid 13:1398–1411
112. Zhou Z, Zhang X, Guo Q et al (2021) Analyzing power and dynamic traffic flows in coupled power and transportation networks. Renew Sustain Energy Rev 135:1–12
113. Sun G, Li G, Xia S et al (2020) ALADIN-based coordinated operation of power distribution and traffic networks with electric vehicles. IEEE Trans Ind Appl 56:5944–5954
114. Abdalrahman A, Zhuang W (2019) PEV charging infrastructure siting based on spatial-temporal traffic flow distribution. IEEE Trans Smart Grid 10:6115–6125
115. Qiu H, Dou S, Shang H et al (2021) Charging station planning based on the accumulation prospect theory and dynamic user equilibrium. Complex Intell Syst 9:2521–2539
116. Lai S, Qiu J, Tao Y et al (2023) Pricing strategy for energy supplement services of hybrid electric vehicles considering bounded-rationality and energy substitution effect. IEEE Trans Smart Grid 14:2973–2985
117. Di X, Liu HX, Pang J et al (2013) Boundedly rational user equilibria (BRUE): mathematical formulation and solution sets. Transp Res Part B Methodol 57:300–313

118. Han K, Szeto WY, Friesz TL (2015) Formulation, existence, and computation of boundedly rational dynamic user equilibrium with fixed or endogenous user tolerance. Transp Res Part B Methodol 79:16–49
119. Zhao T, Yan H, Liu X et al (2022) Congestion-aware dynamic optimal traffic power flow in coupled transportation power systems. IEEE Trans Ind Inform 19:1833–1843
120. Yang H, Xu Y, Guo Q (2024) Dynamic incentive pricing on charging stations for real-time congestion management in distribution network: an adaptive model-based safe deep reinforcement learning method. IEEE Trans Sustain Energy 15:1100–1113
121. Xie S, Hu Z, Wang J (2020) Two-stage robust optimization for expansion planning of active distribution systems coupled with urban transportation networks. Appl Energy 261:1–17
122. Alizadeh M, Wai H-T, Chowdhury M et al (2017) Optimal pricing to manage electric vehicles in coupled power and transportation networks. IEEE Trans Control Netw Syst 4:863–875
123. Manshadi SD, Khodayar ME, Abdelghany K et al (2018) Wireless charging of electric vehicles in electricity and transportation networks. IEEE Trans Smart Grid 9:4503–4512
124. He F, Yin Y, Zhou J (2013) Integrated pricing of roads and electricity enabled by wireless power transfer. Transp Res Part C Emerg Technol 34:1–15
125. Alizadeh M, Wai H-T, Goldsmith A et al (2019) Retail and wholesale electricity pricing considering electric vehicle mobility. IEEE Trans Control Netw Syst 6:249–260
126. Zhou Z, Zhang X, Guo Q et al (2020) Integrated pricing framework for optimal power and semi-dynamic traffic flow problem. IET Renew Power Gener 14:3636–3643
127. Xie S, Wu Q, Hatziargyriou ND et al (2022) Collaborative pricing in a power-transportation coupled network: a variational inequality approach. IEEE Trans Power Syst
128. Lv S, Wei Z, Chen S et al (2021) Integrated demand response for congestion alleviation in coupled power and transportation networks. Appl Energy 283:1–12
129. Lv S, Chen S, Wei Z et al (2022) Power-transportation coordination: toward a hybrid economic-emission dispatch model. IEEE Trans Power Syst 37:3969–3981
130. He F, Yin Y, Wang J et al (2013) Sustainability SI: optimal prices of electricity at public charging stations for plug-in electric vehicles. Netw Spat Econ 16:131–154
131. Lv S, Sun G, Wei Z et al (2024) Optimal pricing and energy sharing of EV charging stations with an augmented user equilibrium model. IEEE Trans Power Syst 39:4336–4349
132. Qian T, Shao C, Li X et al (2020) Enhanced coordinated operations of electric power and transportation networks via EV charging services. IEEE Trans Smart Grid 11:3019–3030
133. Cui Y, Hu Z, Duan X (2021) Optimal pricing of public electric vehicle charging stations considering operations of coupled transportation and power systems. IEEE Trans Smart Grid 12:3278–3288
134. Sohet B, Hayel Y, Beaude O et al (2021) Hierarchical coupled driving-and-charging model of electric vehicles, stations and grid operators. IEEE Trans Smart Grid 12:5146–5157
135. Moradipari A, Alizadeh M (2020) Pricing and routing mechanisms for differentiated services in an electric vehicle public charging station network. IEEE Trans Smart Grid 11:1489–1499
136. Zhang Y, Zhou Y, Jiang C et al (2021) Plug-in electric vehicle charging with multiple charging options: a systematic analysis of service providers' pricing strategies. IEEE Trans Smart Grid 12:524–537
137. Park S, Cho K, Hoefter G et al (2022) Electric vehicle charging management using location-based incentives for reducing renewable energy curtailment considering the distribution system. Appl Energy 305:1–13
138. Kapoor A, Patel VS, Sharma A et al (2022) Centralized and decentralized pricing strategies for optimal scheduling of electric vehicles. IEEE Trans Smart Grid 13:2234–2244
139. Acharige SSG, Haque ME, Arif MT et al (2023) Review of electric vehicle charging technologies, standards, architectures, and converter configurations. IEEE Access 11:41218–41255
140. Zheng Y, Wang Y, Yang Q (2023) Two-phase operation for coordinated charging of electric vehicles in a market environment: from electric vehicle aggregators' perspective. Renew Sustain Energy Rev 171:1–14
141. Wang M, Li X, Dong C et al (2023) Day-ahead optimal bidding for a retailer with flexible participation of electric vehicles. IEEE Trans Smart Grid 14:1482–1494

142. Liu W, Chen S, Hou Y et al (2022) Trilevel mixed integer optimization for day-ahead spinning reserve management of electric vehicle aggregator with uncertainty. IEEE Trans Smart Grid 13:613–625
143. Shojaabadi S, Talavat V, Galvani S (2022) A game theory-based price bidding strategy for electric vehicle aggregators in the presence of wind power producers. Renew Energy 193:407–417
144. Sun Y, Chen Z, Li Z et al (2019) EV charging schedule in coupled constrained networks of transportation and power system. IEEE Trans Smart Grid 10:4706–4716
145. Chen Y, Zheng Y, Hu S et al (2024) Optimal operation of fast charging station aggregator in uncertain electricity markets considering onsite renewable energy and bounded EV user rationality. IEEE Trans Ind Inform 20:13384–13395
146. Gupta V, Konda SR, Kumar R et al (2018) Multiaggregator collaborative electric vehicle charge scheduling under variable energy purchase and EV cancelation events. IEEE Trans Ind Inform 14:2894–2902
147. Tookanlou MB, Kani SAP, Marzband M (2021) A comprehensive day-ahead scheduling strategy for electric vehicles operation. Int J Electr Power Energy Syst 131:1–17
148. Duan X, Hu Z, Song Y (2021) Bidding strategies in energy and reserve markets for an aggregator of multiple EV fast charging stations with battery storage. IEEE Trans Intell Transp Syst 22:471–482
149. Al Wahedi A, Bicer Y (2020) Development of an off-grid electrical vehicle charging station hybridized with renewables including battery cooling system and multiple energy storage units. Energy Rep 6:2006–2021
150. Cui P, Zhang J, Liu Y et al (2023) Comprehensive analysis of clean fuel vehicle life cycle environment under multiple fuel scenarios. Energy 275:1–9
151. Ruffini E, Wei M (2018) Future costs of fuel cell electric vehicles in California using a learning rate approach. Energy 150:329–341
152. Watabe A, Leaver J, Ishida H et al (2019) Impact of low emissions vehicles on reducing greenhouse gas emissions in Japan. Energy Policy 130:227–242
153. Peterson MB, Barter GE, West TH et al (2014) A parametric study of light-duty natural gas vehicle competitiveness in the United States through 2050. Appl Energy 125:206–217
154. Sun K, Li K, Zhang Z et al (2022) An integration scheme of renewable energies, hydrogen plant, and logistics center in the suburban power grid. IEEE Trans Ind Appl 58:2771–2779
155. Najafi A, Homaee O, Jasinski M et al (2023) Integrating hydrogen technology into active distribution networks: the case of private hydrogen refueling stations. Energy 278:1–12
156. Xu C, Wu Y, Dai S (2020) What are the critical barriers to the development of hydrogen refueling stations in China? A modified fuzzy DEMATEL approach. Energy Policy 142:1–14
157. Chen Y, Zheng Y, Hu S et al (2024) Risk-averse energy dispatch for hybrid energy refueling stations considering boundedly rational mixed user equilibrium and operational uncertainties. Appl Energy 376:1–18
158. Bhandari R, Shah RR (2021) Hydrogen as energy carrier: techno-economic assessment of decentralized hydrogen production in Germany. Renew Energy 177:915–931
159. Jiang J, Zhang L, Wen X et al (2022) Risk-based performance of power-to-gas storage technology integrated with energy hub system regarding downside risk constrained approach. Int J Hydrog Energy 47:39429–39442
160. Tang O, Rehme J, Cerin P (2022) Levelized cost of hydrogen for refueling stations with solar PV and wind in Sweden: on-grid or off-grid? Energy 241:1–14
161. Mansouri SA, Nematbakhsh E, Jordehi AR et al (2023) An interval-based nested optimization framework for deriving flexibility from smart buildings and electric vehicle fleets in the TSO-DSO coordination. Appl Energy 341:1–28
162. Tabar VS, Tohidi S, Ghassemzadeh S (2023) Risk-constrained day-ahead planning of an energy hub integrated with the on-site hydrogen fueling station and active battery swapping infrastructure considering high level of renewable energies and load redistribution. J Energy Storage 72:1–17

163. Gan W, Shahidehpour M, Guo J et al (2021) Two-stage planning of network-constrained hybrid energy supply stations for electric and natural gas vehicles. IEEE Trans Smart Grid 12:2013–2026
164. Shi M, Huang Y (2023) Dynamic planning and energy management strategy of integrated charging and hydrogen refueling at highway energy supply stations considering on-site green hydrogen production. Int J Hydrog Energy 48:29835–29851
165. Cai P, Mi Y, Ma S et al (2023) Hierarchical game for integrated energy system and electricity-hydrogen hybrid charging station under distributionally robust optimization. Energy 283:1–15
166. Mehrjerdi H (2019) Off-grid solar powered charging station for electric and hydrogen vehicles including fuel cell and hydrogen storage. Int J Hydrog Energy 44:11574–11583
167. Zhang J, Sun K, Li C et al (2023) MPC-based co-optimization of an integrated PV-EV-hydrogen station to reduce network loss and meet EV charging demand. eTransportation 15:1–9
168. Fang X, Wang Y, Dong W et al (2023) Optimal energy management of multiple electricity-hydrogen integrated charging stations. Energy 262:1–11
169. Zheng W, Li J, Shao Z et al (2023) Optimal dispatch of hydrogen/electric vehicle charging station based on charging decision prediction. Int J Hydrog Energy 48:26964–26978
170. Zhang C, Chen Z, Guo Q (2024) Research on optimal operation strategy of regional multi-heterogeneous energy charging station considering bidirectional satisfaction and traffic network coupling. Sustain Energy Grids Netw 37:1–9
171. Xu X, Hu W, Liu W et al (2021) Risk-based scheduling of an off-grid hybrid electricity/hydrogen/gas/refueling station powered by renewable energy. J Clean Prod 315:1–15
172. Zhou S, Zhuang Y, Gu W et al (2018) Operation and economic assessment of hybrid refueling station considering traffic flow information. Energies 11:1–20
173. Tostado-Véliz M, Arévalo P, Jurado F (2021) A comprehensive electrical-gas-hydrogen microgrid model for energy management applications. Energy Convers Manag 228:1–13

Chapter 3
Research Challenges and Book Organization

Abstract This chapter provides a comprehensive exploration of the critical challenges and advanced solutions associated with the optimal planning, scheduling, and market participation of plug-in electric vehicles (PEVs) within active distribution networks and competitive electricity markets, while also addressing the emerging complexities of power-transportation coupling. The research is motivated by the significant uncertainties in PEV charging behaviors, which pose threats to grid safety and economics. The book is organized to first tackle these foundational issues through three primary research domains. Initially, it addresses the need for accurate spatiotemporal charging demand forecasting and the coordinated planning of charging facilities with distributed energy sources, proposing novel methodologies like time-electricity quantity analysis and Markov models to capture the stochastic nature of PEV usage. Subsequently, the focus shifts to the operational strategies of electric vehicle aggregators in electricity markets, where it delves into risk-averse bidding strategies, the design of agency models to secure charging flexibility from users, and the management of multi-entity competition, utilizing frameworks like Stackelberg games and conditional value at risk to balance profitability and uncertainty. Finally, the book explores the cutting-edge field of fast charging station optimization within coupled power-transportation networks, explicitly incorporating users' bounded rationality into traffic flow and charging behavior models. It introduces innovative concepts such as bounded rationality dynamic user equilibrium models and three-layer game-theoretic pricing strategies to guide station operations and enhance grid interaction. Furthermore, the scope extends to the coordinated operation of multi-energy integrated service stations, proposing risk-averse dispatch methods for clusters offering electric, hydrogen, and gas refueling services. Through these interconnected parts, the book systematically develops and validates a suite of analytical, game-theoretic, and stochastic optimization methods, ultimately providing a robust theoretical and practical framework for enabling a secure, efficient, and economically viable integration of electric vehicles into future low-carbon, sustainable energy and transportation systems.

Q. Yang et al., *Integration of Massive Electric Vehicles to Smart Grid*, Smart Energy Systems, https://doi.org/10.1007/978-981-92-0557-8_3

3.1 Research Challenges

Firstly, for the research field of optimal planning and scheduling of plug-in electric vehicles in active distribution networks, although electric vehicles have outstanding advantages in energy conservation and environmental protection, their driving and charging needs are influenced by various factors, e.g., as user habits and subjective willingness, resulting in significant uncertainty in their charging load in both time and space scales. This will bring many thorny issues to the safe and economic operation of the power grid, as well as the construction of electric vehicle charging facilities. Therefore, it is necessary to accurately predict the charging demand of large-scale electric vehicles, understand the spatiotemporal characteristics of electric vehicle charging load, so that power grid operators can estimate the charging load of electric vehicles in advance and prevent potential risks it may bring. On one hand, by formulating a reasonable guidance and scheduling mechanism, the charging load of electric vehicles can be optimized and scheduled, so that electric vehicles can be charged in an orderly manner, thereby avoiding various negative effects caused by the disorderly charging of electric vehicles. Even grid-connected electric vehicles can be used as a power source to feed electricity back to the grid, thereby reducing users' charging costs and improving the operation status of the grid. On the other hand, by reasonably planning the charging facilities for electric vehicles and considering various factors, e.g., transportation system and power grid topologies and demand, convenient and economical charging services for electric vehicles can be provided with the assurance of the safety and economic operation of the power grid. Only by simultaneously solving the charging scheduling strategy and planning of charging facilities for electric vehicles can we increase the penetration rate of electric vehicles, address the problems brought by large-scale electric vehicles, and maximize the advantages of electric vehicles.

(1) Electric vehicle charging demand forecasting: existing literature still has some shortcomings in predicting the charging load of electric vehicles: (1) The vast majority of literature only considers the scenario of electric vehicles charging at home after a day's journey, i.e. the conventional charging load. If the daily driving range of the electric vehicle exceeds its maximum range, emergency charging at a fast charging station is needed to replenish electricity. Therefore, the fast charging load of electric vehicles is also an important component of electric vehicle charging, but it has not been considered in detail or even involved in most literature; (2) Most of the existing literature used survey data, e.g., travel time and daily mileage of electric vehicles, as benchmarks for predicting their charging behavior. However, these data cannot calculate whether electric vehicles have a demand for fast charging. Only by considering the changes in battery power of electric vehicles throughout the day can the fast charging load of electric vehicles be accurately obtained. This work has not yet been seen in existing literature, and finally, (3) there is currently no literature on the transition between various states of large-scale electric vehicle operation or charging, nor is there any literature reporting on the intrinsic relationship between various

states of electric vehicles and battery capacity. At present, the only literature that uses Markov models to describe the travel and charging behavior of electric vehicles has not yet fully and systematically addressed the modeling of electric vehicle charging behavior.

(2) Planning of electric vehicle charging facilities: There are still some limitations and areas that need improvement in the existing literature on the collaborative planning of electric vehicle charging facilities and distributed power sources: (1) The vast majority of literature only considers one constraint of the distribution network or transportation network, but electric vehicle charging facilities are both a public transportation infrastructure and a power load. When planning, attention should be paid to their two different attributes, and the coupling relationship between the distribution network and transportation network needs to be considered simultaneously, so that the planning results can meet multiple constraints of the distribution network and transportation network; (2) considering that electric vehicles have two different charging modes with conventional and fast charging characteristics, and their corresponding charging facilities are also different, it is necessary to plan two different charging facilities simultaneously to meet the different charging needs of electric vehicles. However, most current literature only considers one of these charging modes. Thirdly, with the increasing penetration rate of distributed power sources, distributed power sources can provide clean energy for electric vehicles, and electric vehicles can also improve the utilization efficiency of renewable energy. Therefore, it is necessary to coordinate the planning of electric vehicle charging facilities and distributed power sources. However, the literature reported so far has not yet established a systematic and complete theory and model of collaborative planning; and finally (3) if we need to simultaneously consider the location and capacity of electric vehicle charging facilities and distributed power sources, as well as the various constraints and uncertainties of the coupling between the distribution network and the transportation network, it will make the problem an extremely complex and difficult to solve optimization problem, and the complexity and time required for its solution will exponentially increase.

(3) Scheduling strategy of electric vehicle charging and discharging: There are still some limitations and areas that need improvement in the existing literature on the scheduling of electric vehicle charging and discharging behavior: (1) Most literature only considers the conventional charging scheduling of electric vehicles, and generally focuses on scheduling the conventional charging load of electric vehicles on a time scale, while there is little literature on the fast charging load of electric vehicles; (2) Most of the existing scheduling methods are based on recent scheduling strategies. As the penetration rate of distributed power sources gradually increases, their uncertainty will seriously affect the accuracy of scheduling methods, and high-precision real-time optimization scheduling strategies are still relatively rare; and finally, (3) most literature generally considers the collaboration between electric vehicles and traditional units or distributed power sources for V2G to provide auxiliary services. After the large-scale application of energy storage technology in power grid frequency

regulation and peak shaving, it is necessary to establish a more realistic mathematical model for frequency regulation and peak shaving; and finally, (4) the current research on applying game theory to V2G strategies for electric vehicles still simplifies many practical situations and has not yet been able to fully describe the actual situation of providing auxiliary services for V2G in electric vehicles.

Secondly, for the research field of optimal operational and charging scheduling strategy of electric vehicle aggregators in competitive electricity markets, existing research has proposed numerous solutions for the bidding strategies and charging scheduling strategies of electric vehicle aggregators. However, most of the work is based on the premise that electric vehicle aggregators have obtained the authority to regulate electric vehicle charging. Under the framework of the electricity market, there is a lack of in-depth research on the interest relationship between electric vehicle aggregators and electric vehicle users. As a key player in vehicle network interaction, obtaining charging flexibility for electric vehicles is also one of the key issues that electric vehicle aggregators need to consider. In addition, electric vehicle aggregators also face many uncertain factors in the process of participating in electricity spot trading, which will bring huge challenges to the market operation of electric vehicle aggregators. Based on the literature review, the following challenges need to be addressed.

(1) Adaptability of market rules for electric vehicle aggregators: under the framework of the spot market, electric vehicle aggregators must adapt to the trading rules of the electricity market, ensure effective coordination between bidding and real-time charging regulation, and improve the economic efficiency of electric vehicle participation in electricity trading. Unlike other load aggregators, electric vehicle aggregators need to deal with uncertainties from both the electric vehicle and electricity markets in order to develop a day-ahead bidding strategy. In addition, during the real-time charging scheduling phase, electric vehicle aggregators need to respond quickly based on real-time clearing electricity prices and electric vehicle charging demands to reduce their operating costs.
(2) Risk management in bidding for electric vehicle aggregators: unlike current electricity prices, real-time market-clearing prices often have significant volatility. If the goal of minimizing expected costs alone is not sufficient to cope with the high uncertainty of real-time market-clearing results, electric vehicle aggregators may suffer huge losses in certain extreme real-time clearing scenarios. Therefore, electric vehicle aggregators must consider multiple potential real-time clearance scenarios during the bidding stage and be able to control their operational risks within a certain range of confidence.
(3) Interaction between electric vehicle aggregators and EV users: the willingness of electric vehicle users to participate in electricity trading is also one of the issues that need to be considered when implementing vehicle network interaction. Due to differences in charging preferences among electric vehicle users, a single agency model may not be able to meet the demands of large-scale electric

vehicle users to participate in the electricity market. Therefore, based on the actual charging scenarios of electric vehicle users, a cooperation model between electric vehicle aggregators and electric vehicle users that meets different user needs can be designed, and a reasonable benefit distribution mechanism between the two parties can be studied on this basis.

(4) Multi-entity competition among electric vehicle aggregators: In a completely open and competitive market environment, when the dispatchable electric vehicle resources are limited, there may be scenarios where multiple electric vehicle aggregators compete with each other. The game mode of one master and multiple followers is no longer applicable to this scenario. In the case of multiple proxy entities competing with each other, electric vehicle aggregators not only need to consider the interaction relationship with electric vehicle users, but also need to balance the interests of other independent aggregators. This will bring greater challenges to the bidding and charging pricing of electric vehicle aggregators.

Thirdly, for the research field of optimization dispatch methods for fast charging stations considering power-transportation coupling and users' bounded rationality, although existing research has conducted in-depth exploration on core issues such as model construction, price setting, power regulation, online transactions, and multi energy collaboration of fast charging stations, and provided important theoretical support and methodological references for the optimization and scheduling of fast charging stations in electric transportation coupling networks, there are still several technical bottlenecks that urgently need to be further overcome. The following challenges are considered worth further research effort:

(1) Electric transportation coupling system model: most existing studies are based on absolute rational single-period static transportation network models, which are difficult to accurately characterize the spatiotemporal dynamic behavior of electric vehicle users during their travel and charging processes. Therefore, it is necessary to develop a multi-period static or dynamic traffic network model that considers user bounded rationality, to better characterize the behavioral characteristics of electric vehicle users and provide more accurate data support for estimating the spatiotemporal distribution of charging demand. This improvement will contribute to the overall optimization of the power transportation coupling network and the efficient operation of fast charging stations. In addition, the modeling of fast charging stations directly affects the coupling characteristics between traffic flow and power flow, but current charging station models based on road segment functions are difficult to accurately describe the conversion process between traffic flow and charging load. Therefore, it is necessary to design an accurate model that can comprehensively reflect the entire charging process of electric vehicle traffic flow, including entry, queuing, charging, and departure.

(2) Pricing-guided optimal scheduling of fast charging stations: Currently, although fixed pricing and node marginal pricing methods are simple and feasible, they limit the autonomous pricing space of fast charging station operators and

cannot fully utilize the guiding role of charging price signals in the spatial and temporal distribution of traffic flow and power flow. The existing optimization pricing methods assume that electric vehicle users have absolute rationality in their response to charging prices. However, this assumption may result in retail charging price schemes that fail to achieve expected profits or effectively improve the operation of the power transportation coupling network. Therefore, it is necessary to conduct an in-depth analysis and exploration of the impact of user bounded rationality on the optimization pricing of fast charging stations. In existing research, market regulatory constraints mainly focus on regulating electricity prices, without effectively regulating the charging fees of electric vehicle users. Therefore, it is necessary to establish a more comprehensive and direct regulatory mechanism that not only guarantees the profits of fast charging station operators but also protects the reasonable rights and interests of electric vehicle users in the monopolistic charging market.

(3) Optimal energy scheduling and power regulation of fast charging stations: the charging scheduling and online power trading issues of fast charging station clusters have not received sufficient attention, especially the aggregator model of fast charging stations with distributed new energy generation and energy storage resources still needs further research. This type of model should handle the impact of multiple uncertain factors, e.g., user demand, the electricity market, and new energy generation. In addition, with the expansion of electric vehicle clusters, the requirements for time granularity and computational performance of fast charging station charging power regulation algorithms will become more stringent. The charging scheduling algorithm for individual vehicles and overall vehicle groups still has shortcomings in terms of computational efficiency and accuracy. When estimating users' next-day travel and charging needs, fast charging station aggregators also need to consider the impact of users' bounded rationality.

(4) Multi-energy coordinated operation in fast charging stations, comprehensive energy service stations that couple charging, hydrogen injection, and refueling facilities face challenges such as diversified energy forms, diverse service targets, and multidimensional operational risks. The operation of integrated energy service stations covers various aspects of energy production, conversion, storage, and sales. Addressing the differences in heterogeneous energy flow coupling mechanisms and achieving multi-energy synergy and complementarity among electricity, hydrogen, and gas is a key prerequisite for managing the orderly integration of integrated energy service station clusters into the grid and participating in electricity market transactions. However, widely used stochastic programming methods cannot effectively cope with operational risks in extreme scenarios, and robust optimization methods may lead to overly conservative scheduling strategies. In addition, the differentiated energy replenishment needs of heterogeneous vehicle users are closely related to their travel behavior and decision-making rationality. However, there is currently a lack of a bounded rational heterogeneous traffic flow model that can simultaneously model the

travel and energy replenishment behavior of electric vehicle, hydrogen fuel cell vehicle, and natural gas vehicle users.

3.2 Book Organization

This book contains a total of 14 chapters, and the content of the chapters is briefly described as follows:

Chapter 1: Introduction. This chapter introduced and elucidated the current status of electric vehicle development and its impact on the power and transportation systems. In addition, this chapter analyzed the significance of constructing electric vehicle charging infrastructure and optimizing the scheduling of fast charging stations in building a green, low-carbon, and sustainable integrated transportation and energy system.

Chapter 2: State-of-the-art literature review. This chapter summarizes the current research status at home and abroad, covering the optimization planning and scheduling of electric vehicle network cooperation, the operation mechanism and charging optimization scheduling of electric vehicle aggregators in a competitive electricity market environment, and the latest research progress in the field of fast charging station optimization scheduling considering power transportation coupling and user bounded rationality.

This chapter: Research challenges and book organization. This chapter discussed the limitations of the existing research and briefly described the content of this book.

Chapter 4: Analytical analysis for PEV charging demand characterization. This chapter studies the analysis and prediction methods for large-scale electric vehicle travel and charging demand, proposes a quantitative analysis method for electric vehicle travel characteristics and charging demand based on time electricity quantity, as well as a Markov model for electric vehicles and an invisible Markov model for electric vehicles considering battery power. By utilizing historical data and Monte Carlo simulation methods, we conducted an in-depth analysis and summary of the inherent patterns of changes in the travel and charging status of electric vehicles throughout the day. This enabled us to accurately predict the charging load and battery power distribution of electric vehicles, as well as evaluate the negative impact that large-scale disorderly charging of electric vehicles will have on the power grid.

Chapter 5: Optimal PEV charging demand scheduling in active power distribution systems. This chapter studies the scheduling strategy of electric vehicle charging behavior in active distribution networks and proposes a time–space segmentation scheduling strategy. Based on real-time information in the intelligent charging scheduling system, while ensuring the satisfaction of electric vehicle charging needs, electric vehicles are scheduled to perform regular charging in the optimal time interval on a time scale to achieve peak shaving and valley filling effects. In addition, at the spatial scale, a driving path planning method considering fast charging for electric vehicle users has been proposed, which can schedule electric vehicles to the optimal charging station for fast charging while ensuring their travel needs.

Chapter 6: Optimal participation of electric vehicle load aggregators in demand response programs. In response to the complex spot trading rules and uncertain market trading environment, this chapter proposes a two-stage electric vehicle cluster charging regulation method and a market operation strategy for electric vehicle aggregators. In order to obtain the charging flexibility of electric vehicles, two cooperative agency models between electric vehicle aggregators and electric vehicle users have been proposed, namely the centralized management model and the distributed incentive model. A day-ahead bidding model and real-time charging regulation model for electric vehicle aggregators were established. The recent bidding model integrates factors such as the randomness of electric vehicle charging, the uncertainty of real-time clearing electricity prices, and user response. The real-time charging scheduling model adopts a rolling optimization method to dynamically adjust the charging power of electric vehicles to maximize the economic benefits of electric vehicle aggregators. This model dynamically adjusts the charging power of electric vehicles based on the demand for grid-connected electric vehicle charging and real-time electricity price signals, to optimize the operating costs of electric vehicle aggregators. Finally, the applicability and effectiveness of the proposed method were extensively validated using a certain electricity market in China as the research object.

Chapter 7: Bidding strategy for PEV aggregators in the electricity market. In response to the potential huge losses that electric vehicle aggregators may suffer from risk-neutral day-ahead bidding methods, this chapter proposes a risk-averse day-ahead bidding method based on the risk measurement criterion of conditional value at risk. This method can cope with the uncertainty of real-time clearing scenarios and control the market participation risk of electric vehicle aggregators within a controllable range. In response to the randomness issue of electric vehicle charging, a probability model was used to simulate the charging behavior of the next day's electric vehicle cluster. A risk avoidance decision-making method was used to model potential real-time clearing scenarios in response to the uncertainty issues in the real-time electricity market. Finally, extensive simulation verification was conducted on the effectiveness of the proposed method based on the PJM market. By comparing it with a risk-neutral bidding strategy, the superiority of a risk-averse bidding strategy in risk management was elucidated.

Chapter 8: Coordinated charging of PEV in a market environment. In response to the interactive relationship between electric vehicle aggregators and multiple electric vehicle users, this chapter establishes a Stackelberg game model between electric vehicle aggregators and electric vehicle users, and proposes a collaborative scheduling method for electric vehicle aggregators' bidding and charging under different agency modes. This chapter uses a Stackelberg game model to study the relationship between the pricing strategy of electric vehicle aggregators and the charging demand of electric vehicle users. In addition, the MPEC method is adopted to transform the two-layer game optimization problem into a single-layer mathematical optimization problem, in order to obtain the Stackelberg equilibrium of the game. In addition, the conditional value at risk was incorporated into the bidding and energy management methods of electric vehicle aggregator markets, quantifying the impact of market uncertainty on the operating costs of electric vehicle aggregators. A mathematical model for balancing risk costs and benefits was established to optimize the bidding strategies of electric vehicle aggregators and the charging coordination strategies of electric vehicle clusters in potential real-time clearance scenarios. Finally, the effectiveness of the proposed method was extensively validated based on the PJM market.

Chapter 9: PEV-to-Grid operational strategies in active power distribution systems. This chapter studies the V2G strategy for electric vehicles to discharge from the active distribution network as a power source to the grid, and proposes an optimal charging and discharging strategy based on a master–slave game between electric vehicle users and grid operators. Firstly, the costs of disorderly charging and discharging, non-cooperative game, and cooperative game among electric vehicles under the influence of dynamic electricity prices were analyzed and compared. Secondly, considering the development trend of energy storage application in future grid frequency regulation and peak shaving, a master–slave game model between electric vehicle users and grid operators was proposed, and the equilibrium point between the charging and discharging costs of electric vehicle users and the cost of energy storage equipment used for peak shaving by grid operators was determined.

Chapter 10: Modeling method for power-traffic coupled networks considering bounded rationality of users. In this chapter, a modeling method for electric power transportation coupling networks considering user bounded rationality is proposed to address the issues of incomplete user information perception and non-absolute sensitivity to travel costs in existing models. The main research work includes: proposing a bounded rationality multi-period static user equilibrium model and a bounded rationality dynamic user equilibrium model, characterizing the spatiotemporal distribution characteristics of traffic flow, where the travel and charging behaviors of electric vehicle users satisfy bounded rationality decision conditions, and quantifying their rationality level based on user tolerance. An improved point queue model for fast charging stations has been constructed, which uses differential equations to describe the entry, queuing, charging, and departure actions of electric vehicle traffic flow at fast charging stations, accurately tracking the conversion process from traffic flow to charging load. In addition, the dynamic coupling equilibrium between the bounded rationality dynamic user equilibrium model and the optimal power flow model was studied using fixed point theory, revealing the dynamic coupling mechanism between traffic flow and power flow under user bounded rationality.

Chapter 11: Dynamic pricing strategy for fast charging stations considering bounded user rationality and market regulation. This chapter proposes an optimal dynamic pricing method for fast charging stations based on the three-layer Stackelberg game framework of the vehicle station network, in response to the problems of low flexibility and insufficient market supervision in the static pricing methods of existing fast charging stations. The main content includes: proposing a pricing model based on a three-layer Stackelberg game of vehicle station network, where the distribution network side uses the node marginal electricity price of the optimal power flow model as the wholesale electricity price, the charging station operator side determines the retail charging electricity price to maximize profits, and the electric vehicle user side determines the spatiotemporal distribution of charging load using a bounded rationality dynamic user equilibrium model. The Gauss–Seidel iterative method is used to solve the three-layer game problem. A charging pricing regulatory mechanism has been introduced to ensure that the charging costs of electric vehicle users remain within a reasonable range by setting regulatory coefficients, while incentivizing charging station operators to guide charging load peak shaving and valley filling. This chapter presents an adaptive estimation method for user rational parameters, dynamically adjusting user cost tolerance based on historical behavioral data. In addition, this chapter analyzed the cross-impact of different regulatory coefficients and user rationality coefficients on the operation of fast charging stations, providing a reference for the setting of regulatory coefficients, and also analyzed the promoting effect of user rationality coefficients on the coordinated operation of coupled networks.

Chapter 12: Energy optimization and control strategy for fast charging station aggregators under an electricity market environment. This chapter proposes a simplified dynamic transportation network modeling method and a fast charging station aggregation merchant stochastic energy optimization control method to address the problems of low efficiency in large-scale electric vehicle scheduling and multidimensional uncertainty in operating scenarios when participating in electricity market transactions through fast charging station clusters. The main content includes: proposing a path merging strategy, merging similar paths equivalently based on bounded rationality decision conditions, and transforming the complex traffic network topology structure into a simplified topology structure. A traffic separation strategy was proposed to focus on the electric vehicle traffic flow with charging demand in the transportation network, separate other background traffic flows, and use time-varying path parameters to replace the congestion effect of background traffic flow. This chapter proposed a grouped charging scheduling strategy to bundle electric vehicles with similar charging needs into a unified scheduling group, reducing the dimensionality of decision variables. This chapter adopts the scenario-based stochastic optimization methods and jointly solves the optimal power bidding strategy, charging scheduling strategy, and new energy access strategy for fast charging station aggregators to cope with the uncertainty of user demand, real-time electricity prices, and new energy.

Chapter 13: Risk-averse energy dispatch for integrated energy service stations. This chapter proposes a comprehensive energy service station aggregator risk avoidance energy optimization and regulation method based on Conditional Value at Risk (CVaR) to address the problem of significant differences in heterogeneous user energy replenishment characteristics and high operational risks when participating in electricity market transactions in the electricity hydrogen gas integrated energy service station cluster. The main content includes: proposing a bounded rationality heterogeneous user equilibrium model to uniformly characterize the spatiotemporal distribution of heterogeneous traffic flows and their energy replenishment needs in low-carbon transportation networks, including electric vehicles, hydrogen fuel cell vehicles, and natural gas vehicles. A comprehensive energy service station based on new energy generation and multi-energy synergy and complementarity of electricity, hydrogen gas has been constructed, providing a one-stop energy guarantee for heterogeneous energy replenishment needs of electricity and hydrogen gas in low-carbon transportation networks. A risk avoidance energy optimization regulation strategy for a cluster of comprehensive energy service stations participating in a competitive electricity spot market has been designed. Based on CVaR, the tail risk of profit loss caused by uncertain factors such as heterogeneous user travel demand, real-time electricity prices, and on-site new energy in extreme scenarios is quantified to enhance the system's ability to resist risks.

Chapter 14: Conclusion and Prospect. A comprehensive summary of the research content of this book and a deep outlook on potential future research directions.

Chapter 4
Analytical Analysis for PEV Charging Demand Characterization

Abstract This chapter presents a comprehensive analytical framework for characterizing the charging demand of plug-in electric vehicles (PEVs) under large-scale penetration scenarios, addressing the critical need for accurate load forecasting to ensure the stable and secure operation of power grids. As PEV adoption increases, the associated charging load introduces significant uncertainties and potential risks to grid operations, necessitating a deep understanding of the spatiotemporal patterns of PEV charging behavior. To tackle the challenge of limited real-world data due to low current PEV penetration, this chapter proposes three complementary modeling approaches. First, a quantitative analysis method based on time and electricity consumption is developed, utilizing Monte Carlo simulation to generate large-scale synthetic travel and charging data grounded in existing travel surveys and battery parameters. Second, a Markov model is introduced to capture the state transition probabilities of PEVs over time, enabling the prediction of vehicle state distributions, e.g., driving, parking, and charging—across different time intervals throughout the day. Third, a hidden Markov model (HMM) incorporating battery state of charge (SoC) is proposed to reveal the probabilistic relationship between vehicle states and battery energy levels, allowing for the inference of SoC distributions from observable state sequences. These models collectively enable accurate prediction of PEV charging demand and energy status over time. Furthermore, the chapter investigates the adverse impacts of uncoordinated PEV charging on the power grid, including increased peak-valley differences and voltage drops, using real grid load data and IEEE benchmark systems. The findings underscore the importance of developing scheduling mechanisms and infrastructure plans to mitigate these effects, providing essential insights for grid operators and policymakers.

4.1 Introduction

After a significant increase in the penetration rate of electric vehicles in the future, a large number of electric vehicles will be connected to the distribution network for charging, and the huge charging load it brings will have many adverse effects on the

Q. Yang et al., *Integration of Massive Electric Vehicles to Smart Grid*, Smart Energy Systems, https://doi.org/10.1007/978-981-92-0557-8_4

safe and stable operation of the power grid. Therefore, we need to accurately predict the charging demand of large-scale electric vehicles, fully understand the spatiotemporal characteristics of electric vehicle charging demand, and prevent potential risks that may arise in a targeted manner. For example, after accurately estimating the charging load of electric vehicles, grid operators can arrange the operation plan of frequency regulation and peak shaving units in advance or establish a reasonable guidance and scheduling mechanism to optimize and schedule the charging load of electric vehicles, so that electric vehicles can be charged in an orderly manner. It is also possible to fully consider the factors of electric vehicle charging load in the planning and construction of the distribution network, so as to avoid the negative problems caused by electric vehicle charging at the planning level.

However, every electric vehicle has its unique travel and charging behavior characteristics. Even for the same electric vehicle, its daily travel and charging behavior may be influenced by many factors and become irregular. This makes the charging demand of a single electric vehicle have obvious randomness and individual differences. Studying the travel and charging behavior of one or several electric vehicles alone cannot represent the travel patterns and charging demand characteristics of large-scale electric vehicles. However, due to the regularity of people's travel habits over time, as the penetration rate of electric vehicles gradually increases, the travel and charging behavior of large-scale electric vehicles is no longer discrete and random, but rather follows certain distribution rules in both time and space. Therefore, how to accurately analyze and predict the travel and charging behavior characteristics of large-scale electric vehicles in the current situation where the penetration rate of electric vehicles is not high enough has become a current research challenge and hotspot.

This chapter makes the technical contributions as follows: Firstly, a quantitative analysis method for the travel characteristics and charging demand of electric vehicles based on time and electricity quantity has been proposed. On the basis of existing survey data on car travel habits and electric vehicle battery parameters, this method fully analyzes the inherent relationship between the travel and charging needs of electric vehicles and factors such as travel habits and battery parameters. The Monte Carlo simulation method is used to simulate large-scale electric vehicle travel and charging behavior, thereby obtaining a large amount of data on electric vehicle travel and charging behavior. The characteristics of large-scale electric vehicle charging demand are analyzed and summarized, and finally, a more accurate prediction model is obtained. Secondly, a Markov model was proposed to describe the state transition process of electric vehicles. This model uses Markov models to describe the process and probability of changes in the travel and charging status of large-scale electric vehicles throughout the day. After obtaining the state distribution of electric vehicles at a certain moment, the probability distribution of the state of large-scale electric vehicles at any time of the day can be predicted. Thirdly, a hidden Markov model for electric vehicles considering battery capacity has been proposed. This model utilizes a hidden Markov model to describe the inherent probability relationship between the daily travel and charging status of large-scale electric vehicles and their battery power distribution. After obtaining the state distribution of electric vehicles at a certain moment, the probability distribution of battery power of electric vehicles at the same moment can be predicted. Meanwhile, if combined with the Markov model of electric vehicles, it can predict the probability distribution of battery capacity of large-scale electric vehicles at any time of the day.

In addition, this chapter also analyzes the negative impact of large-scale unregulated charging of electric vehicles on the power grid based on the predicted demand for electric vehicle charging.

4.2 Impact Factors of EV Charging Behaviors

There are many factors that affect the charging behavior of electric vehicles, which are mainly classified into the following three categories in this article: battery parameters of electric vehicles, charging power of electric vehicles, and user travel needs of electric vehicles.

Battery parameters for electric vehicles. The maximum capacity, power consumption per kilometer, and range of the battery have an undeniable impact on the driving behavior and charging needs of electric vehicles. In the current situation where battery technology and energy storage technology are not yet mature and perfect, the range of electric vehicles is generally not high compared to traditional internal combustion engine vehicles, and the current setting of electric vehicle charging stations is not dense enough, which will cause serious "range anxiety" for users, directly affecting the driving behavior and charging needs of electric vehicles.

At present, various electric vehicle manufacturers around the world have launched multiple models of pure electric vehicles to the market, and the battery parameters of different styles of electric vehicles vary greatly. This leads to the necessity of analyzing and summarizing the main battery parameter characteristics of various styles of electric vehicles on the market before predicting the travel behavior and charging demand of large-scale electric vehicles, to obtain representative electric vehicle battery parameters for load prediction.

Based on the survey from the European pure electric vehicle market released by the European Union, the popular models of pure electric vehicles on the market can be divided into two categories based on the parameters of the power battery they carry, represented by M1 and N1, respectively. The maximum capacity of the power battery for two categories of pure electric vehicles C_{bat} can be approximately assumed to follow a normal distribution with upper and lower limits $f(C_{bat}) \sim N\left(\mu_{C_{bat}}, \sigma_{C_{bat}}^2\right)$, as shown in Eq. (4.1), where $\mu_{C_{bat}}$ is the mean of the normal distribution, $\sigma_{C_{bat}}$ is the standard deviation of the normal distribution, and its specific values and other battery parameters such as power consumption per kilometer E_c, as well as the market share of the two categories, are shown in Table 4.1.

$$f(C_{bat}) = \frac{1}{\sqrt{2\pi}\sigma_{C_{bat}}} e^{-\frac{\left(C_{bat} - \mu_{C_{bat}}\right)^2}{2\sigma_{C_{bat}}^2}} \tag{4.1}$$

Charging power of electric vehicles. The charging power directly determines the charging time of electric vehicles, and compared with traditional internal combustion engine vehicles, the excessively long charging time is also one of the important reasons affecting the popularity of electric vehicles. In addition, multiple charging standards for electric vehicles can also be applied in parallel on the market. Based on the most widely used charging standard, this article considers the power used in the conventional charging mode of electric vehicles as a constant power of 3.3 kW (220 V/15 A). The power of fast charging mode P_{fch} is considered as a constant power of 50 kW (400 V/125 A).

The travel needs of electric vehicle users. The travel needs of users are the most fundamental factor affecting the driving and charging behavior of electric vehicles, as the primary role of electric vehicles is to meet the travel needs of users. To understand the characteristics of electric vehicle driving and charging behavior, it is necessary to first analyze the travel demand characteristics of car users.

Firstly, according to a survey report released by the UK government on the domestic use of cars, 61% of cars in the UK are mainly used for commuting by office workers. Therefore, in this article, they are named Commuting PEVs (C-PEVs). These types of vehicles have two fixed daily commuting needs, one in the morning and one in the evening, commuting to and from the user's residence and workplace. In addition, another 39% of cars are used by governments, businesses, or unemployed people for other non-commuting purposes, so they are named Other Purpose PEVs (O-PEVs) in this article. The daily travel demand for this type of vehicle is not fixed and it may not be available or there may be multiple trips.

Meanwhile, according to authoritative survey data, e.g., the 2009 National Household Travel Survey (NHTS) released by the US Department of Transportation and the 2011 Beijing Transportation Development Annual Report released by the Beijing Transportation Development Research Center, the distribution of departure and arrival times for commuting electric vehicles in the morning and evening can be obtained. After processing the original survey data using normalization, maximum likelihood estimation, curve fitting, and other methods, it can be found that the departure and arrival times of commuting electric vehicles in the morning and evening all

Table 4.1 PEV classification and battery parameters

Electric vehicle category	M1	N1
Maximum battery capacity C_{bat} (kWh)	$\mu_{C_{bat}}^{M1} = 28.5, \sigma_{C_{bat}}^{M1} = 14.7,$ $C_{\max}^{M1} = 72.0, C_{\min}^{M1} = 10.0$	$\mu_{C_{bat}}^{N1} = 23.0, \sigma_{C_{bat}}^{N1} = 9.5,$ $C_{\max}^{N1} = 40.0, C_{\min}^{N1} = 9.6$
Electricity consumption per kilometer E_c (kWh/km)	$E_c^{M1} = 0.159$	$E_c^{N1} = 0.185$
Market sharems	$ms^{M1} = 89\%$	$ms^{N1} = 11\%$

Table 4.2 PEV temporal travel patterns over a day

Types of electric vehicle applications	Travel time	Mean μ_t (h: min)	Standard deviation σ_t (h)
Commuting electric vehicles	t_{md}	6:52	1.3
		6:52	1.3
	t_{ma}	8:00	3.4
		8:00	3.4
	t_{ed}	16:52	2.3
		16:52	2.3
	t_{ea}	17:29	3.25
		17:29	3.25
Non-commuting electric vehicles	t_d	13:51	5.2
		13:51	5.2
	t_a	17:29	3.25
		17:29	3.25

follow a normal distribution. The mean values of the normal distribution are 6:52, 8:00, 16:52, and 17:29, respectively. The standard deviations σ_t are 1.3, 3.4, 2.3, and 3.25, respectively. Similarly, after processing the survey data on the daily travel time of non-commuting electric vehicles, it can be concluded that their departure time t_d follows a normal distribution with a mean of 13:51 and a standard deviation of 5.2, while their arrival time t_a can be considered the same as that of commuting electric vehicles due to the lack of corresponding data. The above travel time data is shown in Table 4.2.

4.3 Monte Carlo Based EV Travel and Charging Demand Prediction

A thorough analysis and accurate prediction of the charging demand for large-scale electric vehicles is an essential preliminary work to improve the penetration rate of electric vehicles and ensure the safe and stable operation of the power grid. However, at present, the ownership of electric vehicles is still in a relatively low state, and the amount of related travel data is too small. The travel and charging behavior of a small number of electric vehicles has obvious randomness and individual differences, which cannot reflect the characteristics of large-scale electric vehicle travel and charging behavior. Therefore, there is currently no method or model that can well describe the characteristics of large-scale electric vehicle travel and charging behavior, and thus cannot accurately predict the charging demand of electric vehicles.

In response to this difficulty, this section proposes a quantitative analysis method for the travel characteristics and charging demand of electric vehicles. This method uses Monte Carlo simulation to randomly generate a large number of electric vehicle samples that conform to the statistical characteristics of existing travel and charging behavior data, based on existing vehicle travel survey data and electric vehicle parameter data, to analyze the time patterns of electric vehicle travel and charging behavior and make more accurate predictions of electric vehicle charging demand.

Monte Carlo simulation method, also known as the computer random simulation method, is a numerical calculation method based on probability and statistics theory that gradually developed in the mid-1940s due to the invention of electronic computers. The Monte Carlo simulation method is named after the world-renowned gambling city of Monaco, Monte Carlo, which happens to be a probability based statistical calculation method.

When studying nuclear reactor equipment and its protective facilities in the Manhattan Project, it was necessary to solve a six-dimensional integral equation. This complex computational problem once left all staff helpless, but the emergence of Monte Carlo simulation methods became the only way to solve the problem at that time. In fact, the prototype of the Monte Carlo simulation method had already appeared before this. Although the accuracy of the results of this experiment is not very high, it reflects the most basic idea of the Monte Carlo simulation method: using the frequency of events to represent the probability of events. With the continuous development of electronic computer technology, especially the emergence of high-speed electronic computers in recent years, it has become possible for people to conduct simulation experiments on computers using mathematical methods in large quantities and quickly. At present, Monte Carlo simulation methods have been widely used in research fields such as mathematics, physics, management, genetics, and social sciences, and have demonstrated their special advantages.

Generally, Monte Carlo simulation methods can be divided into two categories: one is that the problem being solved itself has inherent randomness, and with the help of computer computing power, this stochastic process can be directly simulated. For example, in physics research, analyzing the transport process of neutrons in reactors, the interaction between neutrons and atomic nuclei is constrained by the laws of quantum mechanics. People can only know the probability of their interaction, but cannot accurately obtain the position of neutrons when interacting with atomic nuclei, as well as the velocity and direction of new neutrons produced by fission. Scientists randomly sample the fission location, velocity, and direction based on their probabilities. By simulating the behavior of a large number of neutrons, the range of neutron transport can be obtained through statistics, which serves as the basis for reactor design. This type is also known as the direct Monte Carlo simulation method. Another type is that the problem being solved can be transformed into a characteristic number of a random distribution, such as the probability of a random event occurring or the expected value of a random variable. By using random sampling methods, the probability of a random event can be estimated based on its frequency of occurrence, or the numerical characteristics of a random variable can be estimated based on the sampled numerical features, and used as the solution to the problem. This method, also known as indirect Monte Carlo simulation, is commonly used to solve complex multidimensional integration problems.

The Law of Large Numbers and Central Limit Theorem in probability theory are the theoretical foundations of Monte Carlo simulation methods. The law of large numbers reflects the property of the sum of a large number of random numbers. If a function *h* is randomly selected in an [*a*, *b*] interval with a uniform probability distribution density *n*, and the function value is calculated for each u_i value $h(u_i)$, the law of large numbers tells us that the sum of these function values divided by the resulting *n* value will converge to the expected value of the function *h*. That is to say, the law of large numbers ensures that after selecting a sufficient number of random samples, the estimated value of the Monte Carlo simulation method will converge to the expected value of the function. To study the degree of convergence and make various error estimates, the central limit theorem is needed. The central limit theorem states that regardless of the probability distribution of the random variable being studied, the sum of the sampled values of several independent random variables always follows a normal distribution. The central limit theorem describes the distribution of function values estimated by Monte Carlo simulation methods when there is a sufficiently large but finite sample size *n*. Based on the above two theorems, the basic principle of the Monte Carlo simulation method can be expressed as follows:

Assuming a function as follows:

$$Y = f(X_1, X_2, \ldots, X_n) \tag{4.2}$$

Among them, the probability distributions of all variablesX_1, X_2, …, X_n are known. The Monte Carlo simulation method uses a random number (or more commonly a pseudo-random number) generator to randomly generate a set of values that conform to the probability distribution of a random variable through direct or indirect sampling based on a known variable probability distribution $X_1, X_2, \ldots, X_n$. Then, the estimated value of a function *Y* is determined according to the relationship Eq. (4.2) with the variable, as shown in Eq. (4.3):

$$y_i = f(x_{1i}, x_{2i}, \ldots, x_{ni}) \tag{4.3}$$

After multiple independent samplings, a set of function values calculated based on the sampled values can be obtained *Y*. According to the central limit theorem, these function values follow a normal distribution $y_1, y_2, \cdots, y_M$. When the sampling frequency reaches a certain number, the Monte Carlo simulation method can provide information such as the probability distribution and statistical characteristics of functions that are close to the true function values *Y*.

According to the above principles, using the Monte Carlo simulation method to solve problems mainly involves the following steps:

(1) Establish a corresponding probability and statistical model for the problem under study, where certain statistical characteristics of the random variables in the model are the solution to the problem, e.g., mathematical expectations, probability distributions.

(2) According to the probability distribution law of the random variables in the model, use a computer to randomly generate several sets of random numbers, that is, sample the model and ensure that the number of generated random numbers (groups) meets the convergence condition.
(3) Repeat the above steps multiple times, that is, conduct multiple simulation experiments, calculate the statistical characteristics of the corresponding sampling values, and find the solution to the problem that needs to be solved.

It should be pointed out that strictly mathematical random numbers can be obtained through existing random numbers. However, in practical applications, random numbers are usually pseudo-random numbers calculated using certain mathematical formulas, and such pseudo-random numbers are not strictly considered random numbers in mathematical terms. However, as long as pseudo-random numbers can pass a series of statistical tests on random numbers, we can use them as true random numbers. This allows for more convenient, economical, and repetitive generation of random numbers. In theory, pseudo-random number generators are required to possess the following characteristics: good statistical distribution properties, long cycle time for pseudo-random number generation and high efficiency. Currently, various methods for generating pseudo-random numbers with different distributions have been proposed. In this work, the random number generation function in MATLAB is used to generate random numbers that conform to the probability distribution of the corresponding variable.

In the analysis and prediction of the characteristics of large-scale electric vehicle travel and charging behavior studied in this chapter, there is currently not enough real data on electric vehicle travel and charging behavior, and the limited amount of electric vehicle data has obvious individual randomness, which cannot represent the travel and charging behavior of a large number of electric vehicles. As a result, there is currently no method or model that can accurately predict the travel and charging behavior characteristics of large-scale electric vehicles. Therefore, based on existing survey data, this section uses the Monte Carlo simulation method to randomly generate the travel and charging behavior of large-scale electric vehicles, obtain a large number of electric vehicle samples that match real data, summarize the inherent rules of electric vehicle travel and charging behavior, and use this to make relatively accurate predictions of electric vehicle charging demand.

4.4 Behavior Analysis of Electric Vehicles Based on a Markov Model

The previous section provided a detailed introduction to a quantitative analysis method for the travel characteristics and charging demand of electric vehicles based on time electricity, which can accurately predict the expected probability of daily travel and charging demand for large-scale electric vehicles. However, the drawback of this method is the lack of description of the transition process between various states of electric vehicles. If there could be a model that describes the probability of state transitions for electric vehicles at various time intervals throughout the day, it would be possible to accurately predict the state distribution of electric vehicles at any future time interval, given the current time interval's state distribution. This can save a lot of computation time and improve the efficiency and accuracy of electric vehicle travel and charging demand prediction, which is of great significance.

Therefore, this chapter proposes a Markov model to describe the state transition probability of electric vehicles. This model utilizes the state transition matrix of the Markov model to describe the process and probability of changes in the travel and charging status of large-scale electric vehicles throughout the day. After obtaining the state distribution of electric vehicles at a certain moment, the probability distribution of the state of large-scale electric vehicles at any subsequent moment can be predicted.

The Markov model was proposed by the Russian mathematician Andrei Markov (1856–1922) and it refers to the Markov property of all state transition processes in the state space of the model, namely "memorylessness": the next state can only be determined by the current state, and previous states in the time series are independent of it, that is, future states and past states are independent of each other. The state in the model can change to another state during a state change process, or maintain the current state. This process of state change is called transition, and the probability of its occurrence is called transition probability. A random process with this state transition property is called a Markov process, and a Markov process with discrete time and state is also called a Markov chain. Its mathematical definition is as follows:

Set I as a state space, $\{S(t), t = 0, 1, 2, \ldots\}$ it is a random sequence defined in that state space. If Eq. (4.4) is satisfied for any non-negative integer t, where i_1, i_2, …, $i_{t+1} \in I$ represents different states,

$$\begin{aligned} &P\{S(t+1) = i_{t+1} | S(1) = i_1, S(2) = i_2, \ldots, S(t) = i_t\} \\ &= P\{S(t+1) = i_{t+1} | S(t) = i_t\} \end{aligned} \tag{4.4}$$

If all conditional probabilities in Eq. (4.4) are meaningful, then the sequence $\{S(t)\}$ is called a Markov chain. Due to the different transition probabilities between states at different times, the random sequence may have different orders, and all existing Markov chains in the state space I form a Markov model.

Thus, the Markov property can be expressed as follows:

If $\{S(t), t = 0, 1, 2, \ldots\}$ set as a Markov chain, for any three consecutive positive integers $l < m < n$, if $S(m)$ they are in the current state, the future state $S(n)$ is only related to $S(m)$ and independent of each other.

In a Markov model, the probability of transitions between states can be described by a state transition matrix, whose elements are transition probabilities p_{ij}, all of which are non-negative real numbers representing probabilities, defined by Eq. (4.5):

$$p_{ij}(t) = P\{S(t+1) = j | S(t) = i\}, \quad i, j \in I \tag{4.5}$$

Here, $p_{ij}(t)$ represents the probability of the system transitioning to a state at the next moment when it is in a state at that moment, also known as the one-step transition probability. Obviously, the transition probability $p_{ij}(t)$ has the property described in Eq. (4.6), that is, the sum of the probabilities of all states from the state i to the next moment is 1.

$$\sum_{j\in I} p_{ij}(t) = \sum_{j\in I} P\{S(t+1) = j|S(t) = i\} = 1 \tag{4.6}$$

Similarly, Eq. (4.7) can be used to define k the step state transition probability, which refers to the probability of the system transitioning to a state through a secondary transition process when it is in a state at t time.

$$p_{ij}^{(k)}(t) = P\{S(t+k) = j|S(t) = i\}, \quad i, j \in I \tag{4.7}$$

The probability of k step state transition $p_{ij}^{(k)}$ also has the same property as previous equations.

The matrix composed of all one-step transition probabilities of the system at time, t as shown in Eq. (4.8), is the one-step state transition matrix at time, $\{S(t)\}$ or simply the state transition matrix $P(t)$.

$$P(t) = P^{(1)}(t) = \left[p_{ij}(t)\right]_{i,j\in I}, \quad m \in T = \{0, 1, 2, \ldots\} \tag{4.8}$$

Similarly, the k step state transition matrix of the system $P^{(k)}$ can be obtained. According to the Chapman-Kolmogorov Equation, Eq. (4.9) holds for any positive integer.

$$P^{(l+n)} = P^{(l)} \cdot P^{(n)} \tag{4.9}$$

From Eq. (4.9), it can be inferred that $P^{(k)} = (P^{(1)})^k$ the step state transition matrix $k$can be determined by a one-step state transition matrix. Therefore, studying the one-step transition matrix of Markov models $P^{(1)}$ is of great significance. After obtaining the one-step transition matrix of the Markov model $P^{(1)}$, the system state after any number of transition steps can be predicted. The prediction method is as follows:

Assuming the system n has an incompatible state, the initial state vector is $S(0)$, as shown in Eq. (4.10), where $S_i(0)$ is the initial probability of the system being in the state i.

$$S(0) = [S_1(0), S_2(0), \ldots, S_i(0), \ldots, S_n(0)] \tag{4.10}$$

After the k step transition, the probability of the system being in a state i is $S_i(k)$, and the state vector after the k step transition is shown as follows.

$$S(k) = [S_1(k), S_2(k), \ldots, S_i(k), \ldots, S_n(k)] \tag{4.11}$$

Therefore, based on the above principle, the state of the system after k step transition can be predicted by Eq. (4.12),

$$S(k) = S(k-1) \times P = S(0) \times P^k \tag{4.12}$$

Among them, P is the state transition matrix of the Markov model, which has the form shown in Eq. (4.13):

$$P = \begin{bmatrix} p_{11} & p_{12} & \cdots & p_{1n} \\ p_{21} & p_{22} & \cdots & p_{2n} \\ \vdots & \vdots & \ddots & \vdots \\ p_{n1} & p_{n2} & \cdots & p_{nn} \end{bmatrix} \tag{4.13}$$

Therefore, the determination of the state transition matrix of the Markov model is crucial for predicting future system states using the Markov model.

In this work, it is considered that electric vehicles have four incompatible states, namely normal charging state, fast charging state, driving state, and parking state. At any time interval, these four states are independent of the initial state and are independent of previous states. Therefore, the state transition process of electric vehicles can also be regarded as a Markov process, i.e. the state transition matrix of the Markov model can be used to describe the process of electric vehicle state transition.

The Markov model of electric vehicles proposed in this chapter can be represented by Fig. 4.1, and its t state transition matrix in the $P(t)$ time interval is shown in Eq. (4.14). As this article divides a day into 48 time intervals, they $p_{ij}(t)$ vary over time, each of which $p_{ij}(t)$ is a 48-dimensional vector.

$$P(t) = \begin{bmatrix} p_{11}(t) & p_{12}(t) & p_{13}(t) & p_{14}(t) \\ p_{21}(t) & p_{22}(t) & p_{23}(t) & p_{24}(t) \\ p_{31}(t) & p_{32}(t) & p_{33}(t) & p_{34}(t) \\ p_{41}(t) & p_{42}(t) & p_{43}(t) & p_{44}(t) \end{bmatrix} \tag{4.14}$$

If the state transition matrix can be determined $P(t)$, after obtaining the initial state distribution of electric vehicles, the future state distribution of electric vehicles can be predicted. Assuming that the initial distribution of electric vehicle states at t_0 time is $\left(S_{nch}^{t_0}, S_{fch}^{t_0}, S_d^{t_0}, S_p^{t_0}\right)$, the distribution of electric vehicle states at the next time can be calculated by the following equation.

$$\left(S_{nch}^{t_1}, S_{fch}^{t_1}, S_d^{t_1}, S_p^{t_1}\right) = \left(S_{nch}^{t_0}, S_{fch}^{t_0}, S_d^{t_0}, S_p^{t_0}\right) \times P(t_0) \tag{4.15}$$

Similarly, the state distribution at any subsequent time t_k can be calculated using the following equation.

$$(S_{nch}^{t_k}, S_{fch}^{t_k}, S_d^{t_k}, S_p^{t_k}) = (S_{nch}^{t_0}, S_{fch}^{t_0}, S_d^{t_0}, S_p^{t_0}) \times P(t_0) \times P(t_1) \times \cdots \times P(t_{k-1}) \tag{4.16}$$

The state transition matrix of the Markov model for this electric vehicle can be represented by the probability of each transition process in the travel and charging behavior data of a large number of randomly generated electric vehicle samples, based on the principles of Monte Carlo simulation. The results are shown in Fig. 4.2. Among them, subgraphs (a)–(d) represent the probability changes of electric vehicles transitioning from normal charging state, fast charging state, driving state, and parking state to the next state during a day.

According to Fig. 4.2, the state distribution of electric vehicles in the t_1 time interval (i.e. 0:00–0:30) is (0.3309, 0.0084, 0.0391, 0.6218). This electric vehicle Markov model can describe the transition process and transition probability between various states of electric vehicles, and can predict the state distribution at any subsequent time after knowing the state distribution of electric vehicles at a certain time. This model reflects the changing patterns of electric vehicle travel and charging behavior, while also improving the flexibility of the entire prediction method. Therefore, it is of great significance for predicting the demand for electric vehicle travel and charging.

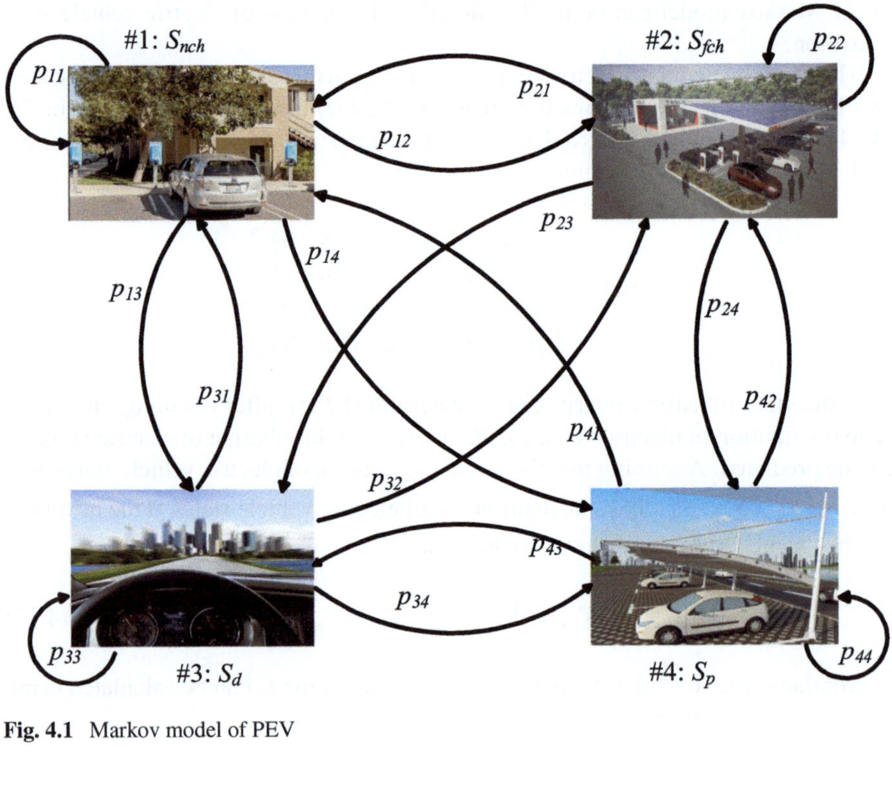

Fig. 4.1 Markov model of PEV

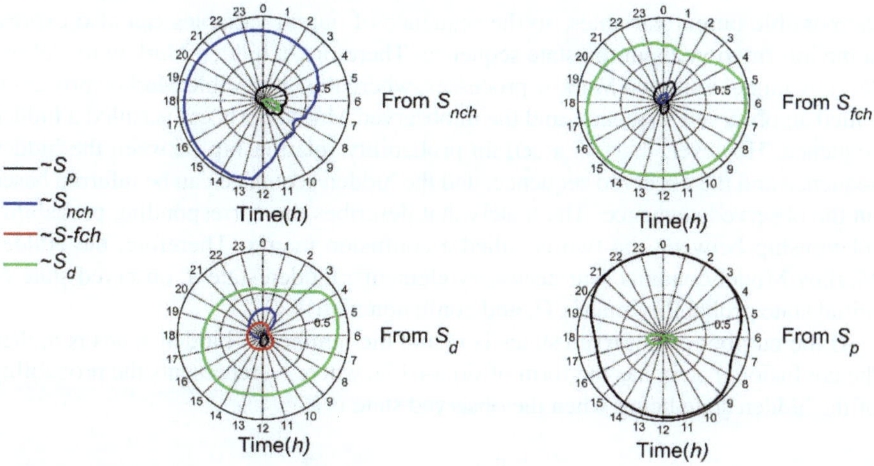

Fig. 4.2 Markov transition matrix of PEV

4.5 Electric Vehicle Energy Prediction Based on a Hidden Markov Model

In practice, the battery capacity will directly affect the travel and charging needs of electric vehicles, as well as the charging time of electric vehicles. Therefore, we not only need to understand the transition rules between the states of electric vehicles, but also need to deeply explore the internal reasons for their state transitions, that is, how battery power affects the transition of electric vehicle states, and quantitatively analyze the probability relationship between the two. This is of great help in further scheduling the charging behavior of electric vehicles.

Therefore, this chapter proposes a hidden Markov model for electric vehicles that considers battery capacity, revealing the inherent probability relationship between the state of large-scale electric vehicles and their battery capacity. After obtaining the state distribution of electric vehicles at a certain moment, the probability distribution of battery power of electric vehicles at the same moment can be predicted. Meanwhile, if combined with the Markov model of electric vehicles, it can predict the probability distribution of battery power for large-scale electric vehicles at any time of the day.

A Hidden Markov Model (HMM) is a stochastic model based on Markov processes, which contains hidden unknown sequences. The Hidden Markov Model was first proposed by American mathematician Leonard E. Baum (1931–2017) and others in the late 1960s. It was later widely used in the field of speech recognition and achieved significant success. It gradually became an indispensable technology for analyzing biological sequences, especially DNA, in bioinformatics.

In a normal Markov model, the state is directly visible to the observer, while in a hidden Markov model, the state is not directly visible, but some output variables affected by the state are visible. Each state has a probability distribution on the possible output variables, so the sequence of output variables can also express some information about the state sequence. Therefore, a hidden Markov model is a Markov model with two Markov processes, where the observable Markov process is called an observed sequence, and the unobserved Markov process is called a hidden sequence. However, there is a certain probability relationship between the hidden sequence and the observed sequence, and the hidden sequence can be inferred based on the observed sequence. The matrix that describes the corresponding probability relationship between the two is called a confusion matrix. Therefore, the Hidden Markov Model contains four necessary elements: hidden state S, observed state O, initial state probability matrix Π, and confusion matrix C.

If the number of observed states is m and the number of hidden states is n, then the confusion matrix has the form of Eq. (4.17), where c_{ij} represents the probability of the hidden state being when the observed state is i.

$$C = \begin{bmatrix} c_{11} & c_{12} & \cdots & c_{1n} \\ c_{21} & c_{22} & \cdots & c_{2n} \\ \vdots & \vdots & \ddots & \vdots \\ c_{m1} & c_{m2} & \cdots & c_{mn} \end{bmatrix} \tag{4.17}$$

If the state vector of the observed state of the system at a certain moment is $O(0)$ as shown in Eq. (4.18), where $O_i(0)$ is the initial probability of the system being in a state.

$$O(0) = [O_1(0), O_2(0), \ldots, O_i(0), \ldots, O_n(0)] \tag{4.18}$$

The state vector of the implicit state of the system at the same time $S(0)$ can be calculated by Eq. (4.19).

$$S(0) = O(0) \times C \tag{4.19}$$

If the Markov state transition matrix of the observed state of the system is P, the implicit state vector of the system after k step transition can be calculated by Eq. (4.20).

$$\begin{aligned} S(k) &= [S_1(k), S_2(k), \ldots, S_i(k), \ldots, S_n(k)] \\ &= O(0) \times P^k \times C(k) \end{aligned} \tag{4.20}$$

Therefore, if the confusion matrix of the Hidden Markov Model can be obtained, the probability distribution of the hidden state can be inferred through observing the sequence. If combined with the Markov state transition matrix of the observation sequence, the probability distribution of the hidden state at any time can be further inferred.

In order to describe the probability relationship between the state of an electric vehicle and its battery level, the principle of the hidden Markov model can be used to define the state of the electric vehicle as the observed state O and the battery level as the hidden state S. Then, the confusion matrix C expresses the probability relationship between the two. In this work, the battery capacity of electric vehicles is divided into three intervals: low, medium, and high: less than 40, 40–60%, and greater than 60%. That is, the hidden state space contains three different states. Therefore, the hidden Markov model of electric vehicles considering battery capacity can be represented by Fig. 4.3. The confusion matrix in the t time interval $C(t)$ is shown in Eq. (4.21), and similarly, $c_{ij}(t)$ it also varies over time. Each one $c_{ij}(t)$ is also a 48-dimensional vector.

$$C(t) = \begin{bmatrix} c_{11}(t)\ c_{12}(t)\ c_{13}(t) \\ c_{21}(t)\ c_{22}(t)\ c_{23}(t) \\ c_{31}(t)\ c_{32}(t)\ c_{33}(t) \\ c_{41}(t)\ c_{42}(t)\ c_{43}(t) \end{bmatrix} \tag{4.21}$$

If the confusion matrix can be determined $C(t)$, the distribution of the electric vehicle battery power interval at a certain time can be predicted after obtaining the state distribution of the electric vehicle. Then, combined with the state transition matrix $P(t)$, the distribution of the electric vehicle battery power interval at any time can be predicted. Assuming that the initial state distribution of electric vehicles at t_0 time is $\left(S_{nch}^{t_0}, S_{fch}^{t_0}, S_d^{t_0}, S_p^{t_0}\right)$, the distribution of power intervals at t_0 time can be calculated by Eq. (4.22).

$$\left(SoC_1^{t_0}, SoC_2^{t_0}, SoC_3^{t_0}\right) = \left(S_{nch}^{t_0}, S_{fch}^{t_0}, S_d^{t_0}, S_p^{t_0}\right) \times C \tag{4.22}$$

The distribution of electricity levels at any time thereafter t_k can be calculated using Eq. (4.23).

$$\begin{aligned}(SoC_1^{t_k}, SoC_2^{t_k}, SoC_3^{t_k}) &= (S_{nch}^{t_0}, S_{fch}^{t_0}, S_d^{t_0}, S_p^{t_0}) \\ &\times P(t_0) \times P(t_1) \times \cdots \times P(t_{k-1}) \times C(t_k)\end{aligned} \tag{4.23}$$

The confusion matrix of the hidden Markov model for this electric vehicle can also be represented by the matching frequency between the states of a large number of randomly generated electric vehicle samples and three power intervals, based on the principle of the Monte Carlo simulation method. The results are shown in Fig. 4.4. Among them, subgraphs (a)–(c) represent the probability changes of battery power in three different SoC ranges when the electric vehicle is in normal charging state, fast charging state, driving state, and parking state during a day.

It should be pointed out that since this article assumes fast charging when the electric vehicle's battery level is below 20%, and its state is determined at the beginning of each time interval and remains unchanged throughout the entire time interval, in

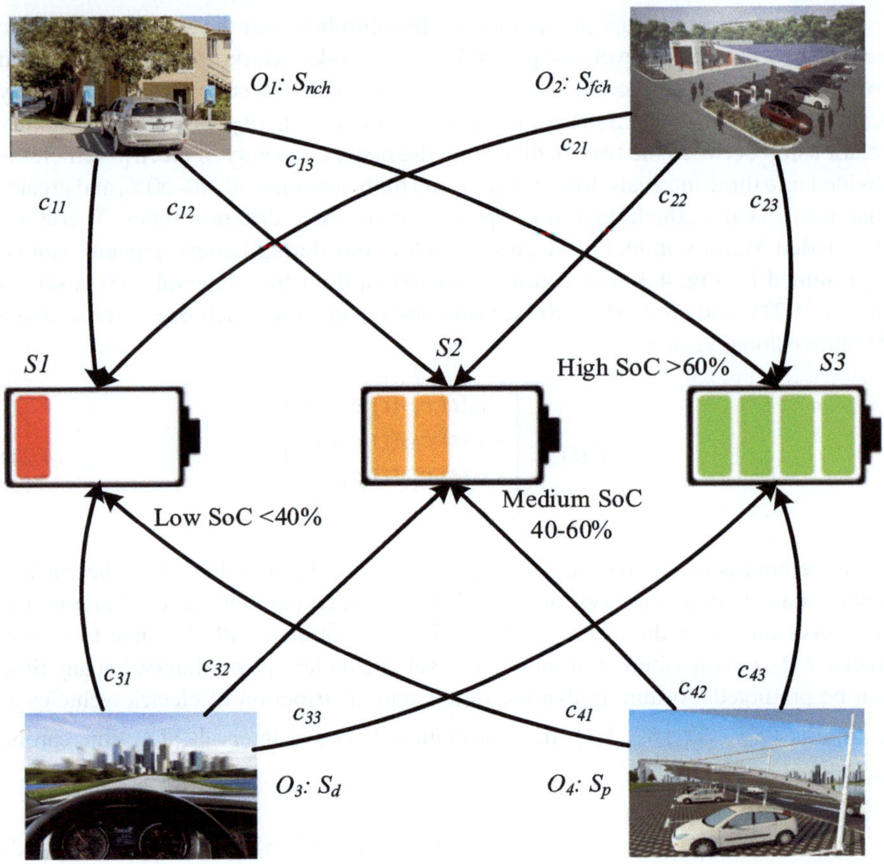

Fig. 4.3 Hidden Markov model of PEV with SoC

Fig. 4.4b, when the electric vehicle is in fast charging mode, its battery level must be below 40%, that is, the probability of the battery level being in the range of less than 40% is always 1. Additionally, it should be noted that in the experiments of this chapter, the battery capacity of electric vehicles was divided into three intervals. In future practical applications, more detailed power intervals can be set as needed to improve the accuracy of the model.

By using the confusion matrix of the hidden Markov model for electric vehicles $C(t)$ and the state transition matrix of the Markov model $P(t)$, it is possible to predict the interval distribution of electric vehicle battery power at the current time and any subsequent time after obtaining the state distribution of the electric vehicle at a certain time. Compared with the electric vehicle Markov model described earlier in this chapter, the electric vehicle hidden Markov model considering battery power proposed in this section further reveals the probability relationship between the electric vehicle state and its battery power. It can predict the operation and charging demand of electric vehicles while obtaining the distribution of their battery power.

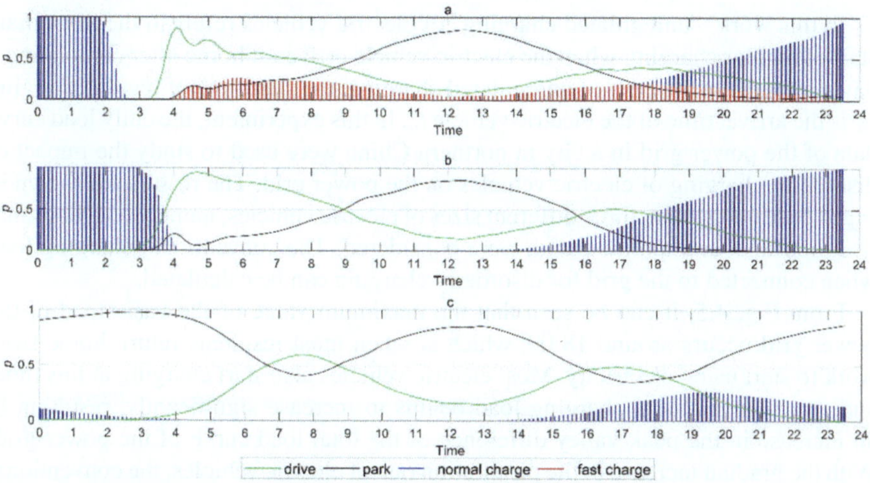

Fig. 4.4 Confusion matrix of the PEV hidden Markov model with SoC

This is of great significance for understanding the essential laws of electric vehicle charging behavior and formulating more reasonable electric vehicle charging and discharging scheduling strategies.

4.6 Impact of EV Charging on Power Grid

According to the probability of the conventional charging state of electric vehicles, it often happens that the conventional charging load of electric vehicles gradually increases from the evening and reaches its maximum value around 21:30. Nearly 40% of electric vehicles will perform conventional charging at this time, which coincides with the late peak period of traditional loads in the power grid. It can be imagined that if a large number of electric vehicles are connected to the distribution network for disorderly charging, a new peak of electric vehicle charging load will be generated at night, which will overlap with the evening peak of the power grid load. This may have a significant impact on the safe and stable operation of the power grid, resulting in many adverse effects. Therefore, it is necessary to study the impact of large-scale electric vehicle disorderly charging on the power grid and evaluate the current number of electric vehicles that the power grid can accept. If the future penetration rate of electric vehicles exceeds the current number of electric vehicles that the power grid can accept, grid operators need to prepare in advance, e.g., formulating reasonable electric vehicle charging scheduling strategies to transfer the charging load of electric vehicles to low load periods, planning the location and capacity of electric vehicle charging facilities reasonably, or upgrading and transforming the existing distribution network.

In this work, "unregulated charging" of electric vehicles refers to the immediate start of regular charging when the electric vehicle arrives at home after completing a day's journey. In the case of unregulated charging, the start time of regular charging t_{sc} is the arrival time of the electric vehicle t_a. In this experiment, the daily load curve data of the power grid in a city in northern China were used to study the impact of disorderly charging of electric vehicles on the power grid. The results are shown in Fig. 4.5. Among them, three different sizes of electric vehicles, namely one thousand, five thousand, and ten thousand, were considered. The conventional charging load when connected to the grid for disorderly charging can be calculated.

From Fig. 4.5, it can be seen that the maximum value of the basic load of the power grid occurs around 18:00, which is when most residents return home from work to start using electricity. Most electric vehicles also start charging at this time, and their conventional charging load begins to increase significantly, resulting in an increase in the peak valley difference of the total load curve of the power grid. With the gradual increase in the penetration rate of electric vehicles, the conventional charging load of electric vehicles will become a new peak in the electricity demand of the power grid, further widening the peak valley difference of the power grid. The peak valley difference rate of power grid load can be defined as Eq. (4.24) to describe the peak valley difference characteristics of the power grid. According to calculations, when the number of electric vehicles connected to the grid for conventional charging increases from one thousand to ten thousand, the peak valley difference rate ξ increases from 60.56 to 81.32%. This will bring a serious impact to the power grid, causing a significant drop in node voltage and putting power equipment in a severely unbalanced operating state.

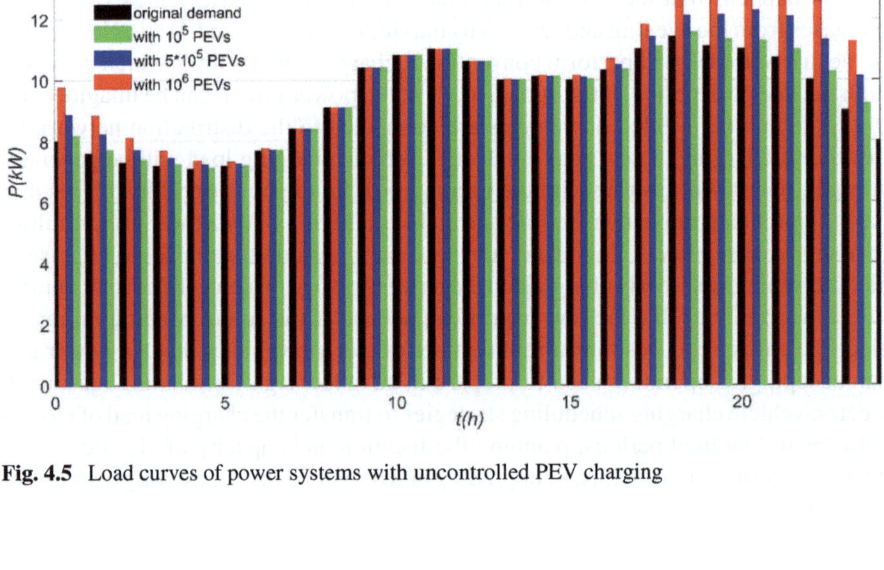

Fig. 4.5 Load curves of power systems with uncontrolled PEV charging

$$\xi = \frac{P_{\max} - P_{\min}}{P_{\min}} \times 100\% \tag{4.24}$$

To further investigate the impact of large-scale electric vehicle disorderly charging on grid voltage, this work selected an IEEE 53-bus distribution network as shown in Fig. 4.6, where nodes # 101, # 102, and # 104 are substations with a total load of 45.67 MW/22.12 Mvar, a maximum voltage of 13.8 kV, and a maximum capacity of 50 MVA. The red nodes are the nodes for conventional charging of electric vehicles, and constant low-power generators are also connected to several nodes to provide certain voltage and power support.

Here, the power system power flow can be calculated using Matpower in MATLAB. It is convenient and fast to calculate power flow parameters such as active and reactive power, as well as node voltage in the distribution network. When the number of electric vehicles connected to the distribution network is 10,000, the

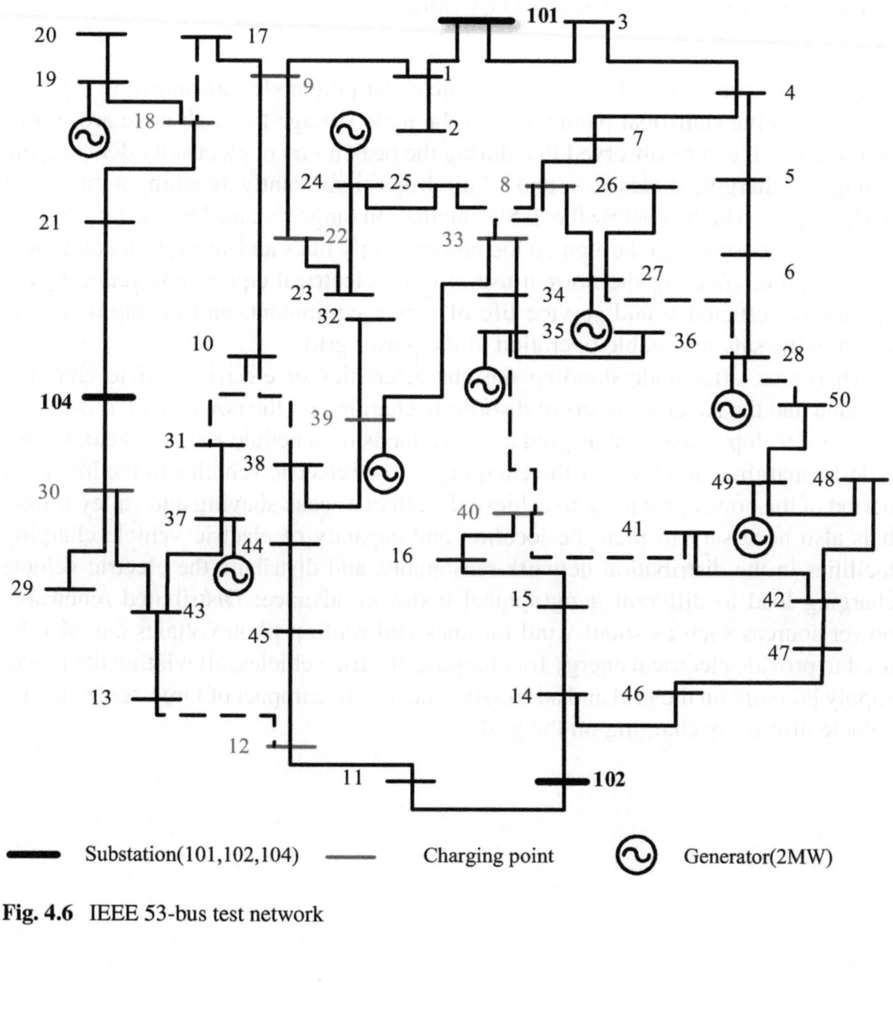

Fig. 4.6 IEEE 53-bus test network

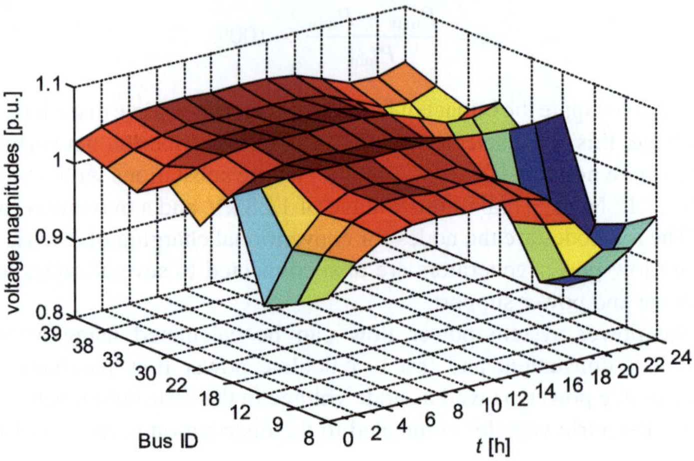

Fig. 4.7 Nodal voltage with uncontrolled PEV charging

daily voltage changes of the 9 electric vehicle charging nodes are shown in Fig. 4.7, and the specific statistical parameters of the node voltage for each node are shown in Table 4.3. It can be observed that during the peak hours of electricity demand, the voltage of charging nodes 8, 9, and 12 all drops significantly, reaching a minimum of 0.846 p.u., which is below the $\pm$ 5% fluctuation range required by the power grid. This will impose a huge burden on the power supply lines and substation equipment in the region, affecting the normal use of other electrical equipment, reducing the operational efficiency and service life of power equipment, and posing a serious threat to the safe and stable operation of the power grid.

Therefore, after understanding the characteristics of electric vehicle charging demand and the adverse effects of disorderly charging on the power grid, it is necessary to develop a reasonable guidance mechanism, schedule electric vehicles for orderly charging, and transfer the charging load of electric vehicles to the low peak period of the power grid load, to achieve the effect of peak shaving and valley filling. It is also necessary to plan the location and capacity of electric vehicle charging facilities in the distribution network reasonably, and distribute the electric vehicle charging load to different geographical nodes in advance; Distributed renewable power sources such as small wind turbines and rooftop photovoltaics can also be used to provide electrical energy for charging electric vehicles, alleviating the power supply pressure on the grid and addressing the negative impact of large-scale electric vehicle disorderly charging on the grid.

Table 4.3 Nodal voltage profiles with uncontrolled PEV charging

node number	Mean node voltage	Standard deviation of node voltage	Worst-case scenario
8	0.996	0.039	0.928
8	0.996	0.039	0.928
9	0.986	0.062	0.874
9	0.986	0.062	0.874
12	0.976	0.071	0.846
12	0.976	0.071	0.846
18	1.047	0.003	1.041
18	1.047	0.003	1.041
22	1.015	0.030	0.963
22	1.015	0.030	0.963
30	1.037	0.013	1.015
30	1.037	0.013	1.015
33	1.020	0.027	0.972
33	1.020	0.027	0.972
39	1.027	0.016	0.999
39	1.027	0.016	0.999
40	1.039	0.013	1.016
40	1.039	0.013	1.016

4.7 Summary

With the gradual popularization of electric vehicles, the charging load of electric vehicles will have a certain adverse impact on the power grid. Therefore, it is necessary to develop a reasonable guidance mechanism to dispatch the charging load of electric vehicles to the low load of the power grid for charging, and it is particularly important to plan the charging facilities of electric vehicles reasonably. The prediction of large-scale electric vehicle charging demand is the foundation of these works. In the current situation where the penetration rate of electric vehicles is still low, it becomes difficult to accurately predict the travel and charging demand of electric vehicles.

In response to this issue, this chapter utilizes existing survey data to deeply analyze and summarize the inherent laws of electric vehicle travel and charging state changes. Three analysis methods or models are mainly proposed: a quantitative analysis method based on time electricity quantity for electric vehicle travel characteristics and charging demand, a Markov model describing the state transition process of electric vehicles, and an electric vehicle hidden Markov model considering battery power. This accurately describes the inherent laws of large-scale electric vehicle travel and charging demand, and can accurately predict electric vehicle travel and charging demand, as well as understand the probability distribution of large-scale electric vehicle battery power. This will play a supporting role in the planning of electric vehicle charging facilities and the scheduling of charging behavior in the future. In addition, this chapter also analyzed the negative impact of large-scale electric vehicle disorderly charging on the power grid using actual data.

Chapter 5
Optimal PEV Charging Scheduling in Active Power Distribution Systems

Abstract This chapter proposes a hierarchical optimization planning method for multiple electric vehicle (PEV) charging facilities and distributed renewable power sources (DGs) in active power distribution systems, explicitly considering the coupling between transportation and distribution networks. With the increasing integration of renewable energy and the growing demand for electric vehicle charging, the inherent uncertainty of DG output and the complex interdependencies between the two networks pose significant challenges to system planning and operation. To address these issues, the K-means clustering algorithm is first employed to analyze historical DG output data, extracting representative daily output curves that capture the stochastic characteristics of renewable generation. Subsequently, a multi-objective collaborative optimization model is established for conventional charging parking lots, fast charging stations, and DGs, incorporating various operational constraints from both networks. A hierarchical solution approach is adopted to sequentially determine the optimal siting and sizing of each facility: the locations of conventional parking lots are based on existing residential and commercial nodes; fast charging station locations are optimized to minimize total travel distance for PEV users; the capacity of fast charging stations is then determined using a queuing theory-based economic model to maximize annualized profit; finally, the allocation of DGs is optimized to minimize a composite index of active power loss, reactive power loss, and node voltage fluctuation in the distribution network. The proposed method is validated on IEEE 53-bus and 123-bus distribution systems coupled with a transportation network. Results demonstrate that the hierarchical planning significantly reduces network losses and voltage deviations compared to scenarios with only PEV charging facilities, effectively mitigating the adverse impacts of uncoordinated charging. Moreover, most DGs are optimally located near charging facilities to enhance renewable energy utilization and provide local power support. This approach simplifies the complex nonlinear planning problem, balances economic and operational objectives, and exhibits good scalability for application in larger, more complex networks.

Q. Yang et al., *Integration of Massive Electric Vehicles to Smart Grid*, Smart Energy Systems, https://doi.org/10.1007/978-981-92-0557-8_5

5.1 Introduction

In recent years, with the gradual consumption of traditional fossil fuels, people have begun to use renewable energy to replace fossil fuels, thereby achieving sustainable development and reducing pollutant emissions. Therefore, in the past decade, renewable energy technologies such as wind turbines and photovoltaics have flourished, and more and more renewable energy sources have been put into use. Compared with traditional thermal power plants, renewable power sources use clean and renewable energy such as wind and light to generate electricity, which can reduce the consumption of fossil fuels, greenhouse gas and pollutant emissions, and is of great significance for global sustainable development and environmental protection. At the same time, renewable power sources do not require large equipment and a large amount of space, and can be connected to the distribution network in a small-scale and decentralized manner, that is, distributed renewable power sources. For example, distributed photovoltaics can be connected to the roofs of thousands of households to work, thereby saving a lot of land and construction costs.

It should be noted that the advantage of electric vehicles in reducing pollutant gas emissions is also based on the use of renewable energy sources to charge electric vehicles. If electric vehicles use the electricity generated by traditional thermal power plants, although they do not produce polluting exhaust emissions during operation, because thermal power plants use fossil fuels, e.g., coal for power generation, if the electricity consumption during operation is converted into the same amount of polluting gases emitted by thermal power plants, the energy is converted twice, and considering efficiency and losses, it may even emit more pollutants than traditional internal combustion engine vehicles. Therefore, encouraging electric vehicles to absorb renewable power sources as much as possible is an important part of reducing carbon emissions and pollutant emissions in the transportation system. Only by combining electric vehicle charging facilities with renewable power sources can the advantages of energy conservation and emission reduction of electric vehicles be fully utilized.

However, the output of renewable power sources is severely limited by natural conditions, such as external factors, such as wind speed and light intensity, which will directly affect the output curve of renewable power sources. However, due to the limitations of current relevant technologies, it is not yet possible to accurately predict the output of renewable power sources. Meanwhile, it is also impossible to completely control the output of renewable power sources to remain constant. The output curve may experience significant fluctuations due to sudden changes in weather conditions, which will have a serious impact on the safe and stable operation of the power grid. Therefore, how to solve the uncertainty of renewable power output has become a research difficulty.

In addition, after considering the conventional charging parking lots for electric vehicles and distributed renewable power sources, the constraints of the transportation network and distribution network have become more complex. It is a very complex and tricky problem to simultaneously solve the site selection and capacity determination of multiple electric vehicle charging facilities and distributed renewable power sources in the planning, and to satisfy the multiple constraints of both in the coupling of the transportation network and distribution network. In response to this issue, this chapter proposes a hierarchical optimization planning method for multiple charging facilities and distributed renewable power sources of electric vehicles, considering the coupling of transportation and distribution networks. The main contents include the following parts:

Firstly, the K-means clustering algorithm is used to summarize and extract statistical features from historical data of renewable power output, to obtain representative daily output curves; Secondly, considering the coupling of transportation and distribution networks, a collaborative optimization planning model for conventional charging parking lots, fast charging stations, and distributed power sources of electric vehicles with multiple optimization objectives was established, considering the various constraints of both; and finally, a hierarchical approach was adopted to gradually solve the optimal location and capacity of various charging facilities and distributed renewable power sources for electric vehicles. While greatly simplifying the complexity of the problem, it ensured the convenience and economy of electric vehicle charging facilities, as well as the safe and stable operation of the distribution network.

5.2 Distributed Generation Modeling

The concept of distributed generation (DG) was officially defined and promoted by the Public Utility Management Policy Act of the United States government in 1978. It refers to small independent power sources installed in a decentralized manner near users, with a power generation of several megawatts to tens of megawatts, which are different from traditional centralized power supply methods. The types of energy generated include renewable energy and non-renewable energy. Among them, non-renewable energy distributed power sources mainly include small internal combustion engines, micro steam turbines, and fuel cells. Intermittent energy generation mainly includes wind power generation, solar photovoltaic cell generation, and biomass energy generation. The distributed power sources mentioned in this chapter will specifically refer to distributed renewable power sources.

At present, the most widely used distributed power sources are small wind turbines and small solar photovoltaic panels. This section will mainly analyze the working principles and output characteristics of these two types of distributed power sources.

As one of the most popular renewable generators, wind turbines are mainly composed of impellers, transmission mechanisms, generators, inverters, speed-limiting safety mechanisms, towers, and other parts. When the wind drives the impellers of the wind turbine to rotate, torque is generated, which can be input to the transmission mechanism through the hub. The generator then converts the mechanical energy in the transmission system into electrical energy through the electromagnetic induction principle, and finally outputs it through the inverter.

According to aerodynamics, the mechanical power output of the fan P_{WT} (W) can be expressed as Eq. (5.1) under ideal conditions.

$$P_{WT} = \frac{1}{2}\rho R^2 v^3 C_P \tag{5.1}$$

Among them, ρ is the air density (kg/m^3); R For the radius of the fan impeller (m); v is the wind speed (m/s); C_P is the wind energy utilization coefficient. For fixed pitch wind turbines, it is related to the tip speed ratio, while for variable pitch wind turbines, it is related to the tip speed ratio and pitch angle.

According to Eq. (5.1), the mechanical power output of the fan is proportional to the cube of the wind speed. The commonly used empirical fitting wind speed distributions include the Weibull distribution, the Rayleigh distribution, the Pearson n-type distribution, etc. Among them, the Weibull distribution is the most widely used, and its density function is shown in Eq. (5.2).

$$f(v) = \frac{k}{\lambda}\left(\frac{v}{\lambda}\right)^{k-1} e^{-\left(\frac{v}{\lambda}\right)^k} \tag{5.2}$$

Among them, $k > 0$ is the shape parameter of the Weibull distribution and $\lambda > 0$ is the proportionality parameter, both of which are related to the meteorological and geographical characteristics of the region.

In addition, wind turbines with different power generation principles and even grid connection methods have different output power curves, which can generally be divided into asynchronous generators and synchronous generators. Among them, synchronous generators require less excitation power during operation, and the terminal voltage and output reactive power can be adjusted through the excitation system. However, their transient stability is poor, and the generator speed is the same as the impeller speed and will vary with wind speed. Asynchronous generators themselves do not have excitation devices and need to absorb reactive power from the outside to generate excitation current. Their speed depends on the frequency of the power grid and only varies within a small range near the synchronous speed. Therefore, asynchronous generators can be directly connected to the grid without the need for synchronous regulation devices, and only the output characteristics of active power need to be considered. Thus, the active output power of wind turbines is greatly uncertain due to the influence of geographical location, natural environment, and wind turbine parameters.

On the other hand, the photovoltaic cells are devices that convert solar energy into electrical energy based on the principle of semiconductor materials' photovoltaic effect. By connecting multiple photovoltaic cells in series and parallel, a photovoltaic cell group, also known as a photovoltaic panel, is formed. Among them, the photovoltaic effect refers to the phenomenon when sunlight shines on a photovoltaic cell, the semiconductor material in the cell absorbs the light energy and generates an electromotive force. If a load is connected to the external circuit, it will form an electric current.

The photovoltaic power generation system mainly includes photovoltaic cell groups, DC combiner boxes, DC distribution cabinets, grid-connected inverters, control systems, and other equipment. First, the photovoltaic cell groups are exposed to light to generate DC power, which is then converted into AC power through a DC converter and finally filtered and boosted before being integrated into the power grid. According to whether it is integrated into the power grid, photovoltaic power generation systems can also be divided into independent power supply photovoltaic power generation systems and grid-connected photovoltaic power generation systems.

The output power of photovoltaic cells is influenced by many factors, such as temperature, light intensity, light receiving area of the photovoltaic panel, photoelectric conversion efficiency, and load size, and generally exhibits a nonlinear relationship. In practical situations, due to the uncontrollable intensity of light and the possibility of significant fluctuations, the output of photovoltaic power sources also has great uncertainty.

Meanwhile, temperature also has a significant impact on the output power of photovoltaic power sources, and even under the same temperature and light intensity conditions, photovoltaic power sources can operate at different voltage states. However, its output power is maximum only at a specific voltage. Therefore, in order to always operate the photovoltaic power source at the maximum output power point and output as much electrical energy as possible, it is necessary to perform Maximum Power Point Tracking (MPPT) control on it. Currently, the commonly used MPPT methods include constant voltage tracking, disturbance observation, and conductance increment.

The control method for grid-connected inverters of photovoltaic power sources generally adopts the voltage source current control method, which controls the output current of the inverter to enable it to track the changes in grid voltage in real time, thereby achieving the goal of grid-connected operation of photovoltaic power sources. Since photovoltaic power sources can maintain their power factor at 1 when connected to the grid, only the active power of the photovoltaic power source can be considered.

5.3 Uncertainty Analysis of Distributed Power Generation

With the increasing penetration rate of distributed power sources in the distribution network, the randomness and uncertainty of their output pose more difficult challenges to the safe and stable operation of the power grid. This requires us to achieve more accurate prediction of the output of distributed power sources on the one hand, and fully consider the uncertainty of distributed power source output when planning the distribution network, improve the robustness of the distribution network, and ensure that the distribution network can operate normally when the output of distributed power sources is within the common fluctuation range. This work adopts the K-means clustering algorithm to solve the uncertainty of distributed power generation.

Clustering analysis originated from taxonomy, which can divide all research objects into different subsets, so that research objects in the same subset have some similar attributes or characteristics, while research objects belonging to different subsets have completely different attributes or characteristics. Cluster analysis is a commonly used unsupervised learning method that has been widely applied in fields such as machine learning, artificial intelligence, data mining, and pattern recognition. The most fundamental implementation method is the K-means clustering algorithm.

The basic idea of the K-means clustering algorithm can be simply described as follows: for a given sample set, the sample set is divided into clusters according to the distance between samples, so that the distance between samples in the clusters is as close as possible, and the distance between clusters is as large as possible. That is, the sample set is divided into clusters by the size of the distance between sample positions, and all samples with closer distances will be divided into the same cluster.

Assuming the clustering of the sample set is divided into clusters $\{C_1, C_2, \ldots, C_k\}$ and considering Euclidean distance, the goal of the K-means clustering algorithm is to minimize the sum of squared distances between all samples and their centroids SSE, as shown in Eq. (5.3), i.e., Sum of the Squared Error (SSE).

$$SSE = \sum_{i=1}^{k} \sum_{x \in C_i} \|x - \mu_i\|^2 \tag{5.3}$$

Among them, μ_i is the mean vector of the cluster C_i, also known as the centroid, which can be calculated by eq. (5.4) m_i and C_i is the number of samples contained in the cluster.

$$\mu_i = \frac{1}{m_i} \sum_{x \in C_i} x \tag{5.4}$$

However, Eq. (5.3) is an NP hard problem that cannot be solved by conventional methods, so the K-means clustering algorithm adopts an iterative approach. Assuming the sample set of input data is $D = \{x_1, x_2, \ldots, x_m\}$, the specific steps of using the K-means clustering algorithm for it are as follows:

Step (1): Set k the value, which is the number of clusters to be clustered into. Generally speaking, a reasonable k value can be selected based on prior experience with the data. If there is no prior knowledge, a reasonable k value can be obtained through cross-validation.

Step (2): Select data from the input data sample set as the initial centroid of each cluster k, form a centroid vector $\{\mu_1, \mu_2, \ldots, \mu_k\}$, and each element represents the centroid of the cluster in sequence$\{C_1, C_2, \ldots, C_k\}$;

Step (3): For all input data x_i, calculate the Euclidean distances between each centroid as shown in Eq. (5.5), and assign them to the cluster represented by the centroid closest to them;

$$d_{ij} = \left\| x_i - \mu_j \right\|^2 \tag{5.5}$$

Step (4): After calculating all the input data, recalculate the C_j centroid of the cluster μ_j as shown in Eq. (5.4) and compare it with the original centroid vector $\{\mu_1, \mu_2, \ldots, \mu_k\}$. If they are not the same, repeat step (3). If there is no change, output the final clustering result$\{C_1, C_2, \ldots, C_k\}$.

To minimize the cluster C_i, it is necessary to continuously update the average value of all samples in the cluster C_i, that is, the centroid μ_i. To demonstrate the feasibility of this method, we can take the derivative in Eq. (5.3), make the derivative equal to zero, and solve for its centroid μ_i, which is Eq. (5.6).

$$\begin{aligned}\frac{\partial}{\partial \mu_i} SSE &= \frac{\partial}{\partial \mu_i} \sum_{i=1}^{k} \sum_{x \in C_i} \|x - \mu_i\|^2 \\ &= \sum_{i=1}^{k} \sum_{x \in C_i} \frac{\partial}{\partial \mu_i} \|x - \mu_i\|^2 \\ &= \sum_{x \in C_i} 2(x_i - \mu_i) = 0\end{aligned} \tag{5.6}$$

By solving the aforementioned equations, it can be proven that *SSE* the minimum value is taken as the average of all samples in the cluster, i.e. on the centroidC_i.

Meanwhile, *SSE* it is a non-convex function, which means that the iterative process may fall into local optima and cannot guarantee that the final result will always be the global minimum. Therefore, the selection of initial centroids has a significant impact on the clustering results and computation time. It is necessary to choose appropriate k centroids, preferably not too close to each other.

In addition, the K-means clustering algorithm determines whether samples can be clustered into the same category based on whether the distance between samples is close enough, and the calculation of sample distance will directly affect the clustering results. In addition to the Euclidean distance mentioned earlier, there are many other distance measures. Several commonly used distance measures are shown below, where x and y are two points in the n-dimensional space.

(1) Euclidean distance, abbreviated as Euclidean distance, is the most common distance calculation method that measures the absolute distance between two points, as shown in eq. (5.7).

$$d(x, y) = \sqrt{\sum_{i=1}^{n} (x_i - y_i)^2} \tag{5.7}$$

(2) Manhattan distance, also known as urban block distance, refers to the sum of the projected distances of a line segment connected by two points on each coordinate axis in a n Cartesian coordinate system, as shown in Eq. (5.8).

$$d(x, y) = \sum_{i=1}^{n} |x_i - y_i| \tag{5.8}$$

(3) Cosine distance, using the cosine value of the angle between two vectors in vector space as a measure of the distance between two objects, as shown in Eq. (5.9).

$$d(x, y) = \cos\theta = \frac{\sum_{i=1}^{n} x_i y_i}{\sqrt{\sum_{i=1}^{n} x_i^2}\sqrt{\sum_{i=1}^{n} y_i^2}} \tag{5.9}$$

(4) Chebyshev distance refers to the maximum difference in coordinates between two points, as shown in Eq. (5.10).

$$d(x, y) = \max_{1 \le i \le n} |x_i - y_i| \tag{5.10}$$

In addition to the four common distance measures mentioned above, there are many other methods for calculating distance. When using the K-means clustering algorithm, it is necessary to analyze the specific problem being studied and choose an appropriate distance calculation method to obtain the optimal clustering results.

The disadvantages of the K-means clustering algorithm are also obvious. Firstly, the selection of initial centroids mentioned earlier will greatly affect the clustering results and speed. Secondly, the number of clusters k needs to be manually determined, and in practice, it is often impossible to determine the number of clusters in advance. Finally, due to the need to calculate the mean of samples within the cluster, the K-means clustering algorithm is particularly sensitive to outlier data. Therefore, when using the K-means clustering algorithm, different numbers of clusters and initial centroids can be selected to repeat the experiment until stable results are obtained.

In this chapter, the K-means clustering algorithm will be used in MATLAB to analyze the uncertainty of the output curve of distributed power sources. The statistical characteristics of the output curve of distributed power sources will be obtained from historical data, and representative daily output curves of distributed power sources will be obtained.

The data used in this chapter is the wind turbine output data of a wind farm located in Zhejiang Province, China, from 2015 to 2017. Sampling is conducted every 15 minutes, with a total of 96 sampling points per day. Firstly, some preprocessing was required on the raw data to remove dates containing missing data, resulting in 662 complete wind turbine daily output curves. In addition, as the daily operating capacity of the wind farm is not the same, this chapter normalized the original data and represented the final output value of the wind turbine by the ratio of the actual output power of the wind turbine to the daily operating capacity. Therefore, all wind turbine outputs clustered are a value between 0 and 1.

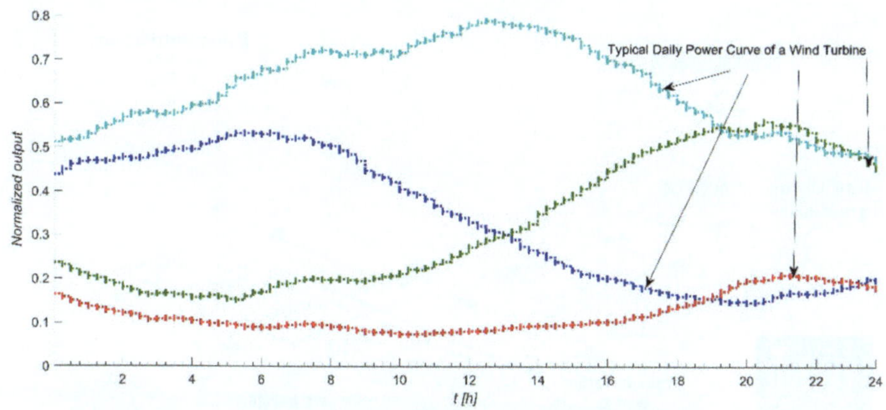

Fig. 5.1 Representative daily curves of wind turbines by K-means clustering

In the K-means clustering of this chapter, each wind turbine's daily power generation profile can be regarded as a 96-dimensional point, and these 662 points can be clustered according to Euclidean distance. After conducting multiple experiments with different k values and initial centroids, it was found that the clustering results were relatively stable at $k = 4$ that time, and the differences between each cluster were also significant, as shown in Fig. 5.1. This can be explained as the local wind speed characteristics being related to the seasons, $k = 4$ corresponding precisely to the different wind speed sizes and variation characteristics of the four seasons of spring, summer, autumn, and winter.

5.4 Collaborative Planning for EV Charging Facilities and DGs

In actual cities, the transportation network and power distribution network are often coupled and inseparable, as shown in Fig. 5.2. The two will affect each other, making it difficult to separate and consider them separately in research, which increases the complexity of the problem. This chapter considers that transportation networks and power distribution networks can be connected together through some common nodes, and these nodes are likely to be important transportation hubs or power load nodes in the city. Thus, a coupling relationship between the transportation network and the power distribution network can be established.

This chapter mainly studies the collaborative planning problem of multiple charging facilities for electric vehicles and distributed power sources that meet various constraints in the coupling of transportation and distribution networks. Electric vehicles have two completely different charging modes corresponding to charging facilities, namely conventional charging parking lots and fast charging stations.

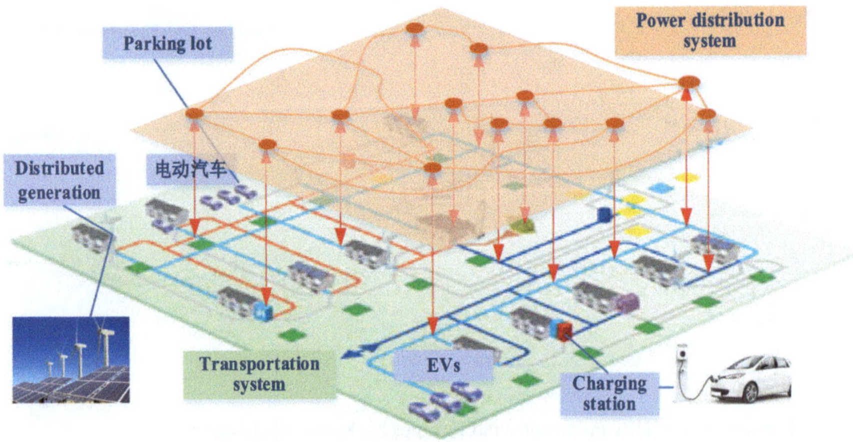

Fig. 5.2 Illustration of transportation-power network couplings

Meanwhile, both types of charging facilities are subject to various constraints imposed by transportation and distribution networks. Under such complex conditions, to collaboratively plan the optimal location and capacity of various optimization objects, e.g., conventional parking lots, fast charging stations, and distributed power sources for electric vehicles, so that the optimization results can meet the changing needs of electric vehicles in the transportation network, minimize the impact on the operation of the distribution network, and balance economy, this problem has become an extremely complex nonlinear problem that is difficult to solve using traditional optimization algorithms.

In response to this difficulty, this chapter adopts a hierarchical method, as shown in Fig. 5.3, to gradually solve the optimal location and capacity of various charging facilities and distributed power sources for electric vehicles. This not only ensures the convenience and economy of electric vehicle charging but also ensures the safe and stable operation of the distribution network, while greatly simplifying the complexity of the problem. Among them, this hierarchical method divides the optimal location and capacity problem of electric vehicle conventional parking lots, fast charging stations, and distributed power sources under the coupling of the transportation network and distribution network into the following steps. In each step, optimization is carried out separately for its respective optimization objectives, and the optimization results are carried into the subsequent steps, so as to conveniently obtain the final planning results while satisfying various constraints and optimization objectives.

Based on the hierarchical solution method described in Fig. 5.3, the collaborative planning problem of multiple charging facilities and distributed power sources for electric vehicles, considering the coupling of transportation and distribution networks, can be modeled and solved according to the following process:

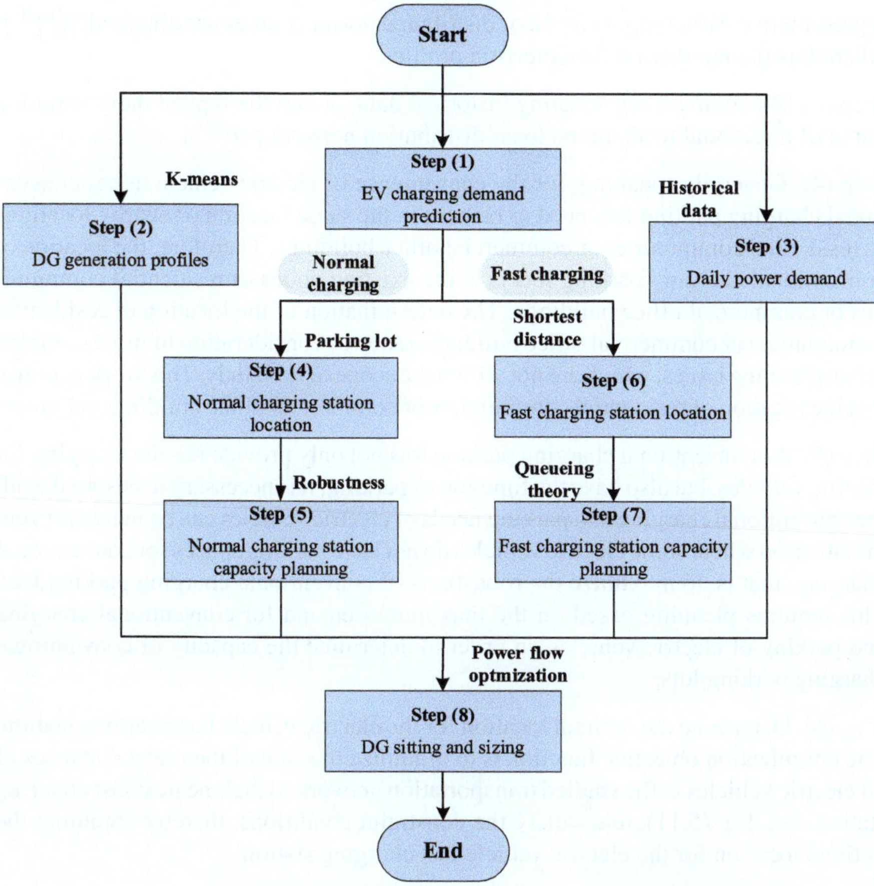

Fig. 5.3 Hierarchical planning solution of PEV charging facilities and DGs

Step (1): Based on the quantitative analysis of the electric vehicle travel statistics and charging demand, the daily variation curves of conventional and rapid charging of electric vehicles can be obtained.

Here, it is assumed that commuting electric vehicles can also receive regular charging in the commercial area after arriving at the workplace in the morning. Therefore, this problem requires the planning of regular charging parking lots in both residential and commercial areas. The conventional charging probabilities of electric vehicles in residential and commercial areas, as well as the daily variation curves of fast charging loads $f\left(P_n^{PEV}\right)$, n can be obtained using the same method, representing the three categories of electric vehicle charging loads mentioned above.

Step (2): The K-means clustering algorithm is used to analyze the uncertainty of the output curve of distributed power sources. The statistical characteristics of the output curve of distributed power sources are obtained from historical data, and representative daily output curves of distributed power sources are obtained $f\left(P_k^{DG}\right)$, where k is the number of characteristic profiles.

Step (3): By summarizing existing historical data, obtain the typical daily variation curve of traditional loads in the local distribution network $f(P^{BD})$.

Step (4): Generally speaking, for the convenience of electric vehicle users, conventional charging parking lots need to be built in the same location or nearby locations as residential communities or commercial office buildings. Therefore, the location of conventional charging parking lots is at the existing nodes of residential communities or commercial office buildings. The determination of the location of residential communities or commercial office buildings requires consideration of more complex urban planning issues, which are not within the scope of this study. This work assumes that the location of residential communities or commercial office buildings is known.

Step (5): As conventional charging parking lots not only provide regular charging for electric vehicles, but also have the function of parking, it is necessary to ensure that all the conventional charging and parking needs of electric vehicles can be met to prevent the situation where some electric vehicles do not have parking spaces for conventional charging, that is, to maximize the robustness of conventional charging parking lots. This requires planning based on the maximum demand for conventional charging and parking of electric vehicles, in order to determine the capacity of conventional charging parking lots.

Step (6): Determine the optimal location for the electric vehicle fast charging station. The optimization objective function is to minimize the sum of the travel distances of all electric vehicles in the studied transportation network to their nearest fast charging station, i.e., Eq. (5.11), and satisfy the constraint conditions, thereby obtaining the optimal location for the electric vehicle fast charging station.

$$\min D = \sum_{l=1}^{n_L} N_l^{PEV} d_l^{\min}, \quad l = 1, 2, \ldots n_L \tag{5.11}$$

Here, it is assumed that the candidate construction locations for electric vehicle fast charging stations are nodes of the original transportation network. However, in reality, electric vehicle fast charging stations may not necessarily be built on nodes of the transportation network, but may also be built on roads between nodes. To increase the granularity of the transportation network and its similarity with the real situation, the additional nodes that are inserted every unit distance in the original transportation network's routes will be selected as candidate construction locations for electric vehicle fast charging stations, along with the nodes of the original transportation network. The number of candidate construction node locations for electric vehicle fast charging stations n_L is the original number of transportation network nodes n_T plus the number of additional nodes. Meanwhile, assuming that the electric vehicles on each road are evenly distributed at each original or additional traffic node, the number of electric vehicles at each node N^{PEV} can be calculated.

Step (7): Using the economic model of electric vehicle fast charging stations based on queuing theory, based on the location of fast charging stations obtained in step (6), with the optimization objective of maximizing the annualized profit of electric vehicle fast charging stations as shown in Eq. (5.12), solve for the optimal number of charging stations l and waiting positions$w$.

$$\max E_l = 365 \sum_{t=1}^{48} \mathrm{Re}(t) - I_{cs} \tag{5.12}$$

It should be noted that Eq. (5.21) does not consider the distance cost converted from the total distance traveled by all electric vehicles within the service range of the fast charging station to charge at the fast charging station. l This is because the distance cost has been minimized in step (6) to obtain the optimal location for the fast charging station. Based on the hierarchical solution method, there is no need to consider the distance cost in step (7).

Step (8): Based on the optimal positions and installed capacities of conventional charging parking lots and fast charging stations for electric vehicles obtained in steps (4)–(7), as well as the typical daily output curves of distributed power sources and traditional load curves of the power grid obtained in steps (2) and (3), continue to configure the positions and capacities of distributed power sources to MOI minimize the multi-objective optimization variables shown in Eq. (5.13).

$$\min MOI = \min\left\{\frac{1}{48}\sum_{t=1}^{48}\left(\sigma_1 PLI_t + \sigma_2 QLI_t + \sigma_3 VDI_t\right)\right\} \tag{5.13}$$

Among them, multi-objective optimization variables MOI are defined as the weighted sum of the average values of three power flow indicators in each time interval of a day: active power loss, reactive power loss, and node voltage fluctuation. To unify dimensions in multi-objective optimization, the three indicators in Eq. (5.21) can be expressed as the ratio between the data after configuring electric vehicle charging facilities and distributed power sources and the original distribution network data; σ_1, σ_2 and σ_3 are weight factors for multi-objective optimization.

$$PLI_t = \frac{PL^t}{PL_0^t} = \frac{\sum_{h=1}^{k}\sum_{i-j=1}^{N_L}\frac{p_{ij,h,t}^2+q_{ij,h,t}^2}{V_{i,h,t}}R_{ij}}{\sum_{i-j=1}^{N_L}\frac{p_{ij,t}^2+q_{ij,t}^2}{V_{i,t}}R_{ij}} \tag{5.14}$$

$$QLI_t = \frac{QL^t}{QL_0^t} = \frac{\sum_{h=1}^{k}\sum_{i-j=1}^{N_L}\frac{p_{ij,h,t}^2+q_{ij,h,t}^2}{V_{i,h,t}}X_{ij}}{\sum_{i-j=1}^{N_L}\frac{p_{ij,t}^2+q_{ij,t}^2}{V_{i,t}}X_{ij}} \tag{5.15}$$

$$VDI_t = \frac{VD^t}{VD_0^t} = \frac{\sum_{h=1}^{k}\sum_{i-j=1}^{N_L}\frac{R_{ij,h,t}p_{ij,h,t}+X_{ij,h,t}q_{ij,h,t}}{V_{j,h,t}}}{\sum_{i-j=1}^{N_L}\frac{R_{ij,t}p_{ij,t}+X_{ij,t}q_{ij,t}}{V_{j,t}}} \tag{5.16}$$

Among them, PLI_t, QLI_t VDI_t, and respectively represent the ratio of active power loss, reactive power loss, and node voltage fluctuation of the distribution network after configuring electric vehicle charging facilities and distributed power sources to the corresponding data of the original distribution network during the time interval. When calculating the original power flow data of the distribution network, that is, on the denominator of each equation, since only one daily variation curve of the traditional load of the power grid was selected in step (3), only one calculation is needed to obtain the original power flow index of the distribution network during the t time interval. When calculating the power flow indicators of the distribution network after configuring electric vehicle charging facilities and distributed power sources, due to the use of the K-means clustering algorithm to extract k daily output curves of distributed power sources, it is necessary to comprehensively consider each characteristic output curve.

The constraints of the objective functions (5.13)–(5.16) are shown in Eqs. (5.17)–(5.25).

Equation (5.17) represents the active power balance at each node in the distribution network, where the active power at each node is provided by the grid and distributed power sources, and consumed by traditional loads, electric vehicle charging, and line losses. Similarly, Eq. (5.18) represents the reactive power balance at each node in the distribution network. Eqs. (5.19)–(5.22) define the node voltage and branch current, as well as their upper and lower limit constraints. Equation (5.23) defines the upper limit of the total penetration rate of distributed power sources. Equation (5.24) defines the upper limit of the capacity of distributed power sources $P^{DG}_{i,\max}$ that can be accessed on a node i. Equation (5.25) defines the relationship between reactive power, active power, and power factor output by distributed power sources.

$$\sum_{l=1}^{NL}\left(S_{il}\times p_{ij,h,t}\right)+P^{SS}_{i,t}+P^{DG}_{i,h,t}=R_{ij}I_{j,h,t}+P^{BD}_{i,t}+\sum_{n=1}^{3}P^{PEV}_{i,n,t} \tag{5.17}$$

$$\sum_{l=1}^{NL}\left(S_{il}\times q_{ij,h,t}\right)+Q^{SS}_{i,t}+Q^{DG}_{i,h,t}=X_{ij}I_{ij,h,t}+Q^{BD}_{i,t}+\sum_{n=1}^{3}Q^{PEV}_{i,n,t} \tag{5.18}$$

$$V^2_{j,h,t}=V^2_{i,h,t}-2\left(R_{ij}p_{ij,h,t}+X_{ij}q_{ij,h,t}\right)+\left(R^2_{ij}+X^2_{ij}\right)I_{ij,h,t} \tag{5.19}$$

$$I_{ij,h,t}V^2_{i,h,t}\geq p^2_{ij,h,t}+q^2_{ij,h,t} \tag{5.20}$$

$$0.95\ pu\leq Vi,h,t\leq 1.05\ pu \tag{5.21}$$

$$\left|I_{ij,h,t}\right|\leq 1.0\ pu \tag{5.22}$$

$$\sum_{i=1}^{NB} P_{i,h,t}^{DG} \le \varphi\left(\sum_{i=1}^{NB} P_{i,t}^{BD} + \sum_{i=1}^{NB}\sum_{n=1}^{3} P_{i,n,t}^{PEV}\right) \tag{5.23}$$

$$0 \le P_{i}^{DG} \le P_{i,\max}^{DG} \tag{5.24}$$

$$Q_{i,h,t}^{DG} = \tan\left(\cos^{-1}\left(PF_{i,t}^{DG}\right)\right) \times P_{i,h,t}^{DG} \tag{5.25}$$

If the minimum value of multi-objective optimization variables can be obtained, *MOI* it can ensure that the comprehensive indicators of network loss and node voltage fluctuation are minimized after connecting electric vehicle charging facilities and distributed power sources to the distribution network, meeting the requirements for safe and stable operation of the distribution network. At the planning level, it solves the impact of uncertainty in electric vehicle charging and distributed power sources on the power grid.

In summary, based on the collaborative optimization planning method proposed in this chapter that considers the coupling of transportation and distribution networks for multiple charging facilities of electric vehicles and distributed renewable power sources, the optimal location and capacity of conventional charging parking lots, fast charging stations, and distributed power sources for electric vehicles can be obtained through a hierarchical solution method. While ensuring that various constraints of the transportation and distribution networks are met, the economy of planning and the complexity of algorithms are also considered.

5.5 Experimental Assessment and Numerical Results

This chapter verified the scalability of the proposed method in complex networks using IEEE 123-bus power distribution network, as illustrated in Fig. 5.4, with a total of 137 branches. Node 1 is a substation with a total load of 77.44 MW/37.59Mvar, a maximum voltage of 13.8 kV, and a maximum capacity of 100MVA.

Firstly, it can be assumed that residential communities are built or planned to be built near traffic nodes # 3, # 5, # 10, # 11, and # 13, and commercial office buildings are built or planned to be built near traffic nodes # 1, # 4, # 16, # 19, and # 22. This chapter assumes that the transportation network and the distribution network are connected through some important transportation hubs or power load nodes. Therefore, some important nodes in the transportation network and distribution network can be paired one by one, i.e. the traffic nodes with a higher busy index correspond to the distribution nodes with higher power load, thus forming a coupling relationship between the two networks, as shown in Table 5.1. Other traffic nodes and additional nodes not mentioned in Table 5.1 can be assumed to be connected to the nearest distribution node.

According to the planning model and hierarchical solving steps, the optimization planning of conventional charging parking lots, fast charging stations, and

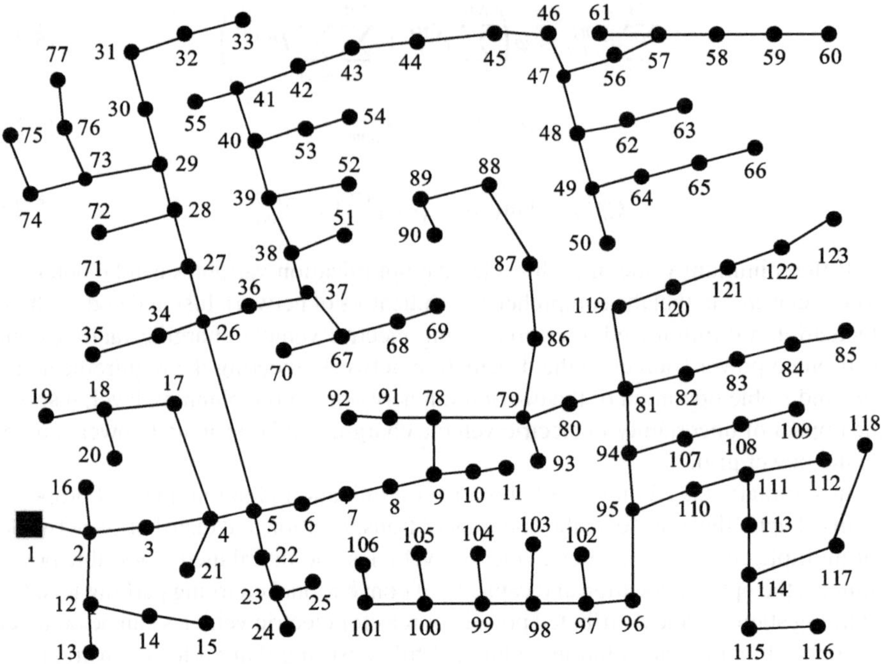

Fig. 5.4 IEEE 123-bus distribution network

Table 5.1 Transportation-power network couplings

Transportation nodes	#1	#2	#3	#4	#5	#10	#11	#13	#16	#17	#19	#22
53 power nodes	#19	#9	#10	#18	#24	#39	#8	#35	#49	#43	#40	#12
123 power node	#41	#29	#73	#26	#47	#37	#49	#79	#119	#111	#95	#2

distributed power sources for electric vehicles can be carried out in this example, thereby obtaining the optimal location and capacity under various constraints of the transportation and distribution networks.

Step (1): Using the quantitative analysis method for electric vehicle travel and charging demand, obtain the daily variation curves of the conventional charging probability and fast charging probability of electric vehicles in residential and commercial areas, as shown in Fig. 5.5.

Step (2): The representative daily output curve of distributed power sources $f\left(P_k^{DG}\right)$ has been obtained using the K-means clustering algorithm;

Step (3): The representative traditional load daily variation curve of the power grid $f(P^{BD})$ still uses the daily load curve data of a city in northern China in Sect. 2.6, as shown in the basic load curve of a city in Northern China.

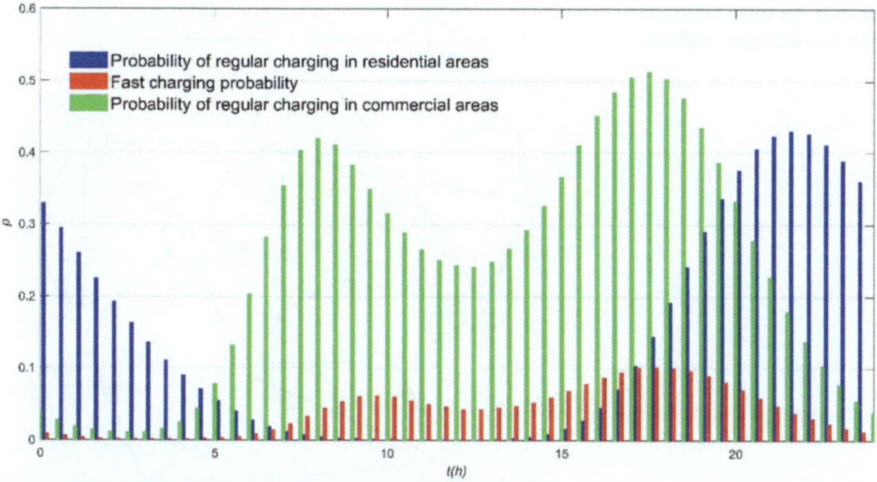

Fig. 5.5 Temporal PEV charging demand over a day

Step (4): According to the assumption in this article, the location of the conventional charging parking lot for electric vehicles is the location of residential areas or commercial office buildings. Therefore, the conventional charging parking lot for electric vehicles is located at traffic nodes # 3, # 5, # 10, # 11, and # 13, as well as # 1, # 4, # 16, # 19, and # 22;

Step (5): Assuming that the total number of electric vehicles in the studied transportation network N^{PEV} is 1000, the expected number of electric vehicles in the three charging categories at any time interval can be obtained. Assuming that the number of electric vehicles served by each conventional charging parking lot is the same, it can be inferred from the optimization objective of maximizing system robustness in the proposed method that the maximum value of each charging probability in a day needs to be used as the capacity planning standard for the parking lot. Therefore, the capacity of each parking lot in the residential area is 82 conventional charging piles, and the capacity of each parking lot in the business district is 49 conventional charging piles.

Step (6): Assuming that the unit distance in the transportation network is 1 km and the distance that electric vehicles can travel with the lowest remaining power is 14 km, the electric vehicle flow analysis model based on the gravity space interaction model can be used to obtain the minimum number of electric vehicle fast charging stations that minimize the objective function Eq. (5.13) and satisfy the constraint condition Eqs. (5.3)–(5.8). The optimal positions are the original traffic node # 2, the original traffic node # 17, and the additional nodes located between the original traffic nodes # 14 and # 22, with a distance of 1 unit from the original traffic node # 22, as shown in Fig. 5.6.

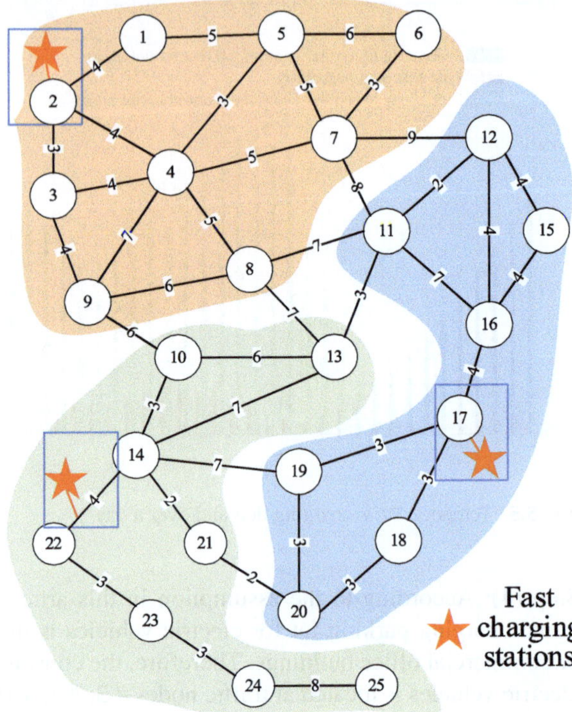

Fig. 5.6 Optimal sitting of PEV fast charging stations

Step (7): Based on the location of the electric vehicle fast charging station obtained in step (6), the arrival rate of each fast charging station can be obtained from the previous equations. Then, based on the queuing theory model, the optimal number of charging stations and waiting positions that maximize the optimization objective function of each fast charging station and satisfy the constraint conditions can be determined, as shown in Table 5.2.

Step (8): Assuming the maximum penetration rate of distributed power sources is 50%, the power factor is 0.9, the rated power of each minimum module of distributed power sources is 0.2 MW, the upper limit of distributed power source power on each distribution node is 2 MW, and the natural conditions of all distribution nodes are the same, all distributed wind turbines operate at the same wind speed. That

Table 5.2 Optimal sizing of PEV fast charging stations

Location of fast charging stations for electric vehicles	Number of charging stations	Number of waiting positions
Original traffic node # 2	14	7
Original traffic node # 17	20	8
Additional nodes near # 22	15	7

is, the daily output curve of each distributed wind turbine is the same. Based on the above assumptions, the optimization problem can be solved using the YALMIP optimization toolkit in MATLAB. The optimal location and capacity of distributed power generation are shown in Table 5.3. The total installed capacity of distributed power generation in the 53-bus distribution network is 19.6 MW, with a penetration rate of 43.6%.

At the same time, Table 5.4 shows the comparison of the average values of three power grid flow indicators before and after the configuration of electric vehicle charging facilities and distributed power sources over a day. It can be found that when only the electric vehicle charging facilities are connected to the original distribution network, the average active power loss, reactive power loss, and node voltage fluctuation of the distribution network have significantly increased. This reflects that the disorderly charging of large-scale electric vehicles mentioned earlier will have a considerable negative impact on the safe and stable operation of the distribution network. After using the hierarchical planning method proposed in this chapter and configuring electric vehicle charging facilities and distributed power sources, all three power grid flow indicators showed a significant decrease, and even operated better than the original distribution network. This indicates that the distributed power sources configured using this planning method can effectively provide power support for the distribution network and electric vehicle charging facilities, thereby alleviating the impact of electric vehicle charging on the power grid and optimizing the operation status of the distribution network. Therefore, the effectiveness of this method is demonstrated.

In addition, to verify the scalability of the hierarchical planning method proposed in this chapter in complex networks, experimental analysis will also be conducted in

Table 5.3 Optimal allocation of DGs in the 53-bus distribution network

Distribution Node (#)	8	9	10	12	18	19	21	24	30	35	38	39	40	42	43	46	49
Capacity (MW)	2.0	1.0	1.4	2.0	0.4	1.4	1.2	1.0	0.8	1.6	0.6	0.6	1.0	0.2	1.6	1.8	1.0

Table 5.4 Power profiles in the 53-bus distribution network

Average value of power grid flow indicators	Original distribution network	Only connected to electric vehicles	After connecting electric vehicles and distributed power sources
Active power loss (kW)	685.40	978.65	427.36
Active power loss (kvar)	583.10	672.34	325.04
Node voltage fluctuation (p.u.)	0.0648	0.1082	0.0430

the IEEE-bus power distribution network. The optimized planning results are shown in Fig. 5.7. Among them, the total installed capacity of distributed power generation is 25.6 MW, with a penetration rate of 28.4%. This indicates that in the 123-bus power distribution network, the number of rechargeable electric vehicles that can be accommodated and the upper limit of distributed power generation capacity can continue to increase, while still ensuring its safe and stable operation.

From Fig. 5.7, it can also be observed that most distributed power sources are optimized and configured near the distribution nodes where electric vehicle charging facilities are located, providing power support for them. This also ensures that the electricity generated by distributed power sources can be consumed by electric vehicles promptly, ensuring that the vast majority of the electricity used for electric vehicle charging is generated through renewable energy generation. This improves the efficiency of renewable energy utilization, reduces resource waste, and reduces emissions of pollutants and greenhouse gases.

Compared with other comprehensive planning methods that consider electric vehicle charging facilities and distributed power sources, the hierarchical planning method proposed in this chapter can decouple and optimize the multiple constraints of transportation and distribution networks, effectively avoiding the optimization

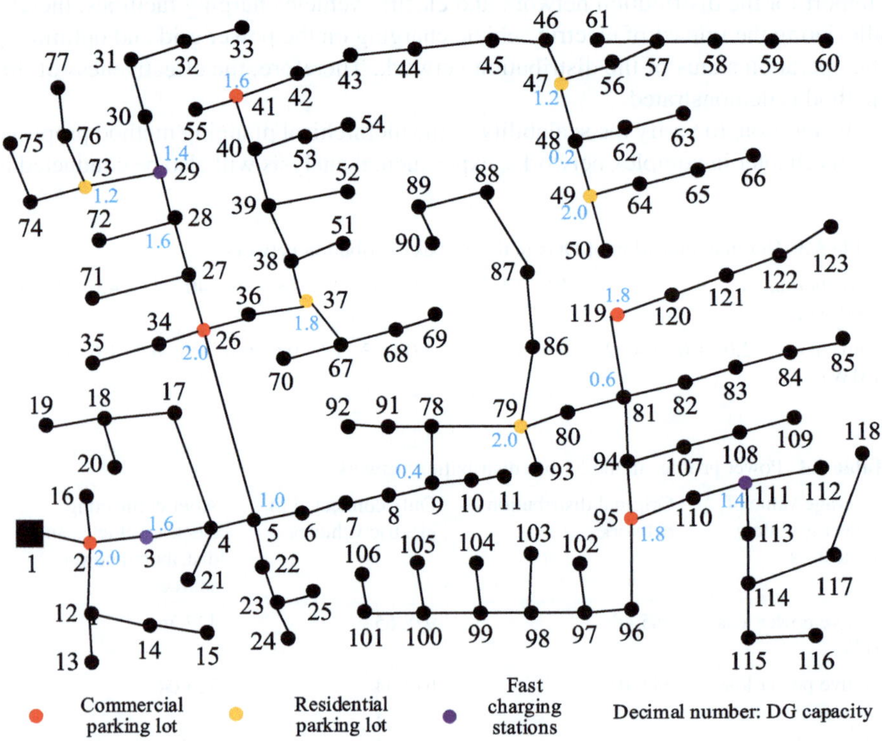

Fig. 5.7 Optimal allocation of PEV charging facilities and DGs in the IEEE 123-bus distribution network

problem from becoming a complex non-convex function that is difficult to solve. Therefore, while ensuring the satisfaction of electric vehicle charging needs and the safe and stable operation of the power grid, electric vehicles can use renewable energy generated by distributed power sources as much as possible for charging to reduce pollution emissions. It also considers the economic costs of electric vehicles and charging stations, significantly reducing the complexity of calculation and solving, and can be easily extended to complex networks.

5.6 Summary

At present, more and more distributed power sources are connected to the distribution network, and the output of distributed power sources is often affected by the natural environment and has strong uncertainty, which will have certain adverse effects on the safe and stable operation of the power grid. At the same time, in order to fully leverage the energy-saving and environmental advantages of electric vehicles, it is also necessary to make use of renewable power sources for charging as much as possible. Therefore, the coordinated planning of electric vehicle charging facilities and distributed power sources is particularly important. However, in the planning process, multiple constraints of the transportation network and distribution network need to be considered. It is necessary to use distributed power sources to meet the changing needs of as many electric vehicles as possible, while also ensuring the stable operation of the distribution network. This makes the planning problem extremely complex and difficult to solve.

In view of this, this chapter proposes a hierarchical optimization planning method for multiple charging facilities and distributed power sources of electric vehicles, considering the coupling of transportation and distribution networks. In the case of coupling between transportation and distribution networks, multiple constraints of both are comprehensively considered. A multi-level and multi optimization objective collaborative optimization planning model for conventional charging parking lots, fast charging stations, and distributed power sources of electric vehicles is established using a hierarchical method and solved layer by layer. While ensuring the convenience and economy of electric vehicle charging facilities, as well as the stability of the distribution network and other constraints, the complexity of problem solving is greatly simplified, and the scalability of the algorithm is improved.

Chapter 6
Optimal Participation of Electric Vehicles Load Aggregators in Demand Response Programs

Abstract In this chapter, we will focus on the optimal participation strategies that electric vehicle load aggregators (EVLAs) can adopt when involved in incentive-based demand response (DR) programs, where EVs would be considered as controllable loads supplying load-reduction services at peak demand times. An MLMF Stackelberg game framework is established, in which DSO, EVLAs and EV users interact hierarchically. In this framework, EVLAs behave like the leaders who choose incentive prices to make the optimal profit through aggregating EV load reduction, whereas EV users play the followers who have to find the optimal amount of load they could reduce to acquire the optimal personal benefit from receiving the incentive, but also having some delay from charging. The existence and uniqueness of the Stackelberg equilibrium (SE) are rigorously proven by using the potential game theory, demonstrating that the non-cooperative game among EVLAs qualifies as an exact potential game with a unique Nash equilibrium (NE). To account for the inherent uncertainties in EV charging behaviors, stochastic models are introduced to characterize those key variables, e.g., plug-in time, charging duration, and energy demand by the use of truncated Chi-square and Gaussian distributions, respectively. Based on the numerical simulation of the community parking lot scenario, three EVLAs are considered to verify the feasibility of the proposed approach. The result demonstrates the significant effect of other influencing elements, including expected DR capacities, available load cutback subject to charger rating, user demand, and competitive equilibrium between EVLAs, on incentive plans for EVLAs. Sensitivity analysis further indicates that incentive prices tend to decrease as the number of EVs or participating EVLAs increases, while the higher DR capacity requirements may lead to increased incentive offers and reduced overall potential function values. This study provides a theoretical and practical foundation for designing an effective DR participation mechanism for EVLAs, highlighting the importance of strategic incentive design and competition management in smart grid environments.

Q. Yang et al., *Integration of Massive Electric Vehicles to Smart Grid*, Smart Energy Systems, https://doi.org/10.1007/978-981-92-0557-8_6

6.1 Introduction

DR programs are engineered to reshape end-user electricity consumption patterns by incentivizing adjustments in usage during those periods of significantly elevated generation/delivery costs or compromised system reliability. These adjustments are triggered by market signals such as time-variable electricity pricing or direct incentive payments. Through strategic load shifting during critical periods, DR initiatives serve three key functions: (1) alleviating grid strains; (2) preventing any outages; (3) cutting down operation costs and greenhouse gas emissions of the power grid.

In the last decade, there has been a lot of interest in EVs as a different kind of flexible load/resource since EVs can provide an energy buffer potentially. The large-scale deployment of Internet-of-Things (IoT) technology allows EVs to become potential load balancing resources with controlled charging power modulating in DR programs. To achieve coordinated EV charging, utilizing EVs to provide DR services is considered a promising solution for EV charging management aided by the G2V and V2G technology.

DR programs generally fall into two categories: price-based methods and incentive-based methods. While price-based DR alters users' consumption habits via different kinds of electrical charges, incentive-based DR leverages constant, sometimes varying rewards for lowering electric loads. In general, incentive-based DR is better on dispatch then price based DR, indicating it will fit well with load aggregation companies that actually want the amount of DR they need to meet their needs. This section investigates the strategies for EVLAs to join incentive-based DR programs. Herein, EVs are treated as controlled electric loads.

In this chapter, incentives are adopted to prompt EVLAs to cut down the charging power of EVs during those higher demand times. EVLAs aggregate and manage the EV cluster to supply DR services to the distribution network. Figure 6.1 shows the operation structure taking EVs as flexible DR resources in the presence of multiple EVLAs. In primary level, the DSO will publish an incentive price as well as a minimum market entry price to all EVLAs. At the secondary level, EVLA decides which load reduction action would be most beneficial and sends those decisions along to the DSO, as well as how they will pay EV users for their electricity reduction capacities. The tertiary layer is to find the optimal load reduction action in terms of what EV users would adopt. It is also supposed that each EV is available for different EVLAs. In that context, the DR service trading interaction between EVLAs and EVs would take the appearance of a multi-leader multi-follower (MLMF) type of Stackelberg game.

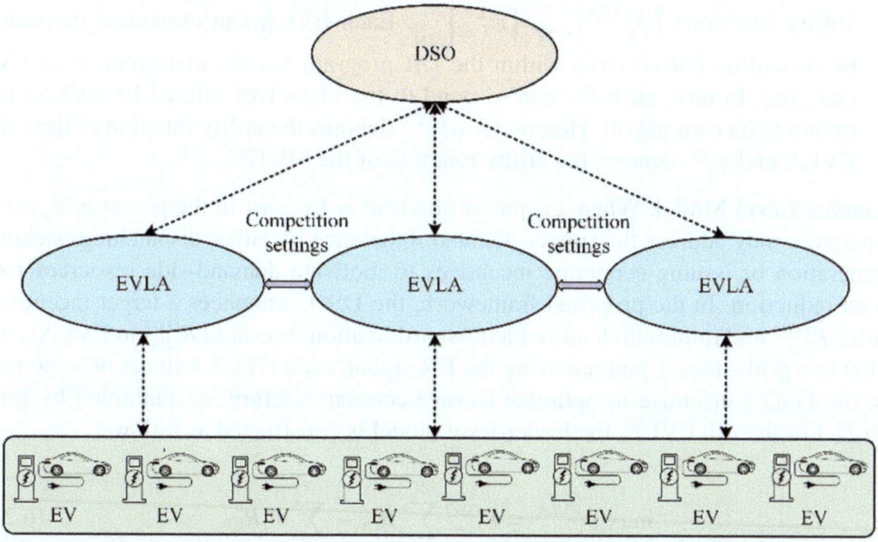

Figure Note: DSO, Distribution System Operator; EVLA, Electric Vehicle Load Aggregators; EV, Electric Vehicle.

Fig. 6.1 Operating structure of utilizing EVs as DR resources with multiple EVLAs

6.2 MLMF Stackelberg Game Formulation

This section aims to establish an MLMF Stackelberg game model to analyze the interactive dynamics between EVLAs and EVs. Specifically, EVLAs function as the leaders by determining incentive prices to secure aggregated load reductions from EVs. Meanwhile, EV users act as the followers, adjusting their charging demand in response to the incentives provided by those EVLAs. Mathematically, the MLMF Stackelberg game can be defined as follows:

$$\Omega = \left\{ \mathbf{M} \cup \mathbf{N}, \{P_i\}_{i \in \mathbf{M}}, \left\{R_{i,j}\right\}_{i \in \mathbf{M}, j \in \mathbf{N}}, \left\{\psi_i^{EVLA}\right\}_{i \in \mathbf{M}}, \left\{\psi_j^{EV}\right\}_{j \in \mathbf{N}} \right\} \quad (6.1)$$

where the game components are described as follows:

- Player sets $\mathbf{M} \cup \mathbf{N}$. $\mathbf{M}$ means the set of EVLAs included in the DR project, and $\mathbf{N}$ means the set of EVs willing to be involved in load reduction. Additionally, M denotes the number of EVLAs, and N denotes the number of EVs.
- Strategy sets $\{P_i\}_{i \in \mathbf{M}}$, $\{R_{i,j}\}_{i \in \mathbf{M}, j \in \mathbf{N}}$. P_i means the incentive price which the i-th EVLA offered to EV users who are willing to reduce their load demand, and $R_{i,j}$ means the load reduction amount adopted by the j-th EV user in response to the incentive provided by the i-th EVLA.

- Utility functions $\{\psi_i^{EVLA}\}_{i\in\mathbf{M}}, \{\psi_j^{EV}\}_{j\in\mathbf{N}}$. Each EVLA can maximize its profit by providing DR services within the DR program via the management of EV charging. In turn, each EV can respond to the incentives offered by EVLAs to optimize its own payoff. Herein, let ψ_i^{EVLA} denotes the utility function of the i-th EVLA and ψ_j^{EV} denotes the utility function of the j-th EV.

Leader-Level Model. When a capacity shortfall is forecast in the power grid, grid operators may address the supply-demand imbalance by either dispatching backup generation or issuing economic incentives to motivate demand-side resources for load reduction. In the proposed framework, the DSO announces a target incentive price P_{Inc}^{DSO} and minimum load-reduction participation threshold R_{Min}^{DSO} to EVLAs, to alleviate grid stress. Upon receiving the DR signal, each EVLA will act in response to the DSO's incentive to optimize its own economic return, as quantified by Eq. (6.2). For the i-th EVLA, the leader-level model is constructed as follows:

$$\max_{P_i} \psi_i^{EVLA} = P_{Inc}^{DSO} \sum_{j=1}^{N} R_{i,j} - \sum_{j=1}^{N} P_i R_{i,j} \tag{6.2}$$

subject to:

$$P_i^{Min} \leq P_i \leq P_i^{Max} \tag{6.3}$$

$$\sum_{j=1}^{N} R_{i,j} \geq R_{Min}^{DSO} \tag{6.4}$$

where Constraint (6.3) states the incentive price is allowed to be above lower bounds and below the upper bounds, and Constraint (6.4) requires that all combined load curtailment contributions from EVs must meet the market participation threshold of DSO.

Follower-Level Model. In the follower-level optimization model, EV users determine their optimal load-reduction strategies by maximizing their individual welfare in response to the incentives provided by EVLAs. Nevertheless, curtailing EV charging power will extend the required charging duration, thereby imposing inconvenience and reduced comfort on users. Consequently, the utility function for the EV user integrates the revenue obtained from incentives and the dissatisfaction cost caused by load reduction. Given the incentive price set by the i-th EVLA, the utility function of the j-th EV user can be formulated as:

$$\psi_j^{EV} = \omega_j P_i R_{i,j} - (1 - \omega_j)\varphi_j^{EV}(R_{i,j}) \tag{6.5}$$

where $\varphi_j^{EV}(R_{i,j})$ denotes the dissatisfaction cost of the EV, which can be calculated as:

$$\varphi_j^{EV}\left(R_{i,j}\right) = \alpha_j R_{i,j}^2 + \beta_j R_{i,j} \tag{6.6}$$

In Eq. (6.5), the first term corresponds to the direct incentive revenue that an EV user receives from EVLAs, whereas the second term reflects the dissatisfaction cost associated with the curtailment of charging loads. ω_j stands for the weighting factor that balances incentive gains and discomfort costs. In Eq. (6.6), the parameters α_j, β_j are user-specific and vary across individual EV users. As illustrated in Fig. 6.2, when α_j remains constant, the utility function reaches its maximum at a lower load-reduction level for a larger β_j. This indicates that an EV user with a larger β_j tends to be more conservative in participating in DR programs. In contrast to the parameter β_j, the parameter α_j exerts a pronounced effect on the utility function of EV users. In general, an EV user with a larger α_j is more sensitive to the magnitude of load reduction than one with a smaller α_j.

Therefore, the follower-level model is formulated as:

$$\max_{R_{i,j}} \psi_j^{EV} = \omega_j P_i R_{i,j} - \left(1 - \omega_j\right)\left(\alpha_j R_{i,j}^2 + \beta_j R_{i,j}\right) \tag{6.7}$$

subject to:

$$0 \leq R_{i,j} \leq R_j^{Max}, \quad \forall i \in \mathbf{M} \tag{6.8}$$

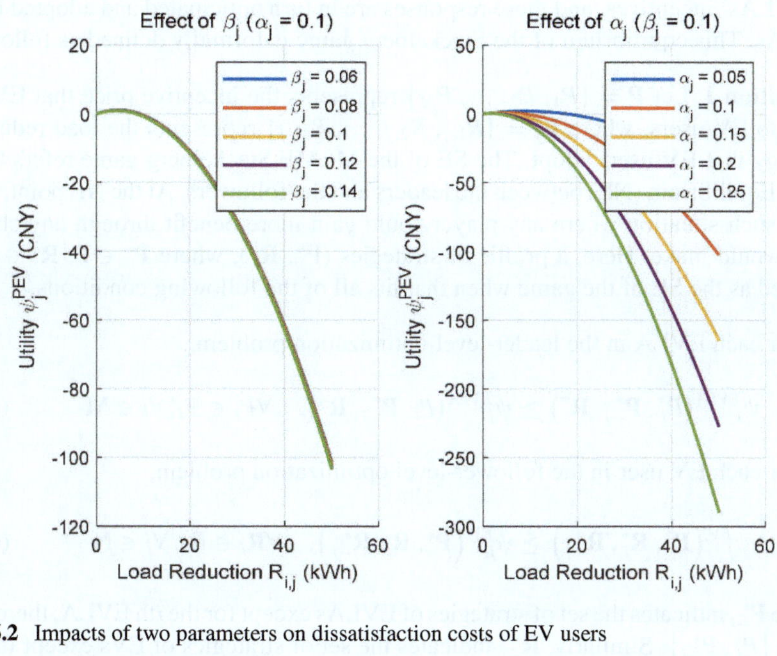

Fig. 6.2 Impacts of two parameters on dissatisfaction costs of EV users

$$R_j^{Max} = p_j^{Rat} \tau_{DR} \tag{6.9}$$

$$\sum_{i=1}^{M} R_{i,j} \le R_j^{Max} \tag{6.10}$$

$$SoC_j^{Int} + \frac{\left(p_j^{Rat} \tau_{GO} - \sum_{i=1}^{M} R_{i,i}\right)\eta_j}{C_j^{Bat}} \ge SoC_j^{Min} \tag{6.11}$$

where Constraints (6.8) and (6.9) specify that the load reduction of an EV engaging in transactions with an EVLA must not exceed the maximum capacity it can provide during DR periods; Constraint (6.10) indicates that the total load reduction of an EV transacting with multiple EVLAs must not exceed its maximum allowable load reduction limit; and Constraint (6.11) represents that the final state of charge (SoC) of the EV's battery pack must not fall below the minimum SoC set by its user.

Definition of Stackelberg Equilibrium. In this game, EVLAs act as dominant decision-makers, with priority over EV users. EV users formulate their vehicle load reduction strategies in response to the incentives provided by EVLAs. Notably, EVLAs will adjust their incentive schemes to either further improve their profits or aggregate a sufficient number of EVs to satisfy the minimum entry requirements for DR markets. The Stackelberg Equilibrium (SE) of the MLMF Stackelberg game is achieved when all EV users adopt strategies that represent their optimal responses to EVLAs' incentives, and these responses are in turn anticipated and adopted by all EVLAs. This equilibrium of the Stackelberg game is formally defined as follows:

Definition 1 Let $\mathbf{P} = \{P_1, P_2, \ldots, P_M\}$ represents the incentive price that EVLAs offer to EV users, where $\mathbf{R}_j = \{R_{1,j}, R_{2,j}, \ldots, R_{M,j}\}$ represents the load reduction strategy that EV users adopt. The SE of the MLMF Stackelberg game refers to the Nash Equilibrium (NE) between the leaders and the followers. At the NE point, there is no such situation where any player could gain more benefit through any change they would make. Here, a profile of strategies $(\mathbf{P}^*, \mathbf{R}^*)$, where $\mathbf{P}^* \in \mathbf{P}$, $\mathbf{R}^* \in \mathbf{R}$, is defined as the SE of the game when that fits all of the following conditions.

For each EVLA in the leader-level optimization problem,

$$\psi_i^{EVLA}\left(P_i^*, \mathbf{P}_{-i}^*, \mathbf{R}^*\right) \ge \psi_i^{EVLA}\left(P_i, \mathbf{P}_{-i}^*, \mathbf{R}^*\right), \quad \forall P_i \in \mathbb{P}_i, \forall i \in \mathbf{M} \tag{6.12}$$

For each EV user in the follower-level optimization problem,

$$\psi_j^{EV}\left(\mathbf{P}^*, \mathbf{R}_j^*, \mathbf{R}_{-j}^*\right) \ge \psi_j^{EV}\left(\mathbf{P}^*, \mathbf{R}_j, \mathbf{R}_{-j}^*\right), \quad \forall \mathbf{R}_j \in \mathbb{R}_j, \forall j \in \mathbf{N} \tag{6.13}$$

where $\mathbf{P}_{-i}^*$ indicates the set of strategies of EVLAs except for the ith EVLA; therefore, $\mathbf{P}^* = \left\{P_i^*, \mathbf{P}_{-i}^*\right\}$. Similarly, $\mathbf{R}_{-j}^*$ indicates the set of strategies of EVs except for the

jth EV, and $\mathbf{R}^* = \left\{\mathbf{R}_j^*, \mathbf{R}_{-j}^*\right\}$. It should be noted that Condition (6.12) ensures the equilibrium among M EVLAs and Condition (6.13) ensures the equilibrium among N EVs.

6.3 Existence and Uniqueness Analysis of Stackelberg Equilibrium

Here, the proof of the existence and uniqueness of SE for that game is manifested via the potential game theory. In this section, the optimal response action of EV users who react to the EVLAs is acquired. And then, the optimal response strategies of the EV users are integrated into the leader-level optimization model to study the NE of the game.

Identify Best Responses of EVs under Multiple Incentives. In the follower-level model, given the incentive offered by the i-th EVLA, the optimal response of the j-th EV can be generated by computing the first-order derivative of ψ_j^{EV} for $R_{i,j}$:

$$\frac{\partial\ \psi_j^{EV}}{\partial\ R_{i,j}} = \omega_j P_i - (1-\omega_j)\big(2\alpha_j R_{i,j} + \beta_j\big) = 0, \quad \forall j \in \mathbf{N} \tag{6.14}$$

Mathematically, setting Eq. (6.14) to zero allows the optimal load reduction of the EV to be expressed as:

$$R_{i,j} = \frac{\omega_j P_i}{2\alpha_j(1-\omega_j)} - \frac{\beta_j}{2\alpha_j}, \quad \forall j \in \mathbf{N} \tag{6.15}$$

In addition, the second-order derivative of ψ_j^{EV} for $R_{i,j}$ can be computed as follows:

$$\frac{\partial^2 \psi_j^{EV}}{\partial\ R_{i,j}^2} = -2\alpha_j(1-\omega_j), \quad \forall j \in \mathbf{N} \tag{6.16}$$

Note that the value of Eq. (6.16) remains negative owing to $\alpha_j > 0$ and $0 < \omega_j < 1$. It implies that ψ_j^{EV} is strictly concave in its feasible region. Consequently, the solution given in Eq. (6.15) is optimal and unique for the EV user the solution presented in Eq. (6.15) is optimal and unique for the EV user seeking to maximize its welfare, as indicated in Eq. (6.7).

Drawing on the best responses of each EV to the incentives provided by EVLAs, the utility function of each EVLA can be reformulated as:

$$\psi_i^{EVLA} = P_{Inc}^{DSO} \sum_{j=1}^{N} \left(\frac{\omega_j P_i}{2\alpha_j (1-\omega_j)} - \frac{\beta_j}{2\alpha_j} \right) - \sum_{j=1}^{N} \left(\frac{\omega_j P_i^2}{2\alpha_j (1-\omega_j)} - \frac{\beta_j P_i}{2\alpha_j} \right) \tag{6.17}$$

Furthermore, the constraints for both leaders and followers can be rewritten as follows:

$$\sum_{j=1}^{N} \left(\frac{\omega_j P_i}{2\alpha_j (1-\omega_j)} - \frac{\beta_j}{2\alpha_j} \right) \geq R_{Min}^{DSO} \tag{6.18}$$

$$0 \leq \frac{\omega_j P_i}{2\alpha_j (1-\omega_j)} - \frac{\beta_j}{2\alpha_j} \leq R_j^{Max} \tag{6.19}$$

$$\sum_{i=1}^{M} \left(\frac{\omega_j P_i}{2\alpha_j (1-\omega_j)} - \frac{\beta_j}{2\alpha_j} \right) \leq R_j^{Max} \tag{6.20}$$

$$\frac{\left(P_j^{Rat} \tau_{GO} - \sum_{i=1}^{M} \left(\frac{\omega_j P_i}{2\alpha_j (1-\omega_j)} - \frac{\beta_j}{2\alpha_j} \right) \right) \eta_j}{C_j^{Bat}} \geq SoC_j^{Min} - SoC_j^{Ini} \tag{6.21}$$

Each EVLA seeks to maximize its profit by leveraging EVs to offer DR services. Nevertheless, the flexible load resources available from EVs are subject to limitations. Accordingly, under Constraints (6.18) and (6.20), each EVLA needs to account for the strategies of other aggregators when formulating its own decisions. Such interdependent competition among multiple EVLAs constitutes a non-cooperative Nash game. As a fundamental analytical tool, Nash Equilibrium plays a crucial role in analyzing the strategic decision-making behaviors of all participants.

Verification of the Game among EVLAs as a Potential Game. The concept of potential games was first introduced by Monderer and Shapley. Over the past decades, such games have attracted considerable research interest due to their favorable properties regarding the existence and attainability of NE. As a category of non-cooperative games, a defining feature of potential games is that key equilibrium information can be encapsulated within a single real-valued function, referred to as the potential function, defined over the strategy space. This function provides an effective framework for analyzing equilibrium characteristics by unifying the objectives of all players into a single expression. Potential games are generally classified into four categories: ordinal potential games, weighted potential games, exact potential games, and generalized ordinal potential games. Importantly, any finite potential game, whether ordinal or exact, guarantees at least one pure-strategy Nash equilibrium. To establish the existence of NE among EVLAs, this section introduces a lemma to verify that the non-cooperative Nash game between EVLAs belongs to the family of potential games. The following section will further prove that this game is specifically an exact potential game.

Theorem 1 *Given a strategic game in which the strategy sets $\mathbb{P}_i$, $i \in \mathbf{M}$, are intervals of real numbers. Assuming that the utility functions are twice continuously differentiable, this game is a potential game if and only if*:

$$\frac{\partial^2\left(\psi_i^{EVLA}-\psi_k^{EVLA}\right)}{\partial P_i \partial P_k}=0, \quad \forall i \in \mathbf{M}, k \in \mathbf{M} \tag{6.22}$$

Proof Firstly, in terms of the variable P_i, the utility function ψ_i^{EVLA} can be rewritten as:

$$\psi_i^{EVLA}=-P_i^2 \sum_{j=1}^N \frac{\omega_j}{2\alpha_j(1-\omega_j)}+P_i \sum_{j=1}^N\left(\frac{P_{Inc}^{DSO}\omega_j}{2\alpha_j(1-\omega_j)}+\frac{\beta_j}{2\alpha_j}\right)-\sum_{j=1}^N \frac{P_{Inc}^{DSO}\beta_j}{2\alpha_j} \tag{6.23}$$

It can be observed that, with reference to Eq. (6.23), the utility function is twice continuously differentiable with respect to P_i. Moreover, the incentive prices of EVLAs are real numbers, which satisfy the basic conditions of Theorem 1.

Furthermore, the difference in utility between two distinct EVLAs can be computed as follows:

$$\begin{aligned}\psi_i^{EVLA}-\psi_k^{EVLA}= & -\left(P_i^2-P_k^2\right) \sum_{j=1}^N \frac{\omega_j}{2\alpha_j(1-\omega_j)} \\ & +\left(P_i-P_k\right) \sum_{j=1}^N\left(\frac{P_{Inc}^{DSO}\omega_j}{2\alpha_j(1-\omega_j)}+\frac{\beta_j}{2\alpha_j}\right)\end{aligned} \tag{6.24}$$

By computing the derivative of Eq. (6.24) with respect to P_i, the following result is obtained:

$$\frac{\partial\left(\psi_i^{EVLA}-\psi_k^{EVLA}\right)}{\partial P_i}=-P_i \sum_{j=1}^N \frac{\omega_j}{\alpha_j(1-\omega_j)}+\sum_{j=1}^N\left(\frac{P_{Inc}^{DSO}\omega_j}{2\alpha_j(1-\omega_j)}+\frac{\beta_j}{2\alpha_j}\right) \tag{6.25}$$

By computing the derivative of Eq. (6.25) with respect to P_k, the following result is obtained:

$$\frac{\partial^2\left(\psi_i^{EVLA}-\psi_k^{EVLA}\right)}{\partial P_i \partial P_k}=0 \tag{6.26}$$

To this end, we have established that the non-cooperative game among multiple EVLAs is indeed a potential game. We now proceed to further verify that this game specifically conforms to an exact potential game through the lens of the potential function theory.

Verification of the Game as an Exact Potential Game. To verify the existence and attainability of NE among multiple EVLAs, Definition 2 is listed as follows:

Definition 2 A non-cooperative strategic game is an exact potential game if there exists a potential function Φ that satisfies the following condition:

$$\begin{aligned}&\psi_i^{EVLA}(P_i, \mathbf{P}_{-i}) - \psi_i^{EVLA}(\hat{P}_i, \mathbf{P}_{-i}) \\ &= \Phi(P_i, \mathbf{P}_{-i}) - \Phi(\hat{P}_i, \mathbf{P}_{-i}), \quad \forall P_i, \hat{P}_i \in \mathbb{P}_i, \forall i \in \mathbf{M}\end{aligned} \tag{6.27}$$

In an exact potential game, any unilateral strategy change by a single player produces a utility variation that exactly matches the corresponding change in the potential function. Importantly, the potential function itself is invariant with respect to player indices. Accordingly, for each player, Φ characterizes how unilateral strategy shifts affect the utilities of all participants in exact potential games.

Theorem 2 *The potential game among EVLAs is an exact potential game if there exists a potential function, i.e.,* Eq. (6.28), *that satisfies Condition* (6.27).

$$\begin{aligned}\Phi(P_i, \mathbf{P}_{-i}) = &-\sum_{j=1}^{N}\frac{\omega_j}{2\alpha_j(1-\omega_j)}\sum_{i=1}^{M}P_i^2 \\ &+\sum_{j=1}^{N}\left(\frac{P_{Inc}^{DSO}\omega_j}{2\alpha_j(1-\omega_j)}+\frac{\beta_j}{2\alpha_j}\right)\sum_{i=1}^{M}P_i - M\sum_{j=1}^{N}\frac{P_{Inc}^{DSO}\beta_j}{2\alpha_j}\end{aligned} \tag{6.28}$$

Proof When the leader changes the incentive strategy from P_i to $\hat{P}_i$, the difference in the leader's utility function (namely, the left of Eq. (6.27)) can be calculated as:

$$\begin{aligned}\psi_i^{EVLA}(P_i, \mathbf{P}_{-i}) - \psi_i^{EVLA}(\hat{P}_i, \mathbf{P}_{-i}) = &-\left(P_i^2 - \hat{P}_i^2\right)\sum_{j=1}^{N}\frac{\omega_j}{2\alpha_j(1-\omega_j)} \\ &+\left(P_i - \hat{P}_i\right)\sum_{j=1}^{N}\left(\frac{P_{Inc}^{DSO}\omega_j}{2\alpha_j(1-\omega_j)}+\frac{\beta_j}{2\alpha_j}\right)\end{aligned} \tag{6.29}$$

To calculate the variation in the value of the potential function, Eq. (6.28) is rewritten as follows:

$$\begin{aligned}\Phi(P_i, \mathbf{P}_{-i}) = &-\sum_{j=1}^{N}\frac{\omega_j}{2\alpha_j(1-\omega_j)}\left(P_i^2+\sum_{k\neq i}^{M}P_k^2\right) \\ &+\sum_{j=1}^{N}\left(\frac{P_{Inc}^{DSO}\omega_j}{2\alpha_j(1-\omega_j)}+\frac{\beta_j}{2\alpha_j}\right)\left(P_i+\sum_{k\neq i}^{M}P_k\right) - M\sum_{j=1}^{N}\frac{P_{Inc}^{DSO}\beta_j}{2\alpha_j}\end{aligned} \tag{6.30}$$

Hence, the term on the right-hand side of Eq. (6.27) can be computed by eliminating the identical components:

$$\Phi\left(P_i, \mathbf{P}_{-i}\right) - \Phi\left(\hat{P}_i, \mathbf{P}_{-i}\right) = -\left(P_i^2 - \hat{P}_i^2\right)\sum_{j=1}^{N}\frac{\omega_j}{2\alpha_j(1-\omega_j)} + \left(P_i - \hat{P}_i\right)\sum_{j=1}^{N}\left(\frac{P_{Inc}^{DSO}\omega_j}{2\alpha_j(1-\omega_j)} + \frac{\beta_j}{2\alpha_j}\right) \tag{6.31}$$

As can be seen, the formulation in Eq. (6.29) is consistent with that in Eq. (6.31). This implies that the utility difference caused by a leader's strategy adjustment is equivalent to the change in the potential function. Thus, the potential game among multiple EVLAs is proven to be an exact potential game.

Verification of the Existence of a Unique NE of the Game. Exact potential games inherently guarantee the existence of at least one NE. Moreover, when the strategy sets are compact and convex, and the potential function is continuously differentiable within the interior of these sets and strictly concave across the entire strategy domain, the NE is unique. The corresponding lemma is formalized in Theorem 3.

Theorem 3 *In an exact potential game, a unique Nash Equilibrium exists if the game meets the following two conditions*: (1) *The strategy set* **P** *is compact and convex*; (2) *The potential function is continuously differentiable on the interior of* **P** *and strictly concave over* **P**.

Proof Here, we only need to demonstrate the uniqueness of the NE, as the exact potential game guarantees the existence of a pure-strategy Nash equilibrium. Regarding the first condition, it can be observed that Constraints (6.18)–(6.21) are linear constraints. Consequently, the strategy set **P** bounded by Eqs. (6.18)–(6.21) is compact and convex.

For the second condition, it is straightforward that the potential function is continuously differentiable, as evidenced by Eq. (6.28). To compute the Hessian matrix of Φ, the second-order partial derivatives of the potential function can be calculated as follows:

$$\frac{\partial^2\Phi(\mathbf{P})}{\partial P_i \partial P_k} = \begin{cases} -\sum_{j=1}^{N}\frac{\omega_j}{\alpha_j(1-\omega_j)}, & i = k \\ 0, & i \neq k \end{cases} \tag{6.32}$$

Hence, the expression for the Hessian matrix of Φ is given as follows:

$$H(\Phi) = \begin{vmatrix} -\sum_{j=1}^{N} \frac{\omega_j}{\alpha_j(1-\omega_j)} & 0 & \cdots & 0 \\ 0 & -\sum_{j=1}^{N} \frac{\omega_j}{\alpha_j(1-\omega_j)} & \cdots & 0 \\ 0 & 0 & \cdots & 0 \\ 0 & 0 & \cdots & -\sum_{j=1}^{N} \frac{\omega_j}{\alpha_j(1-\omega_j)} \end{vmatrix}_{M \times M} \quad (6.33)$$

It is observed that the Hessian matrix is negative definite, since all its diagonal elements are negative owing to $\alpha_j > 0$ and $0 < \omega_j < 1$. This indicates that the potential function is strictly concave on the interior of **P**. Ultimately, it is concluded that there exists exactly one NE solution in the potential game among M EVLAs. Furthermore, the optimal load reduction of EVs responding to EVLAs' incentives is unique. Thus, the MLMF Stackelberg game possesses a unique Stackelberg equilibrium.

Solution of the Proposed Stackelberg Game. The preceding section has established that the non-cooperative game among M EVLAs admits a unique Nash equilibrium. For potential games, reaching such an equilibrium is equivalent to identifying the optimal solution of the potential function over its strategy set. Accordingly, solving the optimization model with objective function (6.28) and Constraints (6.3) and (6.18)-(6.21) will generate the NE point, thereby obtaining the optimal incentive strategy P_i for the i-th EVLA.

Furthermore, the best-response strategies of EVs to given incentives provided by EVLAs can be derived. According to Eq. (6.15), the optimal load reduction strategy for the j-th EV under those incentives of EVLAs can be computed, in the form of $\mathbf{R}_j^* = \left\{R_{1,j}^*, R_{2,j}^*, \ldots, R_{M,j}^*\right\}$. Consequently, the optimal load reduction of EVs can be determined, denoted as $\mathbf{R}^* = \left\{\mathbf{R}_1^*, \mathbf{R}_2^*, \ldots, \mathbf{R}_N^*\right\}$. To this point, the unique SE of the game's strategy profile $(\mathbf{P}^*, \mathbf{R}^*)$ has been established.

6.4 Electric Vehicle Charging Uncertainty

Unlike conventional DR resources such as industrial loads, EV charging behaviors exhibit substantial uncertainty stemming from the spatiotemporal dynamics of EV travel. EV charging is governed by multiple stochastic variables, including arrival and departure times, driving mileage, initial battery SoC, and target energy demand. To maximize their payoffs, EVLAs must fully incorporate the stochastic charging characteristics of EVs into their DR provision decisions. For a large community containing commercial and official buildings, the uncertainties related to EVs are modeled as described below.

For charging stations located in the community, the Chi-square distribution function is adopted to describe the arrival times of EVs during daily commutes. EV Users are assumed to connect their vehicles to electric vehicle supply equipment (EVSE) for charging immediately upon entering the EV parking area. Therefore, the plug-in time of EVs can be expressed by a truncated Chi-square distribution, as formulated in Eq. (6.34):

$$f_A(\tau_{In}) = \frac{1}{2^{n/2}\Gamma(n/2)}(\tau_{In})^{n/2-1}\exp(-\tau_{In}/2), \quad \tau_{In} \in \left[\tau_{In}^{Min}, \tau_{In}^{Max}\right] \tag{6.34}$$

where n represents the freedom degree of this distribution, $\Gamma(\cdot)$ represents the Gamma function, and $\left[\tau_{In}^{Min}, \tau_{In}^{Max}\right]$ represents the truncated region indicating that the plug-in time of EVs falls in between the lower bound and upper bound of the truncated region.

To ensure the logical validity of EVs' plug-out time (i.e., it is greater than the plug-in time), the charging duration of EVs is simulated by using a truncated Gaussian distribution, as expressed in Eq. (6.35). Therefore, the plug-out time of EVs is modeled via Eq. (6.36).

$$f_D(\tau_{Dur}) = \frac{1}{\sigma_{Dur}\sqrt{2\pi}}\exp\left[-\frac{(\tau_{Dur}-\mu_{Dur})^2}{2\sigma_{Dur}^2}\right], \quad \tau_{Dur} \in \left[\tau_{Dur}^{Min}, \tau_{Dur}^{Max}\right] \tag{6.35}$$

where μ_{Dur} and σ_{Dur} represents the mean and standard deviation of the distribution of the variable, and $\left[\tau_{Dur}^{Min}, \tau_{Dur}^{Max}\right]$ represents the truncated region of the random variable, as given in (34).

$$\tau_{Out} = \tau_{In} + \tau_{Dur} \tag{6.36}$$

where τ_{Out} represents the initial time of EV charging, τ_{In} represents the ending time of EV charging, and τ_{Dur} represents the total charging time of EVs.

The number of EVs being charged in each period is calculated based on the connection and disconnection time of EV charging, as formulated in Eq. (6.37):

$$N^t = N^{t-1} + N_{In}^t - N_{Out}^t, \forall t \tag{6.37}$$

where N^t and N^{t-1} denote the number of EVs being charged at time t and t-1, N_{In}^t denotes the number of EVs connecting to EVSE for charging at time t, and N_{Out}^t denotes the number of EVs, which are disconnected from EVSE at time t.

Additionally, the battery capacity of EVs is taken from current mainstream EV models stochastically. Moreover, the energy demand of EVs is modeled by using a truncated Gaussian distribution, as shown in Eq. (6.38).

$$f_P(e_{PEV}) = \frac{1}{\sigma_{PEV}\sqrt{2\pi}}\exp\left[-\frac{(e_{PEV}-\mu_{PEV})^2}{2\sigma_{PEV}^2}\right], \quad e_{PEV} \in \left[0, e_{PEV}^{Max}\right] \tag{6.38}$$

where μ_{PEV} and σ_{PEV} indicate the mean and standard deviation of the random variable e_{PEV}, and e_{PEV}^{Max} indicates the maximum amount that the EV can be charged in DR periods, which is calculated by Eq. (6.9).

6.5 Case Study and Result Analysis

Parameter Settings. The proposed approach is verified via simulations conducted on community parking lots, where three EVLAs oversee electric vehicle charging loads to engage in DR programs. Each EVLA defines its target DR capacity and establishes upper and lower bounds for DR transaction prices, as specified in Table 6.1. The distribution system operator (DSO) initially sets the load reduction incentive price at 2.5 CNY/kWh, with a minimum requirement of 1000 kWh for market participation in DR services. The probability distribution parameters associated with EV variables are listed in Table 6.2, and the number of EVs with distinct charging patterns is illustrated in Fig. 6.3.

The weighting coefficient, reflecting the trade-off between incentive revenue and dissatisfaction cost for EV users, is generated via a discrete uniform distribution over the interval [0.1, 0.9]. The corresponding parameters α_j and β_j are randomly sampled from the sets $\{0.05, 0.1, 0.15, 0.2, 0.25\}$ and $\{0.06, 0.08, 0.1, 0.12, 0.14\}$, respectively. In simulations of parameters associated with user discomfort levels, the following constraint must be enforced to guarantee a positive value:

Table 6.1 Parameters of EVLAs

Electric vehicle load aggregators (EVLAs)	EVLA #1	EVLA #2	EVLA #3
Expected minimum DR capacity (kWh)	1000	1500	2000
Minimum DR transaction price (CNY/kWh)	0.1	0.2	0.3
Maximum DR transaction price (CNY/kWh)	1.5	2.0	2.5

Table 6.2 Probability distribution of EV variables

EV Variables	Plug-in time	Charging duration	Energy demand
Probability distribution	Truncated chi-square distribution	Truncated Gaussian distribution	Truncated Gaussian distribution
Mean	8:00	4 h	10 kWh
Standard deviation	2 h	2 h	5 kWh
Minimum	6:00 AM	0 h	0 kWh
Maximum	4:00 PM	12 h	$p_j^{Rat}\tau_{DR}$ kWh

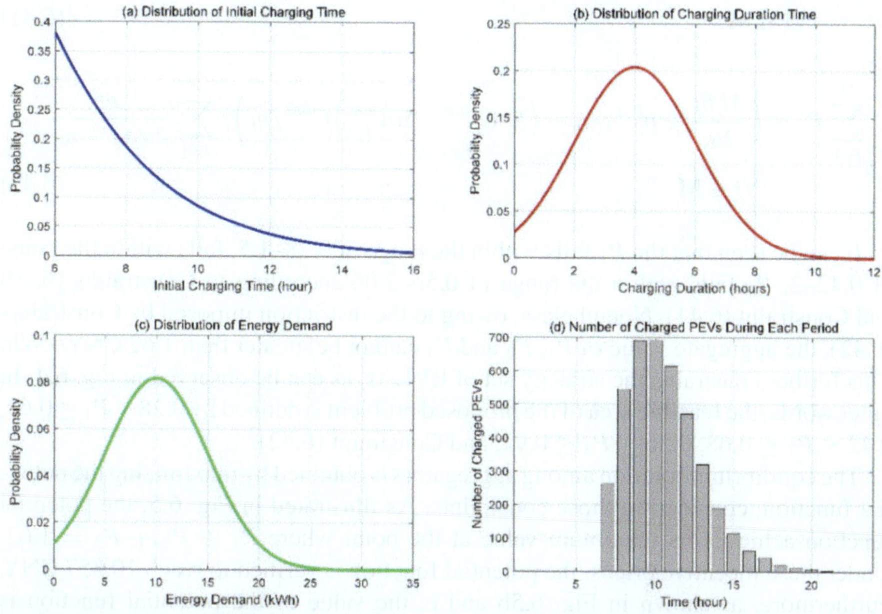

Fig. 6.3 EV charging simulations: **a** distribution of initial charging time; **b** distribution of charging duration; **c** distribution of energy demand; **d** number of EVs being charged in each period

$$R_{i,j} = \frac{\omega_j P_i}{2\alpha_j(1-\omega_j)} - \frac{\beta_j}{2\alpha_j} \geq 0 \iff \beta_j \leq \frac{\omega_j P_i}{1-\omega_j} \leq \frac{\omega_j P_i^{Max}}{1-\omega_j}, \quad \forall j \in \mathbf{N} \tag{6.39}$$

The charging data of EVs is simulated using the platform of MATLAB R2021b, deployed on a workstation with a 2.50 GHz Intel Core i5 processor and 16 GB of random-access memory (RAM). For the quadratic programming (QP) formulation and solution, Lingo 18 × 64 is utilized, with its global solver employed to derive results.

NE Solutions among EVLAs. Grounded in the potential game theory, the NE among EVLAs corresponds to maximizing the game's potential function over its strategy set. The SE between EVs and EVLAs can then be obtained via Eq. (Eq. (6.15)). The optimization model for the NE solutions among EVLAs comprises the potential function (6.28) together with Constraints (6.3) and (6.18)–(6.21). Using these constraints, the feasible range of the incentive price for each EVLA can be reformulated as:

$$P_i \leq \min\left\{P_i^{Max}, 2\alpha_j\left(R_j^{Max} + \frac{\beta_j}{2\alpha_j}\right)(1-\omega_j)/\omega_j\big|_{\forall j \in \mathbf{N}}\right\}, \quad \forall i \in \mathbf{M} \tag{6.40}$$

$$P_i \geq \max\left\{P_i^{Min}, \left(R_{Min}^{DSO} + \sum_{j=1}^{N}\frac{\beta_j}{2\alpha_j}\right)/\sum_{j=1}^{N}\frac{\omega_j}{2\alpha_j(1-\omega_j)}, \beta_j(1-\omega_j)/\omega_j\big|_{\forall j \in \mathbf{N}}\right\},$$

$$\forall i \in \mathbf{M} \tag{6.41}$$

$$\sum_{i=1}^{M} P_i \le \left(\frac{M\beta_j}{2\alpha_j} + p_j^{Rat}\tau_{GO} - \left(SoC_j^{Min} - SoC_j^{Ini} \right) C_j^{Bat} / \eta_j \right) \Big/ \sum_{i=1}^{M} \frac{\omega_j}{2\alpha_j(1-\omega_j)},$$
$$\forall i \in \mathbf{M} \tag{6.42}$$

It can be seen that the P_1 falls within the range of 0.28–1.5, falls within the range of 0.42–2, P_3 falls within the range of 0.56–2.06 according to Constraints (6.40) and Constraint (6.41). Nonetheless, owing to the restriction imposed by Constraints (6.42), the aggregate value of P_1, P_2 and P_3 cannot be greater than 1.62 CNY/kWh. This further constrains the strategy set of EVLAs, as can be observed in Fig. 6.4. In other words, the feasible area of the proposed problem is defined by $0.28 \le P_1 \le 0.64$, $0.42 \le P_2 \le 0.68$, $0.56 \le P_3 \le 0.92$, and Constraint (6.42).

The equilibrium solution among aggregators is obtained by maximizing the potential function, considering those constraints. As illustrated in Fig. 6.5, the potential function achieves its maximum value at the point where $P_1 + P_2 + P_3 = 1.62$. Under these incentive prices, the potential function is verified to reach 10,357 CNY. Furthermore, as shown in Fig. 6.5b and c, the value of the potential function is considerably lower than 10,357 CNY whenever the aggregate incentive price of EVLAs is less than 1.62 CNY/kWh. This observation demonstrates that the optimal

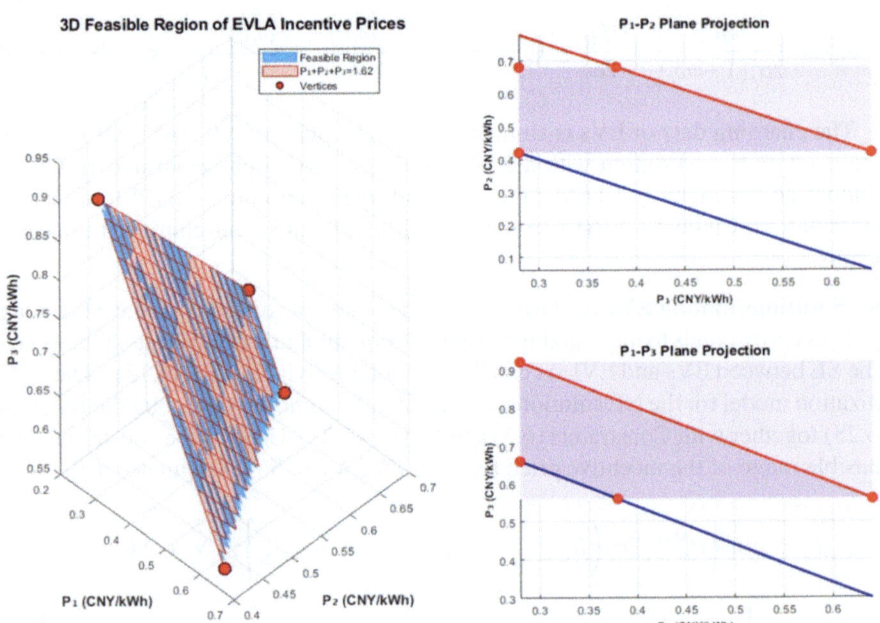

Fig. 6.4 Feasible region of the proposed model using a potential function for NE

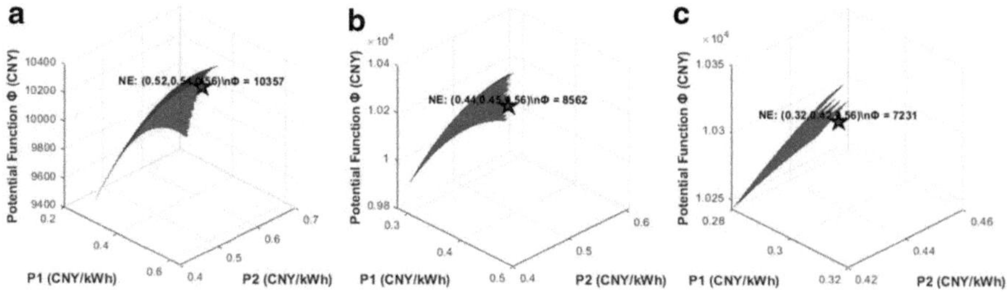

Fig. 6.5 NE of the game among EVLAs with the maximization of the potential function with constraints: **a** $P_1 + P_2 + P_3 = 1.62$; **b** $P_1 + P_2 + P_3 = 1.45$; **c** $P_1 + P_2 + P_3 = 1.30$

NE solution is attained exactly when the total incentive price of EVLAs equals its maximum allowable value. The detailed NE solutions of the game are presented in Table 6.3. At the NE point, the potential function is maximized, implying that the total net profit across all EVLAs is optimized when each EVLA implements its corresponding optimal strategy.

The utility functions of individual EVLAs are examined in Fig. 6.6. As observed, none of the EVLAs can achieve their respective maximum net profits within the feasible region, which is attributed to the restriction imposed by Constraint (6.42). In particular, the utility functions of EVLA #1, EVLA #2, and EVLA #3 would reach 4011.4 CNY, 4210.3 CNY, and 5140.3 CNY, respectively, if their incentive prices were set at the upper bounds. Nevertheless, the corresponding total incentive price of the three aggregators would amount to 2.24 CNY/kWh, which exceeds the threshold of 1.62 CNY/kWh and thus violates Constraint (6.42). Constraint (6.42) essentially captures the limited availability of EV resources, which is governed by the rated charging power of the charging equipment and the energy demand of electric vehicles. This result reveals that the available demand response capacity from EVs exerts a substantial influence on both the profitability and incentive strategies of EVLAs.

Furthermore, the maximum DR transaction price for EVLA #2 is adjusted to 1.2 CNY/kWh, and the minimum expected DR capacity for EVLA #3 is revised to 5000 kWh. Accordingly, the incentive price ranges for the three EVLAs are $P_1 \in [0.28, 2.06]$, $P_2 \in [0.42, 1.2]$, $P_3 \in [1.36, 2.06]$. The results show that the potential function attains a value of 17,018.4 CNY at the NE point. As depicted in Fig. 6.7, each EVLA can maximize its profit at its corresponding optimal incentive price when Constraint (6.42) is neglected. In detail:

Table 6.3 Simulation results for EVLAs' strategies and net profits

Electric vehicle load aggregators (EVLAs)	EVLA #1	EVLA #2	EVLA #3	Total
Optimal DR capacity (kWh)	1647.3	1647.6	2000	
Incentive price (CNY/kWh)	0.51	0.51	0.60	
Net profit (CNY)	3278.5	3278.5	3800	10,357

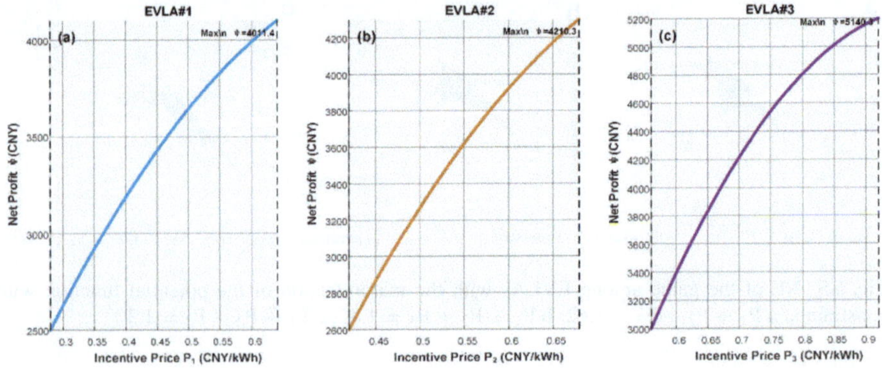

Fig. 6.6 Utility function of EVLAs in its feasible region: **a** EVLA #1; **b** EVLA #2; **c** EVLA #3

- EVLA #1 achieves a maximum net profit of 5690 CNY at an incentive price of 1.29 CNY/kWh;
- EVLA #2 attains a net profit of 5655 CNY at its upper incentive price limit (1.20 CNY/kWh), which serves as its optimal value;
- EVLA #3 realizes an optimal net profit of 5673.4 CNY at its lower incentive price limit (1.36 CNY/kWh).

These findings confirm that, without Constraint (6.42), EVLAs obtain their maximum net profit either at the inflection point of their utility functions or at the upper or lower boundaries of their incentive price intervals.

Sensitivity Analysis Considering Other Factors. To investigate the potential influence of other factors on the incentive strategies of EVLAs, simulations are performed with modified EV scales and expected DR capacities, as presented in Figs. 6.8 and 6.9. As observed in Fig. 6.8a, EVLA #3 provides a higher incentive price compared with EVLA #1 and EVLA #2. This phenomenon arises from the larger DR capacity

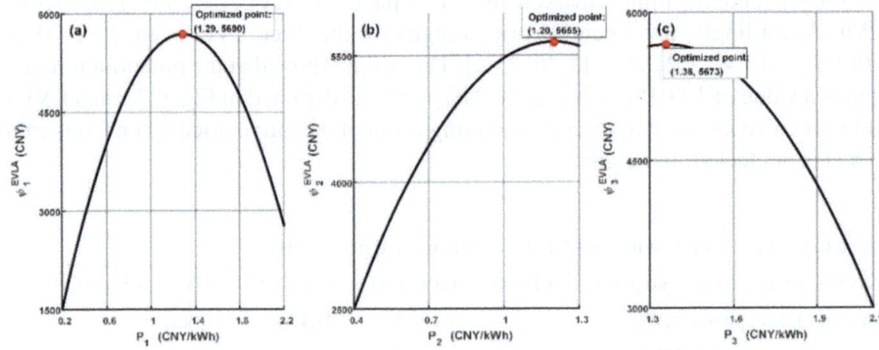

Fig. 6.7 Utility function of EVLAs without Constraints (6.42): **a** EVLA #1; **b** EVLA #2; **c** EVLA #3

demand of EVLA #3 under a small EV fleet size, as elaborated in Table 6.1. Of particular note, the incentive prices of EVLA #1 and EVLA #2 converge to 0.51 CNY/kWh when the EV quantity approaches around 1000. Moreover, when the number of EVs rises to 1200, the incentive prices of all three EVLAs converge uniformly to 0.54 CNY/kWh. The aggregated DR loads of the EVLAs display a comparable variation trend, as depicted in Fig. 6.8b.

This observation reveals that EVLAs are required to increase their incentive prices to procure additional DR capacities when EV-integrated DR resources are limited. By contrast, when DR resources are sufficiently abundant, EVLAs are relieved

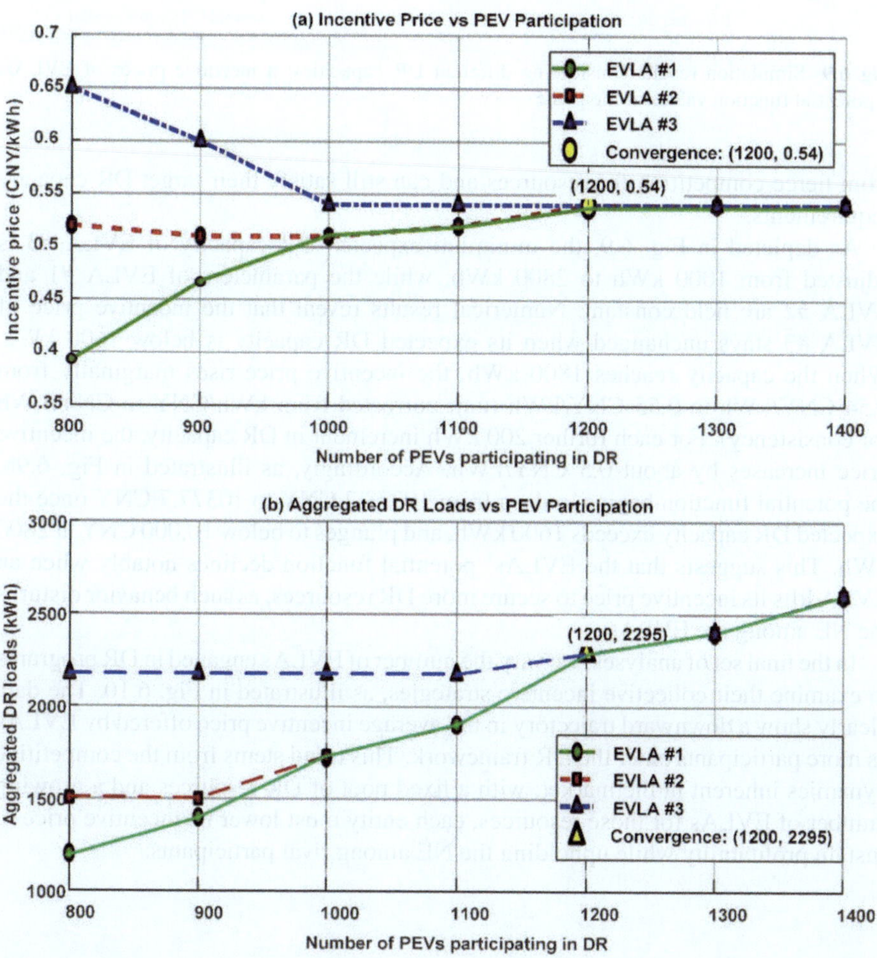

Fig. 6.8 Simulation results considering EV scales: **a** incentive prices of EVLAs; **b** aggregated DR loads of EVLAs

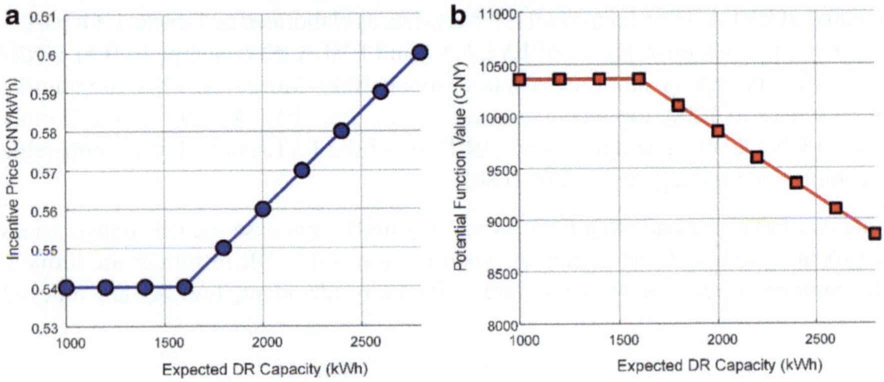

Fig. 6.9 Simulation results considering different DR capacities: **a** incentive prices of EVLAs; **b** potential function values of the game

from fierce competition for resources and can still satisfy their target DR capacity requirements.

As depicted in Fig. 6.9, the minimum expected DR capacity of EVLA #3 is adjusted from 1000 kWh to 2800 kWh, while the parameters of EVLA #1 and EVLA #2 are held constant. Numerical results reveal that the incentive price of EVLA #3 stays unchanged when its expected DR capacity is below 1600 kWh. When the capacity reaches 1800 kWh, the incentive price rises marginally from 0.54 CNY/kWh to 0.55 CNY/kWh (unit corrected from kWh/CNY to CNY/kWh for consistency). For each further 200 kWh increment in DR capacity, the incentive price increases by about 0.5 CNY/kWh. Accordingly, as illustrated in Fig. 6.9b, the potential function begins to drop from 10378.2 CNY to 10377.7 CNY once the expected DR capacity exceeds 1600 kWh, and plunges to below 10,000 CNY at 2800 kWh. This suggests that the EVLAs' potential function declines notably when an EVLA lifts its incentive price to secure more DR resources, as such behavior disturbs the NE among the EVLAs.

In the final set of analyses, we vary the number of EVLAs engaged in DR programs to examine their collective incentive strategies, as illustrated in Fig. 6.10. The data clearly show a downward trajectory in the average incentive price offered by EVLAs as more participants enter the DR framework. This trend stems from the competitive dynamics inherent in the market: with a fixed pool of DR resources and a growing number of EVLAs for those resources, each entity must lower its incentive price to sustain profitability while upholding the NE among rival participants.

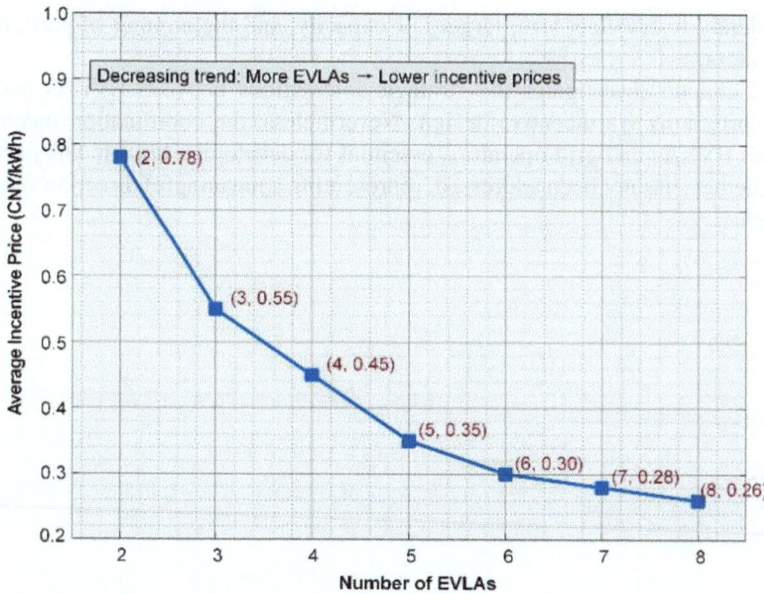

Fig. 6.10 Incentive prices of EVLAs with the scale of EVLAs

6.6 Summary

EVLAs serve as the critical intermediaries connecting EV fleets and grid operators. While EVs provide a flexible resource for DR services in power systems, their reliable participation in DR initiatives requires well-designed incentive mechanisms. To this end, this study proposes an innovative competition model grounded in Stackelberg game theory to formulate incentive strategies for EVLAs when procuring DR capacity from EVs. In the EVLA-EV interaction framework, EVLAs act as the leaders seeking profit maximization by aggregating flexible EV resources, whereas EVs act as followers optimizing their benefits by offering DR services. Using the potential game theory, this work rigorously demonstrates the existence and uniqueness of the SE. Furthermore, stochastic characteristics of EV charging are modeled to capture uncertainties in EV charging behaviors. Comprehensive numerical experiments are performed to verify the effectiveness of the proposed approach.

Results reveal that DR participation strategies of EVLAs are governed by three major factors: (1) The expected DR capacity and DR transaction price ranges established by EVLAs determine the feasible domain of the optimization model, thereby shaping their incentive pricing; (2) The available DR capacity from EVs, limited by charging equipment constraints and energy requirements, considerably restricts the solution space and strongly affects EVLAs' incentive decisions; and (3) Under the Nash equilibrium among competing EVLAs, their optimal incentive prices adapt dynamically to EV fleet size, target DR capacity, and the number of participating aggregators.

This chapter investigates the strategic interactions between EVLAs and EVs in DR programs via incentive design. Nevertheless, the coordination mechanism between EVLAs and grid operators, essential for developing holistic DR participation schemes, has not been addressed, representing a meaningful direction for future research.

Chapter 7
Bidding Strategy for EV Aggregators in the Electricity Market

Abstract Electric vehicle aggregators (EVAs) play an indispensable role in the integration of EV resources and load management. In the electricity spot market, EVAs can leverage the flexibility of EV charging to reasonably schedule EV loads and reduce their operational costs. In general, the electricity spot market consists of the day-ahead and real-time markets. In the day-ahead bidding stage, EVAs ought to submit electricity bids in advance to lock in the EV charging requirement for the next day. As the day-ahead bidding process accounts for the majority of the EVA's operational costs, it is crucially important. Nevertheless, the randomness of EV charging and the uncertainty of spot market operation would pose substantial challenges to the day-ahead bidding decision of EVAs. Most existing studies adopt a scenario-based approach to address uncertainty and derive day-ahead bidding strategies of EVAs to minimize expected costs or maximize expected profits. However, such aggressive bidding strategies may be insufficient to hedge against operational risks caused by market price volatility, exposing EVAs to considerable financial losses. To tackle the above issues, this chapter introduces Conditional Value-at-Risk (CVaR) as a risk metric to cope with the uncertainty in the real-time clearing market, optimize the CVaR of EVA bidding, and mitigate potential market risks. On this basis, a risk-averse day-ahead bidding method for EVAs is proposed. First, a day-ahead bidding model for EVAs under deterministic information is established, and the corresponding nonlinear programming model is linearized. Furthermore, to capture the stochastic nature of EV charging, a probabilistic model is employed to simulate the charging behavior of the EV fleet for the next day. Aiming at the uncertainty of the real-time electricity market, a risk-averse decision-making approach is utilized to model potential real-time clearing scenarios, and a day-ahead bidding model with the objective of minimizing CVaR is formulated. Finally, the proposed risk-averse bidding strategy is compared with the risk-neutral one, and the superiority of the risk-averse bidding strategy in risk management is demonstrated. Results show that higher confidence levels (e.g., 0.95 or 0.99) effectively constrain costs within predefined intervals with minimal variation in optimized CVaR values. Furthermore, the linearized model significantly outperforms the original non-linear formulation in computational efficiency, especially for a large EV fleet. This chapter concludes that the risk-averse bidding approach can provide superior risk management for

Q. Yang et al., *Integration of Massive Electric Vehicles to Smart Grid*, Smart Energy Systems, https://doi.org/10.1007/978-981-92-0557-8_7

EVAs, though future research should address real-time scheduling complexities and computational speed for practical implementation.

7.1 Introduction

Transportation electrification is universally acknowledged as a pivotal policy of many nations worldwide to mitigate carbon emissions and decrease fossil fuel dependence. In contrast to traditional gasoline vehicles, electric vehicles (EVs) present prominent merits, e.g., superior energy efficiency, reduced operational expenses, and enhanced compatibility with renewable energy integration. The proportion of EVs in total vehicle sales has expanded markedly over the past three years, surging from around 4% in 2020 to 14% in 2022, with a sustained upward trend expected in the foreseeable future. The large-scale penetration of EVs is expected to profoundly reshape the overall electricity demand profile, introducing new complexities into the secure and economic operation of modern power systems. As an effective countermeasure, coordinated EV charging schemes have been widely investigated to support the integration of EVs and renewable energy, thereby strengthening the economic efficiency and operational stability of power systems.

Given that individual EVs lack sufficient controllable capacity to qualify for electricity market participation, electric vehicle aggregators (EVAs) have emerged as critical intermediary entities. Representing individual EV users, EVAs serve as a vital interface between end users and power system operators, enabling aggregated participation in electricity transactions. Fundamentally, the core function of the EVA is to consolidate the flexible charging resources of numerous EVs for market engagement, with the dual objectives of minimizing operational costs while fully satisfying the charging requirements of EV users. Furthermore, EVAs can obtain additional economic benefits through energy price arbitrage and the provision of grid-support services by utilizing the EV cluster as flexible demand-side resources. Nevertheless, the optimal day-ahead (DA) bidding decision-making of EVAs is severely challenged by multiple intertwined uncertainties, dominated by the stochastic characteristics of EV charging behaviors and the highly volatile nature of real-time market prices, which necessitate systematic and effective solutions.

A substantial body of existing literature has concentrated on the bidding strategy of EVAs with the expected profit maximization of EVAs using stochastic optimization. Nevertheless, many available methodologies fail to comprehensively quantify and manage the potential financial risks induced by fluctuating market prices. This research limitation may expose EVAs to considerable economic losses under extreme market-clearing conditions. Risk-neutral and risk-averse strategies represent two dominant paradigms for addressing stochastic programming problems. Although a few studies have incorporated Conditional Value-at-Risk (CVaR) into the risk management framework for day-ahead bidding decisions of the EVA, a systematic and comparative analysis of the performance and applicability of these two alternative strategies in solving DA bidding problems remains absent in the current literature.

To bridge this research gap, this chapter develops a risk-averse DA bidding model based on CVaR minimization, which effectively reduces the probability of severe cost deviations for EVAs participating in the day-ahead market. Specifically, the CVaR metric is integrated into the proposed framework to hedge against potential operational risks in the spot electricity market. The risk-averse optimization model is formulated by minimizing the CVaR value across a set of representative real-time market-clearing scenarios. Moreover, the model is equivalently transformed into a tractable linear programming formulation that guarantees computational efficiency and convergence. By using the risk-neutral bidding strategy as a benchmark, comprehensive comparative results indicate that the proposed risk-averse approach exhibits superior performance in terms of operational risk control and overall decision robustness.

7.2 Proposed Risk-Averse Approach for Day-Ahead Bidding

Operating Framework of Market Participation of EVAs. Most electricity spot markets typically adopt a two-settlement system, e.g., the PJM Market, NYISO, and CAISO in the United States, as well as the Guangdong Electricity Market and Shanxi Electricity Market in China. From a temporal perspective, the electricity spot market consists of a day-ahead market and a real-time (or balancing) market. In the day-ahead electricity market, power suppliers (e.g., power plants) submit their generation schedules to the power exchange, while power consumers (e.g., electricity retailers) declare their day-ahead bidding volumes. The power exchange clears the day-ahead electricity price based on the bidding and offering plans of both supply and demand sides, and the resulting day-ahead price is used for the first-settlement of day-ahead bids. In the real-time electricity market, the power exchange clears the real-time energy price according to the day-ahead bidding information and the grid load status obtained from ultra-short-term forecasting, where the real-time price is applied to the second-settlement of deviations between actual and contracted generation or consumption.

Notably, to prevent the arbitrage behavior of electricity retailers, the power exchange imposes penalties on the imbalance energy when the deviation between the day-ahead bidding volume and the actual electricity consumption is excessively large. For instance, in the U.S. PJM market, a penalty is triggered if the imbalance energy exceeds 20% of the day-ahead bidding volume. In China's Guangdong Electricity Market, this penalty threshold is set at 5%. Within the operation framework of electricity markets, specific market access criteria and requirements are imposed. For electricity retailers, the trading capacity in a single transaction must reach a certain threshold, usually at the megawatt-hour level. However, the energy trading capacity

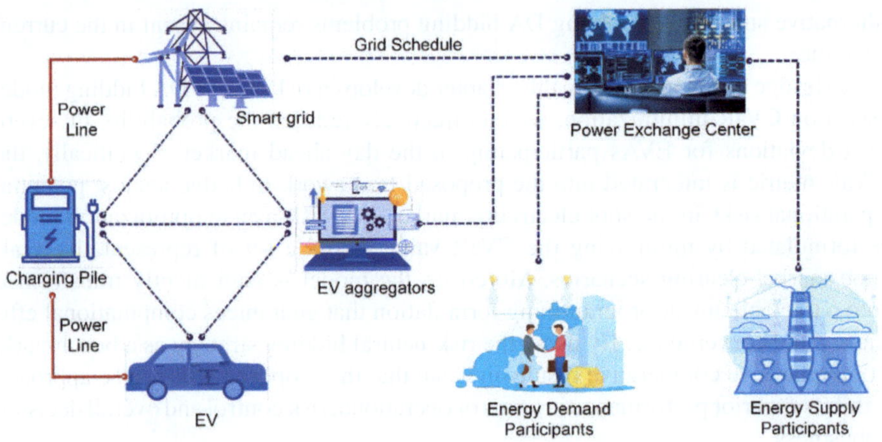

Fig. 7.1 General framework of EVAs participating in electricity markets

of an individual EV is extremely limited and far below the minimum capacity required for market participation. Therefore, individual EVs cannot directly participate in the electricity market. As a special type of electricity retailer, EVAs are mainly responsible for aggregating the EV fleet to achieve scale effects and thus meet the market access threshold, as shown in Fig. 7.1. Under the electricity market framework, EVAs need to conduct day-ahead bidding and real-time power regulation based on multiple factors, including electricity price dynamics, grid operating conditions, and charging characteristics of EV users, to reduce their operational costs.

As an intermediary between EV fleets and the Independent System Operator (ISO), EVAs first sign contracts with EV users to obtain the authority to regulate EV charging and thereby participate in electricity spot transactions. In this study, it is assumed that the EVA has obtained the charging management authority and established two-way communication with EV supply equipment (EVSE), enabling the regulation and control of EV charging time and charging power. In addition, this chapter only considers the coordinated charging of EVs. That is, only the grid-to-vehicle (G2V) mode is considered, excluding the vehicle-to-grid (V2G) mode. Under this assumption, the EVA acts only as a power purchaser in the electricity market, without considering its role as a power supplier. Furthermore, it is assumed that all reasonable bidding demands of the EVA in the day-ahead electricity market will be satisfied.

Within this framework, the EVA participates in both day-ahead and real-time markets to minimize its total operating costs, as formulated in Eq. (7.1). Under prevailing market regulations, financial penalties are triggered whenever energy deviations exceed a predefined tolerance level, defined as a specified proportion of the day-ahead energy commitment. Accordingly, such penalty costs should be explicitly integrated into the formulation of the day-ahead bidding strategy for the EVA.

$$C = \sum_{t \in \Omega_H} \lambda_t^{DA} E_t^{DA} + \sum_{t \in \Omega_H} \lambda_t^{RT} (E_t^{RT} - E_t^{DA}) \tag{7.1}$$

where λ_t^{DA} denotes the DA market price at hour t, E_t^{DA} denotes the quantity of DA bidding of the EVA, λ_t^{RT} denotes the RT market price at hour t, E_t^{RT} denotes the consumed energy of EVs at hour t.

In this chapter, we assume that day-ahead market-clearing prices can be forecasted accurately. Nevertheless, EVAs still confront substantial uncertainties arising from EV charging patterns and real-time market-clearing conditions, which can considerably affect their day-ahead bidding strategies. Such uncertainties are explicitly formulated and comprehensively integrated into the proposed model to support optimal bidding decisions.

The modeling framework for the day-ahead bidding problem of EVAs is illustrated in Fig. 7.2. In the uncertainty modeling stage, the Monte Carlo simulation method is employed to generate a sufficient number of EV charging events, and the scenario-based approach is further adopted to capture potential clearing scenarios in the real-time electricity market. In the formal modeling phase, an original stochastic programming model for day-ahead bidding is first established. This model takes the cost or risk associated with EVA market participation as the optimization objective, subject to constraints including EV charging requirements and market participation limitations. Since the adopted objective function incorporates nonlinear terms as illustrated in Eq. (7.1), the formulated problem belongs to the nonlinear programming (NLP) problem. However, the NLP problem suffers from high computational complexity in the solution process and may even fail to converge to the optimal solution. Moreover, the large number of EV decision variables to be optimized may lead to the curse of dimensionality. To address these issues, dummy variables are introduced to transform the NLP problem into a linear programming (LP) problem. Finally, risk-averse and risk-neutral stochastic optimization approaches are utilized to model and solve the EVA day-ahead bidding problem.

Uncertainty Characterization for Day-Ahead Bidding. In the day-ahead electricity market, EVAs need to consider the dual uncertainties arising from next-day EV charging behaviors and market-clearing prices, so as to formulate optimal day-ahead bidding strategies. How to address the uncertainties from both EVs and the electricity market is critical to reducing the operating cost of EVAs. In this chapter, statistical methods are employed to characterize and model the randomness of EV charging and the uncertainty of market clearing.

Affected by user travel demands, EV charging behaviors exhibit significant randomness, and their charging time and charging demand are difficult to predict. However, for the specific charging scenario or an EV fleet, the charging behaviors generally follow certain statistical distributions. Therefore, based on historical charging data of EVs, statistical methods can be adopted to summarize the corresponding laws or distribution characteristics, thereby simulating the next-day EV charging behaviors. For residential EV charging, the normal distribution is used to describe the grid-connection time and grid-disconnection time of EVs, while the

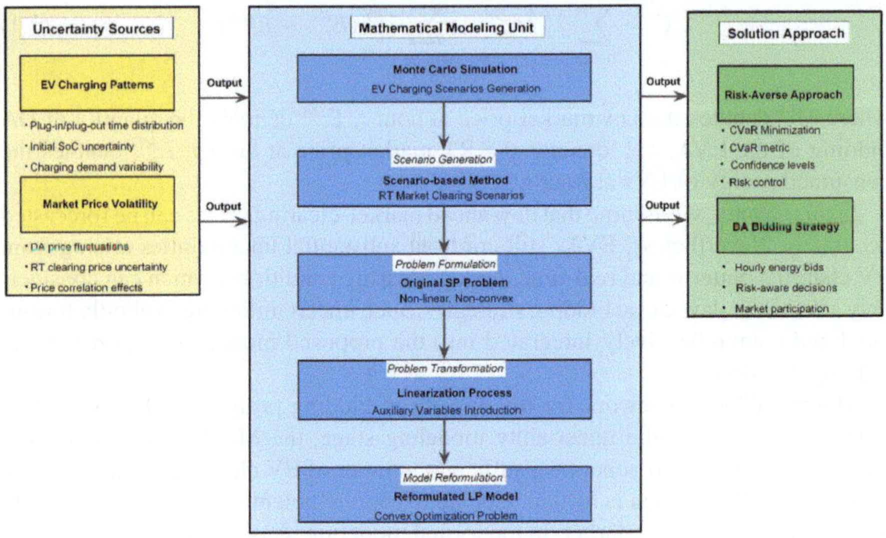

Fig. 7.2 Flowchart of the proposed methodology for DA bidding strategy

continuous uniform distribution is applied to characterize the initial state of charge (SoC) of EVs before charging. Finally, the Monte Carlo simulation method is utilized to generate EV charging events that satisfy given constraints.

The uncertainties in the electricity market mainly stem from day-ahead and real-time market-clearing prices. Day-ahead clearing prices present certain regularities, and relatively accurate forecasts can be achieved based on existing prediction approaches. In contrast, real-time clearing prices are subject to considerable fluctuations, influenced by real-time power supply–demand balance and grid operating conditions. Accordingly, existing studies tend to consider potential real-time clearing scenarios and adopt stochastic programming (SP) to tackle the day-ahead bidding problem of EVAs. For stochastic programming problems, probability functions can be used to describe the occurrence probabilities of real-time clearing scenarios, and the day-ahead bidding strategy can be optimized to minimize the expected cost. Mathematically, this approach is referred to as the risk-neutral approach. Nevertheless, since the risk-neutral method merely targets the minimization of expected cost, EVAs may suffer substantial losses in the electricity market under certain real-time clearing scenarios. For this reason, this chapter introduces the CVaR to handle the uncertainty of real-time electricity prices, and proposes a risk-averse decision-making approach to optimize the day-ahead bidding strategy with the objective of minimizing CVaR. The advantage of the proposed method lies in its ability to constrain the operational risk of EVAs within a certain range under a specified confidence level.

Mathematical modeling via CVaR for uncertainty handling. This study considers two separate timescales corresponding to market transactions and EV charging coordination, respectively. Consistent with the PJM market architecture, hourly time intervals are employed for the financial settlements between the EVA and the independent system operator. For real-time charging dispatch, a finer sub-hourly resolution of 15-min intervals is implemented. In the day-ahead bidding framework, the aggregator seeks to minimize its total operational cost via optimal scheduling of the EV fleet. For each predefined real-time clearing scenario w, the SP formulation governing the day-ahead bidding decision can be constructed as follows:

$$(\textit{Original SP Problem})\; C(w) = \min_{E_t^{DA}} \sum_{t\in\Omega_H} \lambda_t^{DA} E_t^{DA} + \sum_{t\in\Omega_H} \left(\lambda_{w,t}^{RT} \Delta E_{w,t} + C_{w,t}^{Pen}\right) \tag{7.2}$$

subject to:

$$\Delta E_{w,t} = E_{w,t}^{RT} - E_t^{DA}, \forall t \in \Omega_H \tag{7.3}$$

$$E_{w,t}^{RT} = \sum_{n\in\Omega_{EV}} \sum_{k\in[4t-3,4t]} P_{w,n,k}^{EV} \cdot \Delta t, \forall t \in \Omega_H \tag{7.4}$$

$$C_{w,t}^{Pen} = \max\left\{0, \lambda^p\left(\left|\Delta E_{w,t}\right| - \rho E_t^{DA}\right)\right\}, \forall t \in \Omega_H \tag{7.5}$$

$$0 \le P_{w,n,k}^{EV} \le P_n^{\text{Max}}, \forall n \in \Omega_{EV}, \forall k \in \left[\Gamma_n^{In}, \Gamma_n^{Out}\right] \tag{7.6}$$

$$\sum_{k\in[\Gamma_n^{In},\Gamma_n^{Out}]} \left(P_{w,n,k}^{EV} \cdot \Delta\tau\right) \cdot \frac{\eta_n}{Bat_n} = SoC_n^{Out} - SoC_n^{In}, \forall n \in \Omega_{EV} \tag{7.7}$$

$$E_{\text{Min}}^{RT} \le E_{w,t}^{RT} \le E_{\text{Max}}^{RT}, \forall t \in \Omega_H \tag{7.8}$$

$$0 \le E_t^{DA} \le E_{\text{Max}}^{DA}, \forall t \in \Omega_H \tag{7.9}$$

where $\lambda_{w,t}^{RT}$ represents the RT market-clearing price at hour t for scenario w; $\Delta E_{w,t}$ refers to the energy deviation at hour t under scenario w, defined as the difference between the expected energy consumption and the day-ahead bidding energy; $C_{w,t}^{Pen}$ is the penalty cost induced by the energy deviation; $E_{w,t}^{RT}$ represents the consumed energy of EVs at hour t for scenario w; $P_{w,n,k}^{EV}$ represents the expected charging power of EV n at time slot k for scenario w; Δt is the duration of time slot; λ^p represents the penalty coefficient for the energy deviation; ρ is the market penalty threshold, defined as a percentage of the day-ahead bidding energy; P_n^{Max} stands for the maximum charging power of the EV n; η_n denotes the charging efficiency of EV n; Bat_n denotes the capacity of the battery pack of EV n; SoC_n^{Out} represents the grid-disconnection time of EV n; SoC_n^{In} represents the grid-connection time of EV

n; E_{Min}^{RT} represents the minimum consumed energy of all EVs; E_{Max}^{RT} represents the maximum consumed energy of all EVs; E_{Max}^{DA} represents the allowable maximum bidding energy of the electricity market.

Equation (7.2) represents the objective function, which denotes the total operating cost of the EVA participating in the electricity market. It consists of the day-ahead bidding cost, the energy deviation cost, and the penalty cost incurred by energy deviations. Equations (7.3) and (7.4) quantify the deviation between the power consumption of the electric vehicle fleet and the day-ahead bidding quantity. Equation (7.5) formulates the penalty cost resulting from energy deviations. Equation (7.6) imposes the constraint that the charging power of each electric vehicle cannot exceed the maximum operating power of its charger. Equation (7.7) ensures that the charging demand specified by electric vehicle users must be satisfied. Equation (7.8) restricts the total power consumption of the electric vehicle fleet to comply with the physical constraints of the power grid. Equation (7.9) specifies that the hourly bidding quantity of the electric vehicle aggregator must be bounded by the minimum and maximum allowable bidding quantities set by the market.

Notably, Eq. (7.5) contains both a max function and an absolute value function. Consequently, the mathematical programming problem formulated above is a nonlinear programming (NLP) problem, which suffers from computational intractability in large-scale EV charging scheduling. It is therefore necessary to transform this NLP problem into a linear programming (LP) problem to facilitate an efficient solution. The auxiliary variables are introduced to linearize the problem, as presented below:

$$(SLP\ Problem)\ C(w) = \min_{E_t^{DA}} \sum_{t\in\Omega_H} \lambda_t^{DA} E_t^{DA} + \sum_{t\in\Omega_H} \left(\lambda_{w,t}^{RT} \Delta E_{w,t} + \overline{\mu}_{w,t} + \underline{\mu}_{w,t} \right) \tag{7.10}$$

subject to:

$$\Delta E_{w,t} = \Delta E_{w,t}^{[+]} - \Delta E_{w,t}^{[-]}, \forall t \in \Omega_H \tag{7.11}$$

$$\overline{\mu}_{w,t} \geq 0, \overline{\mu}_{w,t} \geq \lambda^p \left(\Delta E_{w,t}^{[+]} - \rho E_t^{DA} \right), \forall t \in \Omega_H \tag{7.12}$$

$$\underline{\mu}_{w,t} \geq 0, \underline{\mu}_{w,t} \geq \lambda^p \left(\Delta E_{w,t}^{[-]} - \rho E_t^{DA} \right), \forall t \in \Omega_H \tag{7.13}$$

$$\Delta E_{w,t}^{[+]}, \Delta E_{w,t}^{[-]} \geq 0, \forall t \in \Omega_H \tag{7.14}$$

and (7.3), (7.4), (7.6)–(7.9).

In the above, $\Delta E_{w,t}^{[+]}$ represents the positive energy deviation at hour t for scenario w, $\Delta E_{w,t}^{[-]}$ represents the negative energy deviation at hour t for scenario w, $\overline{\mu}_{w,t}$ represents the penalty cost induced by the positive energy deviation higher than threshold at hour t for scenario w, $\underline{\mu}_{w,t}$ represents the penalty cost induced by the negative energy deviation higher than threshold at hour t for scenario w.

Regarding this SLP problem, determining the bidding strategy of the EVA is difficult considering the presence of multiple random variables associated with the RT clearing scenario w. In Eq. (7.5), $\boldsymbol{x} = \left\{E_t^{DA}, \forall t \in \Omega_H\right\}$ is the decision vector, and $\boldsymbol{v} = \left\{\lambda_{w,t}^{RT}, \forall t \in \Omega_H, w \in \Omega_{CS}\right\}$ is the random vector. Herein, $\mathbf{v}$ represents the uncertainty in RT market clearing, which can significantly impact the operating costs of EVAs.

$$C(\boldsymbol{x}, \boldsymbol{v}) = \sum_{t \in \Omega_H} \lambda_t^{DA} E_t^{DA} + \sum_{t \in \Omega_H} \left(\lambda_{w,t}^{RT} \Delta E_{w,t} + \overline{\mu}_{w,t} + \underline{\mu}_{w,t}\right) \tag{7.15}$$

For a given day-ahead bidding strategy $\mathbf{x}$, the operating cost of the EVA $C(\boldsymbol{x}, \boldsymbol{v})$ is a distribution function with respect to the random variable $\mathbf{v}$. Assume that the real-time market clearing scenarios, namely the random variable $\mathbf{v}$, follow a certain statistical distribution, with its probability density function denoted by $p(\mathbf{v})$. Therefore, for a fixed day-ahead bidding strategy $\mathbf{x}$, the probability that $C(\boldsymbol{x}, \boldsymbol{v})$ is below a specified threshold ζ can be defined as:

$$\Psi(\boldsymbol{x}, \zeta) = \int_{C(\boldsymbol{x}, \boldsymbol{v}) \le \zeta} p(\boldsymbol{v}) d\boldsymbol{v} \tag{7.16}$$

Based on the formulation given in Eq. (7.15), the Value-at-Risk (VaR) and CVaR of the EVA participating in the day-ahead market can be defined. VaR refers to the maximum potential loss expected to occur over a period at a given confidence level. Although VaR indicates the maximum loss under a specified confidence level, it fails to characterize the distribution of extreme losses beyond this threshold. In contrast, CVaR further quantifies the expected loss exceeding the VaR value, thereby capturing the average loss caused by tail extreme events. By formulating the conditional expectation of tail losses, CVaR overcomes the limitations of VaR in risk measurement. Accordingly, CVaR is particularly suitable for market environments with frequent extreme events. VaR and CVaR can be formally expressed in Eqs. (7.17)–(7.18):

$$(\alpha - VaR) \;\; \zeta_\alpha(\boldsymbol{x}) = \min\{\zeta \in \mathbb{R} | \Psi(\boldsymbol{x}, \zeta) \ge \alpha\} \tag{7.17}$$

$$\begin{aligned} &(\alpha - CVaR) \;\; \phi_\alpha(\boldsymbol{x}) = E[C(\boldsymbol{x}, \boldsymbol{v}) | C(\boldsymbol{x}, \boldsymbol{v}) \ge \zeta_\alpha(\boldsymbol{x})] \\ &= (1 - \alpha)^{-1} \int_{C(\boldsymbol{x}, \boldsymbol{v}) \ge \zeta_\alpha(\boldsymbol{x})} C(\boldsymbol{x}, \boldsymbol{v}) p(\boldsymbol{v}) d\boldsymbol{v} \end{aligned} \tag{7.18}$$

To facilitate the calculation of $\phi_\alpha(\boldsymbol{x})$, function $F_\alpha(\boldsymbol{x}, \zeta)$ are introduced:

$$F_\alpha(\boldsymbol{x}, \zeta) = \zeta + (1 - \alpha)^{-1} \int_{\boldsymbol{v}} (C(\boldsymbol{x}, \boldsymbol{v}) - \zeta)^+ p(\boldsymbol{v}) d\boldsymbol{v} \tag{7.19}$$

where $[q]^+ = \max(q, 0)$.

Hence, $\alpha - CVaR$ of the operating cost of the EVA can be computed by (20):

$$\phi_\alpha(\boldsymbol{x}) = \min_\zeta F_\alpha(\boldsymbol{x}, \zeta) \tag{7.20}$$

By minimizing the CVaR, the EVA can substantially reduce the risk associated with its DA bidding operations. Therefore, the objective function under the risk-averse approach is expressed in Eq. (7.21):

$$\min \phi_\alpha(\boldsymbol{x}) = \min_{\boldsymbol{x}} \min_\zeta F_\alpha(\boldsymbol{x}, \zeta) = \min_{\boldsymbol{x},\zeta} F_\alpha(\boldsymbol{x}, \zeta) \tag{7.21}$$

Here, the integral in $F_\alpha(\boldsymbol{x}, \zeta)$ as shown in Eq. (7.19) can be approximated via discretization. For the potential real-time market-clearing scenarios w, let π_w denote the occurrence probability. The corresponding approximation of $F_\alpha(\boldsymbol{x}, \zeta)$ is expressed in Eq. (7.22):

$$\widetilde{F}_\alpha(\boldsymbol{x}, \zeta) = \zeta + (1 - \alpha)^{-1} \sum_{w \in \Omega_{CS}} \pi_w (C(\boldsymbol{x}, \boldsymbol{v}_w) - \zeta)^+ \tag{7.22}$$

Therefore, Eq. (7.21) can be approximated to Eq. (7.23):

$$\min \phi_\alpha(\boldsymbol{x}) \approx \min_{\boldsymbol{x},\zeta} \widetilde{F}_\alpha(\boldsymbol{x}, \zeta) = \min_{\boldsymbol{x},\zeta} \left\{ \zeta + (1 - \alpha)^{-1} \sum_{w \in \Omega_{CS}} \pi_w (C(\boldsymbol{x}, \boldsymbol{v}_w) - \zeta)^+ \right\} \tag{7.23}$$

Furthermore, taking $z_w = (C(\boldsymbol{x}, \boldsymbol{v}_w) - \zeta)^+$ for linearization, the risk-averse bidding model can be rewritten as:

$$(\textit{Risk-averse Approach}) \ \min \phi_\alpha(\boldsymbol{x}) = \min_{\boldsymbol{x},\zeta} \left\{ \zeta + (1 - \alpha)^{-1} \sum_{w \in \Omega_{CS}} \pi_w z_w \right\} \tag{7.24}$$

subject to:

$$z_w \geq 0, z_w \geq C(\boldsymbol{x}, \boldsymbol{v}_w) - \zeta, \forall w \in \Omega_{CS} \tag{7.25}$$

and (7.3), (7.4), (7.6)–(7.9), (7.11)–(7.14), $\forall w \in \Omega_{CS}$.

The aforementioned stochastic optimization framework incorporates potential real-time market-clearing scenarios to address the uncertainty inherent in real-time dispatch outcomes. Furthermore, this approach accounts for the tail risk induced by real-time market volatility, thereby preventing EVAs from incurring substantial losses in the real-time electricity market. In addition, through linear discretization techniques, the original model is reformulated into an LP formulation. As a typical convex optimization model, the LP formulation satisfies the convexity requirements for both the objective function and constraint conditions. Consequently, it effectively avoids the local optimality pitfalls commonly encountered in conventional nonlinear programming and can leverage mature algorithms such as the interior-point method and simplex method to ensure the rapid convergence to a globally optimal solution. Meanwhile, the convex structure of the LP model significantly reduces computational complexity, rendering it well-suited for the energy bidding problem of large EV fleets.

Benchmark Bidding Strategies. To verify the superiority of the risk-averse day-ahead bidding strategy in terms of risk management, a comparative analysis between the risk-averse bidding strategy and the risk-neutral counterpart is conducted in this study. The risk-neutral decision-making approach takes the maximization of expected utility as its core decision paradigm, in which uncertain parameters are modeled as expected value measures under a probability distribution, making it suitable for decision-making environments with risk indifference. Herein, potential real-time market-clearing scenarios are characterized using a probabilistic model, and the optimization is performed to minimize the expected operating cost of the EVA. The corresponding mathematical formulation is presented as follows:

$$(\textit{Risk-neutral Approach})\ \min \sum_{t\in\Omega_H} \lambda_t^{DA} E_t^{DA} + \sum_{w\in\Omega_{CS}} \sum_{t\in\Omega_H} \left(\lambda_{w,t}^{RT} \Delta E_{w,t} + \overline{\mu}_{w,t} + \underline{\mu}_{w,t}\right) \tag{7.26}$$

subject to constraints (7.3), (7.4), (7.6)–(7.9), and (7.11)–(7.14), $\forall w \in \Omega_{CS}$.

It should be noted that the risk-neutral decision-making approach, which adopts expected utility maximization as its decision criterion, tends to overlook certain extreme real-time clearing scenarios, which may expose the EVA to substantial operational risks. Furthermore, consistent with the linearization method employed in Sect. 7.2, the aforementioned model linearizes the penalty function in the objective function, converting the nonlinear model into an LP formulation and thus reducing the computational complexity of the model.

7.3 Real Case Studies and Performance Analysis

Case Description and Parameter Settings. This chapter focuses on the day-ahead bidding problem of EVAs in the context of residential EV charging stations. In this case study, only the electricity procurement cost of the aggregator participating in the energy market is considered, while the revenue obtained from selling electricity to EV users is excluded. The number of electric vehicles involved in day-ahead energy transactions is set to 1000. It is assumed that residential EVs adopt a nighttime charging pattern. The charging time of EVs (including grid connection time and grid disconnection time) follows a normal distribution. The characteristics of the charging

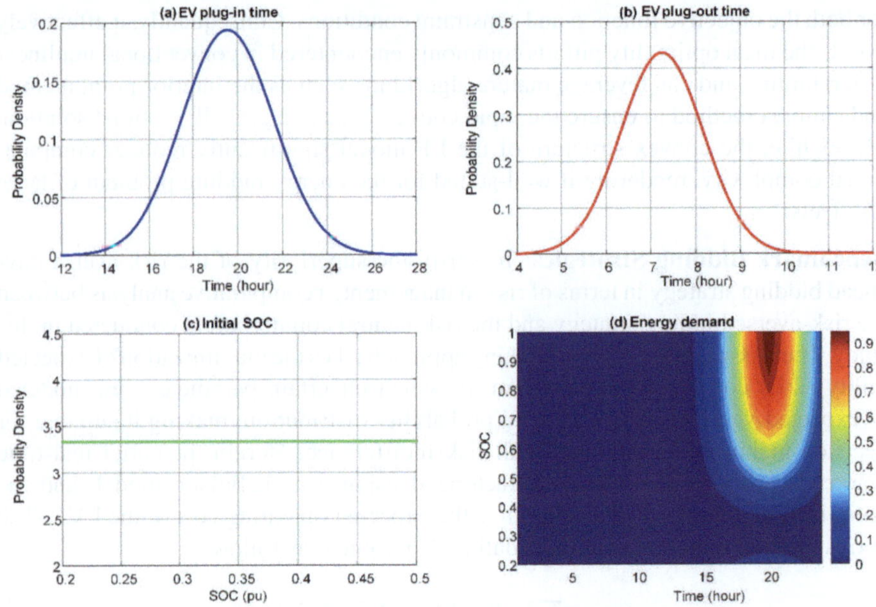

Fig. 7.3 Charging behavior characteristics of EVs in residential areas on weekdays

behavior are illustrated in Fig. 7.3, where the grid connection time of residential EVs obeys the normal distribution $N(19.55, 2.06)$, and the grid disconnection time follows the normal distribution $N(7.25, 0.92)$. In addition, the initial state of charge (SoC) of electric vehicles is assumed to follow a continuous uniform distribution $U(0.2, 0.5)$. The terminal state of charge (SoC) of EVs is set to 100% or the maximum achievable SoC within their grid-connected duration. Furthermore, the charging efficiency of all EVs is assumed to be 0.95. In the residential charging scenario, the maximum operating power of EV chargers is set to either 1.92 kW (Level 1: 16A/120 V) or 19.2 kW (Level 2: 80A/240 V).

In the PJM market, if the deviation between actual energy consumption and the day-ahead bidding volume exceeds 20% of the day-ahead bidding volume, the ISO imposes a financial penalty on electricity retailers. The energy deviation exceeding the penalty threshold is penalized at a rate of 0.15 USD/kWh (i.e., $\rho = 0.15$ \$/kWh). In addition, the hourly maximum bidding volume of the electric vehicle aggregator is set to 3000 kWh. In this study, 100 typical real-time market-clearing scenarios are adopted to characterize the uncertainty of the real-time electricity market, and each scenario is assumed to occur with an equal probability. Finally, the confidence level for calculating CVaR is set to 0.95 (i.e., $\alpha = 0.95$). The proposed model was implemented on a personal computer (PC) configured with an Intel Core i5 central processing unit (CPU) running at 2.50 GHz and 16 gigabytes (GB) of random-access memory (RAM), with Lingo 18.0 × 64 employed as the solution tool.

Result Analysis in a Case Study with Fixed Parameters. This subsection focuses on investigating the inherent correlations between the risk-averse day-ahead bidding strategy and factors such as day-ahead electricity prices, penalty costs, and confidence levels. By taking the parameters set in the previous subsection as model inputs, the optimized day-ahead bidding curve of the EVA can be obtained. The relationship between the bidding volume of the EVA and the day-ahead electricity price is illustrated in Fig. 7.4. It can be observed that the EVA tends to purchase more electricity during periods with lower day-ahead prices. For instance, the bidding volume exceeds 1000 kWh per hour during the 1st–6th hours, 13th–16th hours, and 21st–24th hours, and even approaches the maximum allowable bidding volume (i.e., 3000 kWh per hour) during the 22nd to 24th hours. In contrast, during those periods with higher day-ahead prices, the EVA prefers to purchase less electricity. For example, the bidding volume is less than 500 kWh per hour during the 8th to 12th hours and 17th–19th hours. The underlying reasons for this phenomenon are as follows: First, most EVs adopt a nighttime charging mode, resulting in high electricity demand in the evening. A more in-depth reason is that both day-ahead and real-time electricity prices are relatively lower in the evening. Therefore, to reduce operational costs, the EVA arranges EVs to charge as much as possible during low-price periods and submits more electricity bids in these periods.

To further verify the effectiveness of the proposed method, the penalty costs caused by energy deviations under various real-time clearing scenarios are analyzed herein, as illustrated in Fig. 7.5. It can be observed that the penalty costs induced by positive energy deviations are significantly higher than those caused by negative energy deviations. Specifically, the penalty cost resulting from positive energy deviations is 232.3 USD, while that from negative energy deviations is only 24.3 USD. The penalty

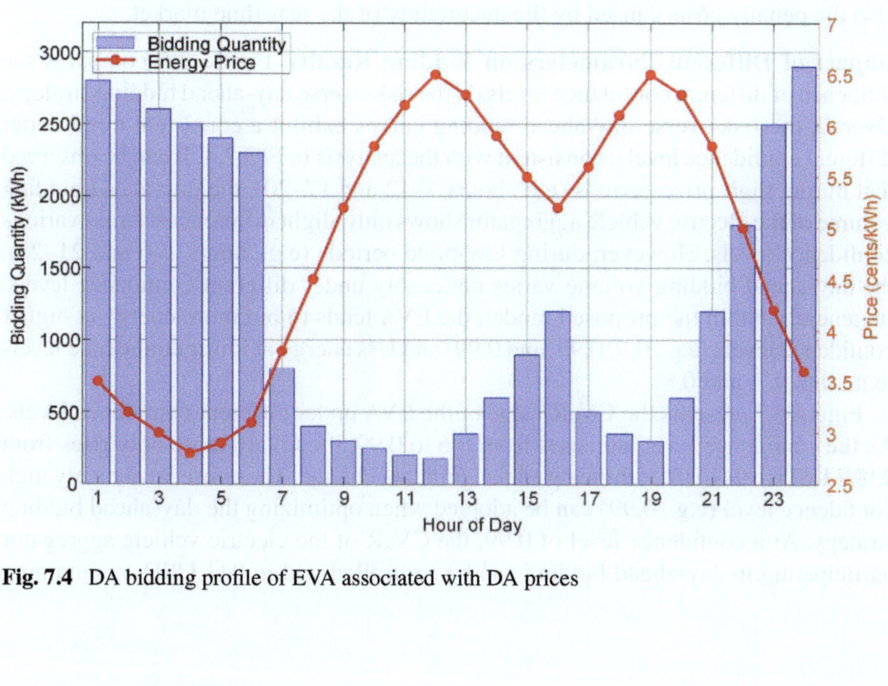

Fig. 7.4 DA bidding profile of EVA associated with DA prices

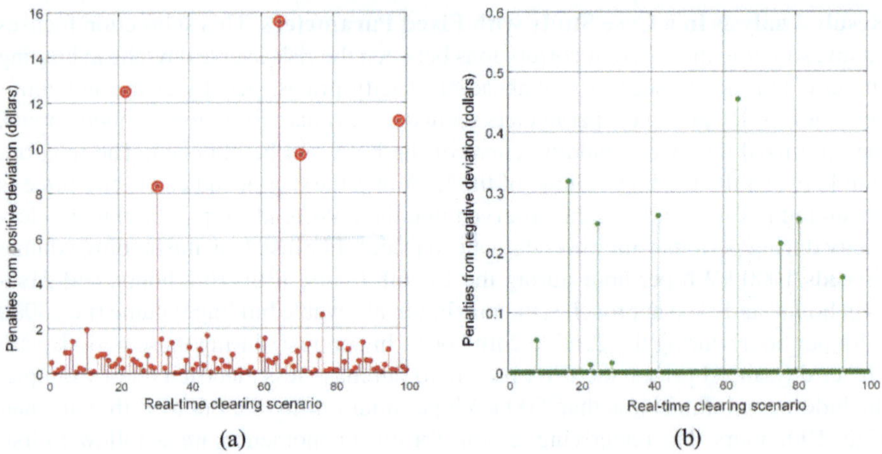

Fig. 7.5 Penalty distribution for 100 RT clearing scenarios: **a** Penalty for positive energy deviation higher than the threshold; and **b** Penalty for negative energy deviation higher than the threshold

costs caused by positive energy deviations are relatively high only under real-time clearing scenarios #22, #31, #65, #71, and #98, and are zero or close to zero under all other real-time clearing scenarios. In contrast, the penalty costs induced by negative energy deviations are zero in most scenarios and only relatively high under the real-time clearing scenario #55. Overall, the optimized day-ahead bidding strategy enables the EVA to bear low penalty costs under all real-time clearing scenarios. It can be concluded that when formulating the day-ahead bidding strategy, the EVA must consider not only the electricity procurement cost in the day-ahead market but also the penalty costs caused by the uncertainty of the real-time market.

Impact of Different Parameters on Bidding Results. Figure 7.6 compares the influence of different confidence levels on the risk-averse day-ahead bidding strategy. Overall, the risk-averse day-ahead bidding curves exhibit a consistent trend under different confidence levels, consistent with the analysis in Fig. 7.4. It can be observed that during high-price periods (e.g., hours 7–12 and 17–20), the day-ahead bidding volume of the electric vehicle aggregator shows only slight differences across various confidence levels. However, during low-price periods (e.g., hours 1–6 and 21–23), the day-ahead bidding volume varies noticeably under different confidence levels. In general, within the proposed model, the EVA tends to bid more energy at higher confidence levels (e.g., 0.9, 0.95, and 0.99) and less energy at lower confidence levels (e.g., 0.6, 0.7, and 0.8).

Figure 7.7 presents the CVaR value of the EVA under different confidence levels. As the confidence level increases from 0.6 to 0.99, the CVaR value only rises from 938.7 USD to 941 USD, showing an insignificant change. Therefore, a relatively high confidence level (e.g., 0.99) can be adopted when optimizing the day-ahead bidding strategy. At a confidence level of 0.99, the CVaR of the electric vehicle aggregator participating in day-ahead bidding can be controlled within 941 USD, maintaining

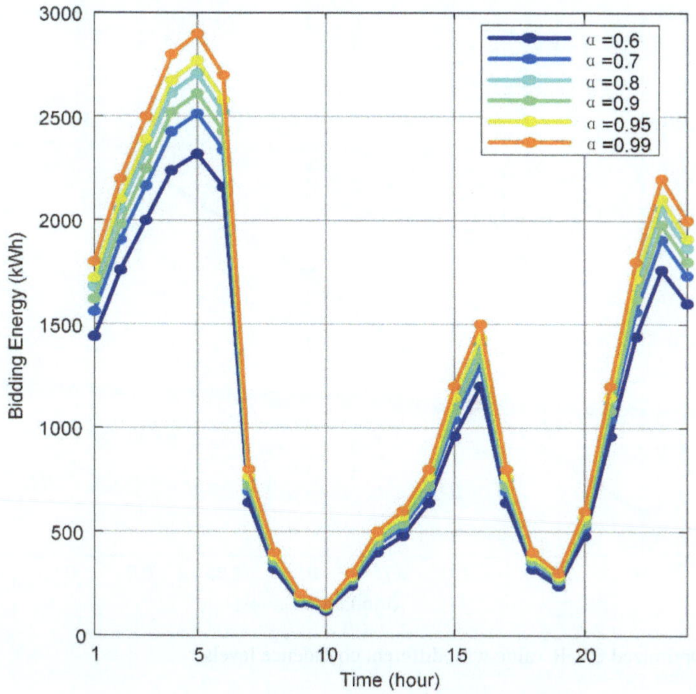

Fig. 7.6 Optimized DA bidding strategy with different confidence levels

its operational risk at a low level. Given the minimal variation in CVaR values across the confidence level range of 0.6–0.99, setting a high confidence level (e.g., 0.99) is advisable. In this case, the probability of the EVA's total operating costs exceeding $941 is very low, signifying reduced risk for the EVA.

In addition, the penalty coefficient for the energy deviations in the electricity market may affect the day-ahead bidding strategy of EVAs. In this subsection, a low day-ahead electricity price profile is adopted, and a sensitivity analysis of the two day-ahead bidding strategies is performed by adjusting the penalty coefficient, as illustrated in Fig. 7.8. It can be observed that when the risk-neutral decision-making approach is employed, different penalty coefficients exert a considerable influence on the day-ahead bidding volume, as shown in Fig. 7.8a. When the penalty coefficient is only 0.05 USD/kWh, the EVA bids at the market upper limit of 3000 kWh during the second hour. This indicates that, under low day-ahead electricity prices, a small penalty coefficient incentivizes the EVA to purchase more electricity to pursue greater arbitrage profits, whereas a large penalty coefficient discourages such speculative behavior. It can therefore be concluded that electricity market operators can regulate the market participation behavior of power retailers by adjusting the penalty coefficient, thereby preventing excessive speculation. In contrast, for the risk-averse bidding strategy, the differences between the day-ahead bidding curves under various penalty coefficients are negligible, as shown in Fig. 7.8b. Accordingly,

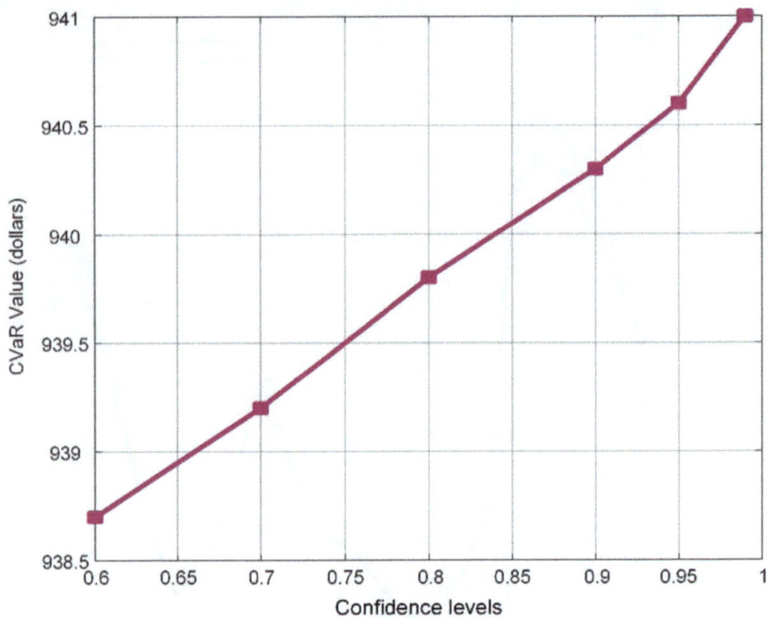

Fig. 7.7 Optimized CVaR value with different confidence levels

it can be concluded that the penalty coefficient has only a minor impact on the risk-averse day-ahead bidding strategy.

Impact of Different Day-Ahead Prices on Bidding Results. This subsection aims to investigate the differences between the risk-averse and risk-neutral decision-making approaches in formulating day-ahead bidding strategies. Figure 7.9 presents a comparison of the results obtained by the two bidding strategies under a low day-ahead electricity price scenario. The right-side ordinate denotes the difference between the real-time clearing price and the day-ahead clearing price, defined as the price spread for concise expression. In addition, a boxplot is used to illustrate the distribution of the price spread between real-time and day-ahead electricity prices.

It can be observed that, in the boxplots for each time interval, the upper quartile (Q3) of the price spread lies above the zero line, indicating that the real-time electricity price exceeds the day-ahead price in most real-time clearing scenarios. When the risk-neutral decision-making approach is adopted, the EVA seeks to minimize its expected operating cost. Driven by the price spread, the aggregator tends to bid for more electricity in the day-ahead market for arbitrage. In contrast, when the risk-averse decision-making approach is employed, the aggregator reduces its bidding volume in the day-ahead market to hedge against the risk that the real-time price may fall below the day-ahead price.

Specifically, at the 8th hour, the bidding volume obtained by the risk-neutral approach is significantly higher than that by the risk-averse approach, which may be attributed to the relatively high real-time electricity price during this period, allowing

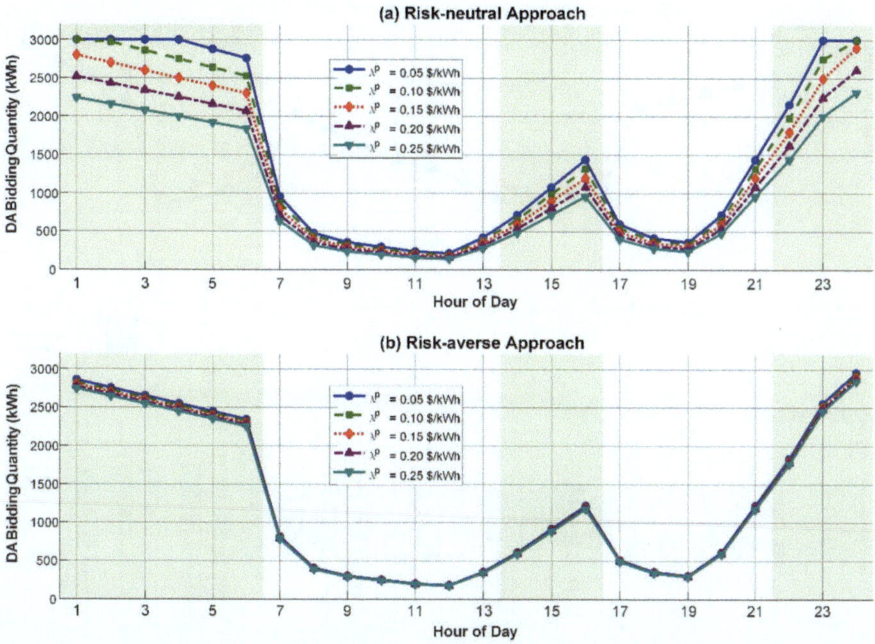

Fig. 7.8 DA bidding strategy with different penalty prices: **a** Risk-neutral approach; **b** Risk-averse approach

the aggregator to achieve considerable arbitrage gains. Accordingly, under the risk-neutral bidding strategy, the EVA prefers to purchase more electricity at this hour.

Furthermore, during the 1st–5th hours and 23rd–24th hours, the risk-neutral approach bids at the upper limit of 3000 kWh per hour. Compared with the risk-averse approach, the risk-neutral method exhibits a more aggressive behavior in formulating the day-ahead bidding strategy. Conversely, to avoid substantial losses in the real-time market, the risk-averse decision-making approach tends to adopt a more conservative day-ahead bidding strategy.

Figure 7.10 illustrates the objective function values of the bidding strategies optimized via the risk-averse and risk-neutral approaches, with the confidence level fixed at 0.95. It is found that across various scales of EV fleets, the costs of the EVA optimized by the risk-averse method are marginally higher than the expected costs derived from the risk-neutral approach. Notably, the cost value associated with the risk-averse strategy serves as the upper limit of potential losses in the majority of scenarios. For example, with 800 EVs, the risk-averse strategy guarantees that the EVA's costs will not surpass $711 in 95% of real-time clearing scenarios. In contrast, while the risk-neutral strategy achieves an expected cost of $702, the actual costs may spike significantly above this figure under certain circumstances. This finding further highlights the advantage of the risk-averse approach in terms of risk management.

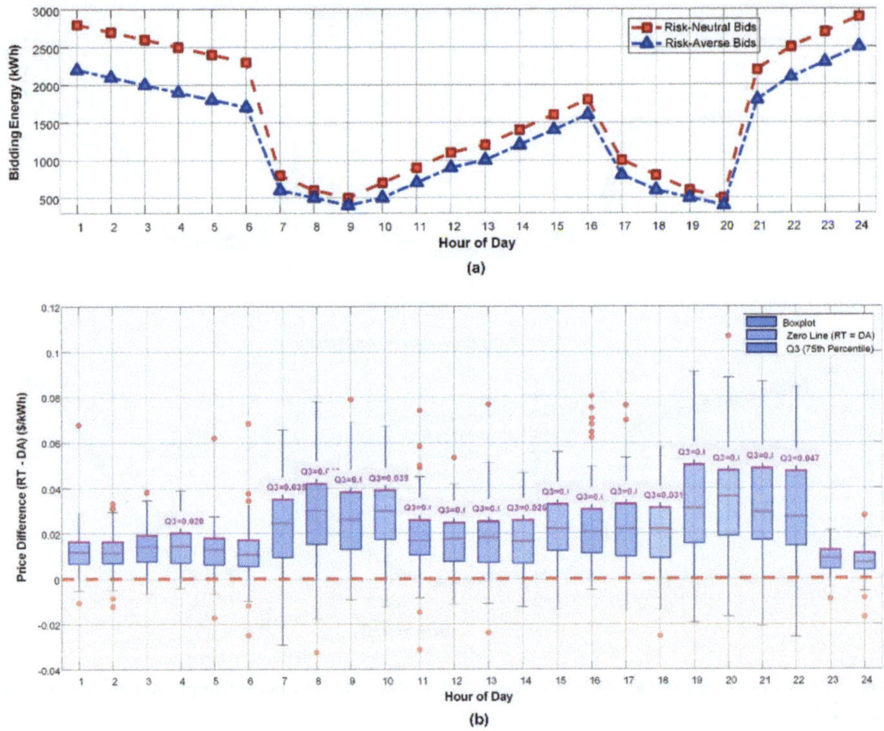

Fig. 7.9 Comparison between risk-neutral and risk-averse approaches for DA bidding in a case with low DA energy prices

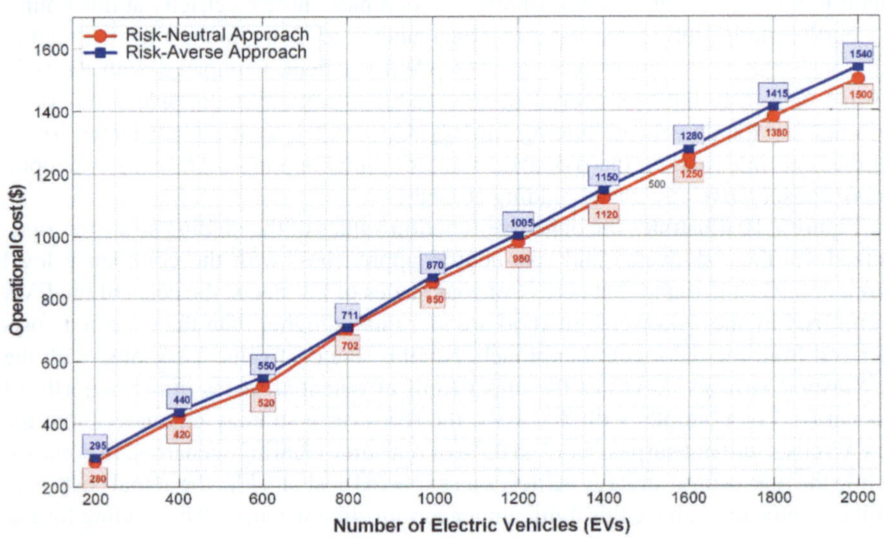

Fig. 7.10 EVA's operational costs when the risk-neutral approach and risk-averse approach are considered

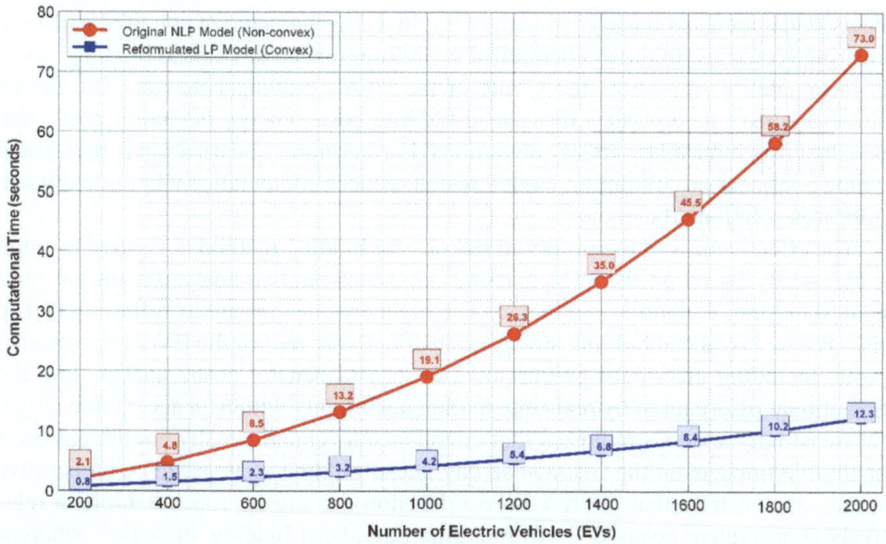

Fig. 7.11 Consumed computational time versus the number of managed EVs

Finally, a comparative analysis of the computational complexity between the constructed LP model and the original NLP model was conducted, as illustrated in Fig. 7.11. It can be observed that the solution time of the linearized risk-averse optimization model is significantly lower than that of the original non-linearized model. This advantage becomes more pronounced, especially when considering large-scale EV fleets. When the number of EVs is less than 1500, the computational time of both models increases linearly with the scale of EVs. However, when the number of EVs exceeds 1500, the computational time of both the linearized LP model and the original NLP model exhibits a rapid upward trend. In terms of computational complexity, the converted LP model has a computational complexity ranging from $O(N^{1.5})$ to $O(N^2)$, whereas the computational complexity of the original NLP model is generally no less than $O(N^3)$. Overall, the reconstructed LP model is superior to the original NLP model in terms of computational complexity.

7.4 Summary

To address the operational risks caused by market-clearing price volatility, this chapter proposes a day-ahead bidding method for EVAs based on CVaR. By introducing CVaR as a risk metric and minimizing CVaR as the objective, the bidding strategy of the EVA is optimized, thereby constraining the risks associated with market participation within a reasonable range. For EV uncertainty modeling, a statistical distribution model is adopted to simulate the charging behavior of EV fleets, and a scenario-based approach is employed to characterize potential real-time clearing scenarios. In the mathematical modeling stage, a day-ahead bidding model is formulated to minimize the CVaR of the EVA's bidding behavior. The model simultaneously incorporates day-ahead bidding costs, energy deviation costs, and penalty costs induced by energy deviations. Furthermore, linearization is performed on the nonlinear programming model, which reduces computational complexity and improves solution efficiency.

The effectiveness of the proposed risk-averse bidding method is verified through a case study based on the PJM market, and a comparative analysis with the risk-neutral bidding method is conducted. The main conclusions are as follows: (1) The risk-averse day-ahead bidding strategy can effectively reduce the EVA's operational costs, including electricity procurement costs and penalty costs, and confine the operational risks caused by real-time market uncertainty within an acceptable range; (2) Sensitivity analysis indicates that confidence level and penalty coefficient exert negligible impacts on the risk-averse day-ahead bidding strategy; (3) Comparative results demonstrate that the risk-averse decision-making approach exhibits a relatively conservative behavior in formulating day-ahead bidding strategies, whereas the risk-neutral approach is more aggressive. Therefore, the risk-averse decision-making method is distinctly superior to the risk-neutral method in terms of bidding risk management.

This chapter primarily explores the risk-averse participation strategy of EVAs in the DA market, leveraging EVs as flexible resources. However, several limitations remain: energy trading interactions between EVAs and EVs (to exploit EV charging flexibility) and real-time adjustments to EVA operational strategies are not addressed. In RT operations, solving time becomes critical, as EVAs require rapid solutions for large-scale EV charging schedules. Consequently, the development of highly efficient scheduling algorithms with significantly reduced computational complexity for EV charging loads is expected to be explored in future work.

Chapter 8
Coordinated Charging Scheduling of EVs in a Market Environment

Abstract This chapter explores the coordinated charging dispatching strategies of electric vehicles (EVs) within a competitive electricity market ecosystem from the viewpoint of EV aggregators, aiming to facilitate large-scale EV integration into power grids. A two-stage energy management system is constructed for EV charging, aligned with the operational guidelines of the Guangdong electricity market. This framework encompasses a day-ahead bidding stage and a real-time charging dispatching stage. During the day-ahead phase, the aggregators formulate optimal energy bids by accounting for various uncertainties inherent to EV charging dynamics, e.g., variability in user arrival/departure timings and initial state of charge, alongside volatilities in market-clearing prices. To hedge against potential financial penalties, this stage integrates projections of future real-time market scenarios. Subsequently, in the real-time phase, a rolling-horizon optimization engine dynamically recalibrates charging protocols to adhere closely to pre-submitted day-ahead schedules, thereby minimizing deviation-related settlement costs. To accommodate the diverse preferences of individual EV users, two distinctive operational modes are engineered: Centralized Management Mode (CMM) and Distributed Incentive Mode (DIM). Under the CMM, the aggregator exerts direct control over charging operations in return for preferential electricity rates. The DIM mode grants users autonomy over their devices, incentivizing them to respond to real-time price signals independently. For simulation purposes, stochastic processes are employed to generate realistic EV charging profiles. Additionally, four distinct real-time price trajectories are modeled by using multi-class support vector machines. Case study results validate that aggregators can generate substantial revenues by capitalizing on the flexibility of EV charging assets, with bidding strategies being primarily dictated by the price differentials between day-ahead and real-time markets. The proposed methodology effectively aligns actual EV load profiles with intended bidding volumes, significantly reduces the computational burden associated with large-scale optimization, and underscores significant market opportunities available to the aggregators in modern energy trading ecosystems.

Q. Yang et al., *Integration of Massive Electric Vehicles to Smart Grid*, Smart Energy Systems, https://doi.org/10.1007/978-981-92-0557-8_8

8.1 Introduction

With the increasing adoption of EVs, the inherent spatial and temporal randomness and dynamics of EV charging activities have posed growing operational challenges to the existing power system. Uncontrolled charging of a large number of EVs may exert adverse effects on the power grid, e.g., overloading feeders and transformers, as well as degrading power quality. To avoid costly upgrades to the current power grid for accommodating those additional EV charging demands, coordinated EV charging has emerged as an economical solution. This approach shifts EV loads to off-peak periods, thereby easing the supply pressure on the grid during peak hours.

The competitive electricity market creates an enabling and practical environment for coordinated EV charging. As a special type of load aggregator, the EV aggregator is the key entity for EVs to participate in energy transactions or ancillary services. In the day-ahead market, EV aggregators face uncertainties from both EVs and the electricity market, including the mobility characteristics of EVs, charging demand, day-ahead clearing prices, and real-time clearing prices. These uncertainties may lead to energy deviations between the day-ahead bidding volume of EV aggregators and the actual electricity consumption of EVs, resulting in high penalty costs. In the real-time market, EV aggregators do not need to modify the bidding decision; instead, they can optimally schedule the charging loads of EVs to achieve optimal market participation. Therefore, formulating optimal day-ahead bidding and real-time charging control strategies is crucial for reducing the operating costs of electric vehicle aggregators. In addition, EV aggregators must integrate sufficient electric vehicle charging transaction capacity to meet the threshold for market participation. Thus, how to develop incentive strategies to capture the flexibility of EV charging is also an issue that EV aggregators need to consider. This chapter introduces the market operation strategies of EV aggregators from three aspects: day-ahead bidding strategies, real-time charging control strategies for EVs, and interaction strategies with EV users.

In the day-ahead market, EV aggregators need to bid for the next day's charging demand, which accounts for most of their operating costs. The randomness of EV charging behavior and volatility of market electricity prices create major challenges for accurate day-ahead bidding. Current research addresses these uncertainties mainly through probabilistic modeling and scenario simulation. To handle price volatility, three decision-making approaches are applied: risk-neutral, risk-averse and robust optimization. Risk-neutral methods are widely used to maximize expected profit but ignore tail risks. Risk-averse approaches balance the profit and the risk, often using the conditional value-at-risk (CVaR) to control potential losses. Robust optimization ensures feasibility under worst-case scenarios but tends to be overly conservative. Overall, the three strategies differ significantly in risk appetite, and risk-averse bidding offers a balanced solution that is neither too aggressive nor too cautious.

During the real-time operation, EV aggregators must respond rapidly to grid dispatch signals, requiring fast and efficient charging and discharging strategies. The time scale for energy market scheduling is around 15 min, while ancillary services, e.g., frequency regulation and demand response. Thus, the computational complexity of optimization algorithms becomes critical for real-time energy management. In the energy market, real-time control helps minimize deviations from day-ahead plans and reduce penalty costs. In ancillary service markets, active charge–discharge regulation improves tracking accuracy for grid signals. The core challenge lies in developing low-complexity, high-efficiency algorithms that can generate optimal charging strategies for large-scale EV fleets within strict time limits.

Most existing studies focus on bidding and real-time scheduling while treating EVs simply as adjustable resources, neglecting users' willingness to provide flexibility. Balancing benefits between EV aggregators and EV users is essential for effective vehicle-grid integration. Current research commonly uses a two-level Stackelberg game framework, where the EV aggregator acts as the leader setting the price or the incentive, and EV users act as the followers deciding the charging power. Multi-level optimization models are also used to coordinate interests among grid operators, EV aggregators, and EV users. The fundamental conflict is that EV aggregators pursue maximum economic benefits, whereas EV users would weigh compensation, battery degradation, and travel flexibility. Therefore, effectively reconciling these conflicting interests is the key to encouraging user participation in electricity markets.

To address complex market rules and multiple uncertainties from both EVs and the market environment, this chapter presents a two-stage charging control framework for the EV cluster to study the optimal operation strategies for EV aggregators in the spot electricity market. Two agent modes are established between EV aggregators and EV users to utilize the charging flexibility of EVs: the centralized management mode (CMM) and the decentralized incentive mode (DIM). On this basis, a stochastic programming model is developed to maximize the expected profit of EV aggregators, taking into account uncertainties in the electricity market and EV charging characteristics, so as to determine the optimal day-ahead bidding strategy. Furthermore, combined with the day-ahead bidding volume and real-time clearing prices, a real-time charging dispatch model is proposed using rolling optimization to dynamically adjust the charging power of EVs, with the target of maximizing the economic benefits of electric vehicle aggregators.

8.2 Problem Formulation of Coordinated EV Charging in Market Frameworks

Operating process of EV aggregators in markets. This chapter investigates the coordinated charging scheduling of EVs in the Guangdong electricity market, one of the first batches of pilot markets in China's power sector reform. The market adopts a security-constrained economic dispatch (SCED) mechanism to determine the marginal clearing price for market settlement.

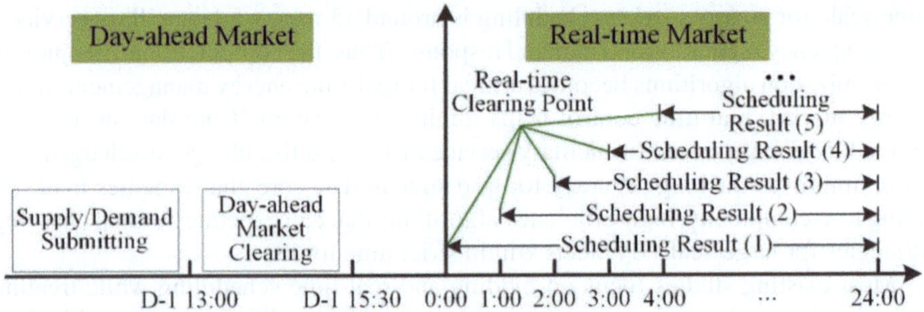

Fig. 8.1 Timeline of EV aggregators participating in energy markets

Figure 8.1 illustrates the participation timeline of EV aggregators in the energy market. In the day-ahead market, EV aggregators are required to submit their power demand bids to the Guangdong Power Exchange Center (GPEC) by 13:00 on the day before the operating day. Meanwhile, power plants must also submit their generation supply information to GPEC. Based on the overall power supply and demand conditions of the grid, GPEC finalizes and releases the day-ahead market clearing results to all participants by 15:30. In the real-time market, GPEC publishes hourly clearing prices based on ultra-short-term load forecasts (e.g., at 15-min intervals) and pre-submitted bidding data from power plants and electricity retailers. Within each real-time clearing interval, EV aggregators dynamically adjust the charging power of connected EVs from the current moment until the end of the operating day, in response to the latest updated clearing prices.

According to the trading rules of the spot market, the market participation process of EV aggregators consists of two stages: day-ahead bidding and real-time charging regulation. Figure 8.2 illustrates the two-stage optimal scheduling process for EV charging under the spot market environment. In the day-ahead bidding stage, EV aggregators formulate day-ahead bidding strategies based on the next-day EV charging demand and market electricity prices, and submit the power demand for the operating day to the power exchange. Meanwhile, power generation enterprises are also required to submit their supply plans to the power exchange. Retail electricity companies usually participate in the electricity market by bidding volume without quoting prices, and accept the market-clearing electricity prices. In contrast, power generation enterprises participate in market competition by bidding both volume and price, which directly affects the formation of market prices. After the power exchange center obtains all information on power supply and demand, it adopts a security-constrained economic dispatch procedure to clear the day-ahead electricity prices. The cleared day-ahead prices will be used for the unified settlement of the day-ahead market.

In the real-time charging scheduling stage, EV aggregators need to formulate real-time EV charging scheduling plans based on the awarded day-ahead power volume and real-time clearing prices, as well as the charging demand of the connected EV fleet, and issue the plans to EV charging facilities for execution. Notably, under the

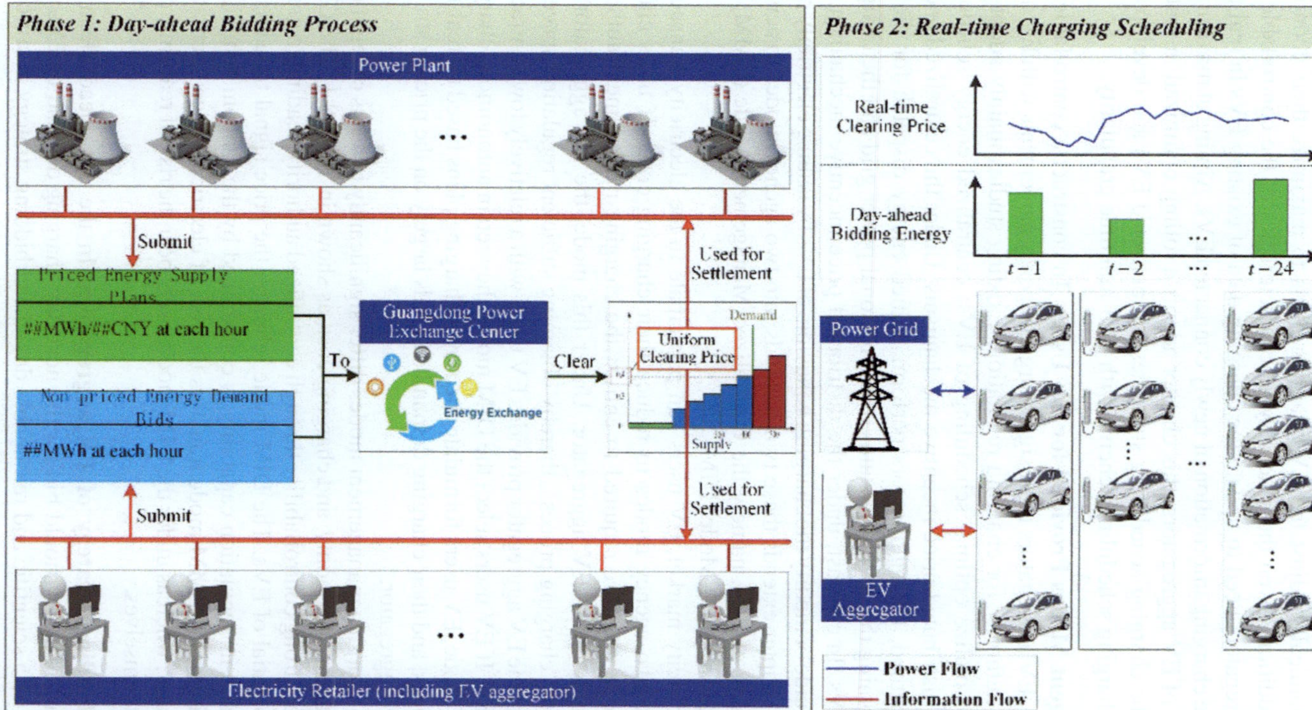

Fig. 8.2 Two-phase operating process of EV aggregators in the market environment

operating rules of this spot market, retail electricity companies are not authorized to modify their bidding volumes in the real-time market. Therefore, EV aggregators can only dynamically adjust the EV charging plans to maximize their economic benefits. In addition, given the time-varying characteristics of the controllable EV fleet, EV aggregators need to update the charging data of existing EVs in real time and collect the charging information of newly connected EVs. Aiming at maximizing the net profit of EV aggregators, this chapter adopts a rolling optimization method to optimize the charging scheduling strategies of connected EVs and develops the optimal EV charging scheduling scheme with a 15-min time granularity.

Charging Agent Modes Provided for EV Users. To construct a controllable EV resource pool, EV aggregators must sign charging agency agreements with EV users to obtain the authority for charging regulation. Existing studies mainly adopt two approaches to realize optimal scheduling of EVs: centralized energy scheduling methods and distributed price incentive mechanisms. Under the centralized energy scheduling method, EV aggregators perform unified energy dispatch for the EV fleet by combining the operating status of the regional power grid and the charging demand of EVs. In contrast, under the distributed price incentive mechanism, EV aggregators guide or modify the charging behaviors of EVs by issuing electricity price signals. This chapter extends these two methods into two agency modes to achieve orderly charging of EVs, namely the Centralized Management Mode (CMM) and the Distributed Incentive Mode (DIM).

In the electricity market, EV users can participate in the electricity transaction through these two agency modes to reduce their charging costs. If users choose the CMM mode, they are required to entrust the charging management authority of their vehicles to the EV aggregator. Under this mode, the EV aggregator can fully control the charging process, thereby obtaining sufficient regulation space. As compensation, the EV aggregator provides EV users with a relatively low contractual electricity price. If EV users select the DIM mode, they retain autonomous charging rights. In this mode, EV users formulate their own charging plans based on real-time electricity prices, and their charging demand depends largely on the pricing strategy set by the EV aggregator.

These two charging management modes differ significantly in terms of regulation approach, charging autonomy, and charging cost, as shown in Table 8.1. The CMM mode achieves strong controllability through centralized authority, leading to greater regulation potential of EVs. The DIM mode relies on the price signal transmission mechanism, and its regulation capability is constrained by the random behavior of users. In addition, the CMM mode transfers the risk of electricity price fluctuations to the aggregator, whereas under the DIM mode, users bear the risk of real-time price volatility by themselves.

Day-ahead bidding strategy of EV aggregators. In the day-ahead electricity market, EV aggregators should bid for the next-day charging demand of EVs. The development of a scientific and reasonable day-ahead bidding strategy is critical to reducing the operating costs of EV aggregators. To this end, on the premise of establishing an agency relationship between EV aggregators and EV users, this section

Table 8.1 Comparison between CMM and DIM

Mode	CMM	DIM
Charging autonomy	No	Yes
Charging cost	Determinate	Indeterminate
Charging flexibility	High	Low
User privacy	Moderately concerned	Marginally concerned

constructs a day-ahead bidding optimization model that integrates multiple factors, including charging randomness, market price volatility, and user response.

As shown in Fig. 8.3, the charging scheduling strategy of EVs is planned over a 24-h daily cycle. Notably, electric energy transactions in the electricity market are usually conducted on an hourly basis. Both day-ahead clearing prices and real-time clearing prices are issued and settled hourly. Therefore, the time granularity of transactions between EV aggregators and the power exchange center is set to 1 h. However, an hourly time granularity is too coarse for EV charging scheduling. On the other hand, an excessively short time granularity for EV charging scheduling would significantly increase the complexity of the real-time charging optimization problem and the frequency of scheduling adjustments. Accordingly, the time granularity for EV charging scheduling is set to 15 min in this study. It can be observed that a 15-min time granularity can not only ensure the accuracy of charging scheduling for EV planning but also avoid high computational costs.

To calculate the operational costs of EV aggregators, define t as the serial number of hours in a day, where $t = 1, 2, \ldots, 24$. The bidding costs of EV aggregators in the day-ahead market can thus be expressed as:

$$C_1 = \sum_{t \in T} \lambda_t^{Day} E_t^{Day} \tag{8.1}$$

where λ_t^{Day} denotes the day-ahead energy price at the t-th hour; and E_t^{Day} denotes the amount of day-ahead bidding energy at the t-th hour.

During the real-time operation, deviations may exist between the actual electricity consumption of EV aggregators and their day-ahead bidding volumes. The energy

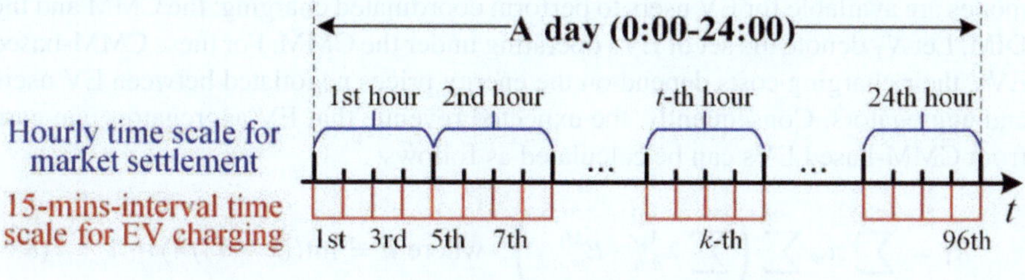

Fig. 8.3 One-day cycle and time scales adopted for EV charging scheduling

deviation will introduce uncertain operating costs to EV aggregators. Therefore, potential real-time clearing scenarios should be incorporated into the day-ahead bidding model. In the Guangdong RT market, the EV aggregator's potential costs associated with imbalance quantities can be expressed as:

$$C_2 = \sum_{w \in W} \pi_w \sum_{t \in T} \left(\lambda_{w,t}^{RT} \cdot \Delta E_{w,t} + C_{w,t}^{Pen} \right) \tag{8.2}$$

where w denotes the index of potential real-time clearing scenarios, π_w denotes the probability of the scenario w occurrence, $\lambda_{w,t}^{RT}$ denotes the real-time price at the t-th hour for the scenario w, $\Delta E_{w,t}$ denotes the deviation between the day-ahead bidding energy and the energy consumed by EVs at the t-th hour for the scenario w, $C_{w,t}^{Pen}$ denotes penalty costs arising from the deviation at the t-th hour for the scenario w.

In Eq. (8.3), $\Delta E_{w,t}$ can be calculated based on (8.4):

$$\Delta E_{w,t} = E_{w,t}^{RT} - E_t^{Day} \tag{8.3}$$

where $E_{w,t}^{RT}$ denotes the consumed energy at the t-th hour for the scenario w.

Furthermore, $C_{w,t}^{Pen}$ can be computed by:

$$C_{\omega,t}^{Pen} = \begin{cases} \left[E_t^{Day} - (1+\varepsilon_0)E_{\omega,t}^{RT}\right](\lambda_{\omega,t}^{RT} - \lambda_t^{Day}), & E_t^{Day} > (1+\varepsilon_0)E_{\omega,t}^{RT} \text{ and } \lambda_{\omega,t}^{RT} > \lambda_t^{Day} \\ \left[(1-\varepsilon_0)E_{\omega,t}^{RT} - E_t^{Day}\right](\lambda_t^{Day} - \lambda_{\omega,t}^{RT}), & E_t^{Day} < (1-\varepsilon_0)E_{\omega,t}^{RT} \text{ and } \lambda_{\omega,t}^{RT} < \lambda_t^{Day} \\ 0, & otherwise \end{cases} \tag{8.4}$$

where ε_0 denotes the allowable bidding deviation rate in Guangdong real-time markets.

Therefore, the total costs of EV aggregators can be calculated as:

$$C = C_1 + C_2 \tag{8.5}$$

In energy markets, EV aggregators can generate revenue by selling energy to EV users. To derive the optimal EV charging strategy, each hour is evenly divided into 4 intervals, resulting in a total of 96 time slots within a day. Let k represent the serial number of the time slots, where $k \in K$ ($K = [1, 96]$). As previously noted, two agent modes are available for EV users to perform coordinated charging: the CMM and the DIM. Let N_1 denote the set of EVs operating under the CMM. For these CMM-based EVs, their charging costs depend on the energy prices negotiated between EV users and aggregators. Consequently, the expected revenue that EV aggregators can earn from CMM-based EVs can be calculated as follows:

$$R_1 = \sum_{w \in W} \pi_w \sum_{k \in K} \left(\sum_{n \in N_1} \lambda_{n,h}^{Agr} \cdot E_{w,n,k}^{M_1} \right), \quad \text{where } h = \text{int}((k-1)/4) + 1 \tag{8.6}$$

where $E^{M_1}_{w,n,k}$ denotes the charging quantity of the n-th EV in the set N_1 at the k-th time slot for scenario w, $\lambda^{Agr}_{n,h}$ denotes the agreement price that EV aggregators sell to the n-th EV in the set N_1 at the h-th hour.

Let N_2 represent the set of EVs adopting the DIM. For these DIM-based EVs, their charging expenses are determined by the real-time market price. Therefore, the revenue that EV aggregators can acquire from DIM-based EVs can be computed using the following formula:

$$R_2 = \sum_{w \in W} \pi_w \sum_{k \in K} \left((1+\theta_0)\lambda^{RT}_{w,h} \cdot \sum_{n \in N_2} E^{M_2}_{w,n,k} \right), \text{ where } h = \text{int}((k-1)/4)+1 \tag{8.7}$$

where $E^{M_2}_{w,n,k}$ denotes the charging quantity of the n-th EV in the set N_2 at the k-th time slot for scenario w, $\lambda^{RT}_{w,h}$ denotes the real-time energy price at the h-th hour for scenario w, θ_0 denotes the service tariff that EV aggregators charge on EVs with the DIM.

In this context, the value of $E^{M_2}_{w,n,k}$ is determined by EV users. Within the demand response program, EV users will curtail their charging volume in response to elevated energy prices. Furthermore, when the energy price escalates to an intolerable threshold, they are inclined to halt EV charging to minimize costs. Building upon this principle, the value of $E^{M_2}_{w,n,k}$ can be mathematically expressed as:

$$E^{M_2}_{w,n,k} = \begin{cases} \phi^{\max}_{n,k} - \varphi_{n,k}\lambda^{RT}_{\omega,h}, & \lambda^{RT}_{\omega,h} \le \phi^{\max}_{n,k} / \varphi_{n,k} \\ 0 & , \lambda^{RT}_{\omega,h} > \phi^{\max}_{n,k} / \varphi_{n,k} \end{cases}, \forall \omega, n \in N_2, k \tag{8.8}$$

where $\phi^{\max}_{n,k}$ indicates the maximum allowable charging energy of the n-th EV in the set N_2 at the k-th time slot; $\varphi_{n,k}$ denotes the response coefficient of the n-th EV at the k-th time slot.

Accordingly, the total revenue generated by EV aggregators can be computed as follows:

$$R = R_1 + R_2 \tag{8.9}$$

Consequently, the expected net profits of EV aggregators under the day-ahead bidding model can be expressed as:

$$P = R - C \tag{8.10}$$

In conjunction with the above equations, the objective function may be formulated as follows:

$$\max \sum_{w\in W}\pi_w\sum_{k\in K}\left(\sum_{n\in N_1}\lambda_{n,h}^{Agr}\cdot E_{w,n,k}^{M_1}\right)+\sum_{w\in W}\pi_w\sum_{k\in K}\left((1+\theta_0)\lambda_{w,h}^{RT}\cdot\sum_{n\in N_2}E_{w,n,k}^{M_2}\right)$$
$$-\left(\sum_{t\in T}\lambda_t^{Day}E_t^{Day}+\sum_{w\in W}\pi_w\sum_{t\in T}\left(\lambda_{w,t}^{RT}\cdot\Delta E_{w,t}+C_{w,t}^{Pen}\right)\right) \tag{8.11}$$

where the mathematical expressions of h and $C_{w,t}^{Pen}$ have been given previously, and

$$\Delta E_{w,t}=E_{w,t}^{RT}-E_t^{Day},\ \text{where}\ E_{w,t}^{RT}=\sum_{n\in N_1}\sum_{k=4t-3}^{4t}E_{w,n,k}^{M_1}+\sum_{n\in N_2}\sum_{k=4t-3}^{4t}E_{w,n,k}^{M_2}, \tag{8.12}$$

subject to:

$$E_{\min}^{Day}\le\sum_{t\in I}E_t^{Day}\le E_{\max}^{Day} \tag{8.13}$$

$$E_{\omega,n,t}^{M_1}\le P_n^{\text{Max}}\,\Delta t,\ \forall n\in N_1 \tag{8.14}$$

$$\sum_{t\in[\Gamma_n^c,\Gamma_n^d]}E_{\omega,n,t}^{M_1}\cdot\frac{\eta_n}{Bat_n}=SoC_n^d-SoC_n^c,\forall\omega,n\in N_1 \tag{8.15}$$

$$E_{\omega,n,t}^{M_1}\ge 0,\forall\omega,n\in N_1,t \tag{8.16}$$

where $E_{\min}^{Day}$ denotes the minimum bidding energy in day-ahead markets, $E_{\max}^{Day}$ denotes the maximum bidding energy in day-ahead markets, P_n^{Max} denotes the maximum charge power of the n-th EV in the set N_1, Δt denotes the length of time slots, Γ_n^c denotes the connection time slot of the n-th EV in the set N_1, Γ_n^d denotes the disconnection time slot of the n-th EV in the set N_1, η_n denotes the charger efficiency of the n-th EV in the set N_1, Bat_n denotes the onboard battery capacity of the n-th EV in the set N_1, SoC_n^d denotes the disconnection battery SoC of the n-th EV in the set N_1, SoC_n^c denotes the connection battery SoC of the n-th EV in the set N_1.

Real-time charging scheduling of EVs. In the real-time charging control stage, EV aggregators need to formulate optimal charging scheduling strategies for EVs by combining the day-ahead awarded power and the charging demand of grid-connected EVs to maximize their economic benefits. Figure 8.4 illustrates the specific procedure for real-time energy regulation of EVs by the EV aggregator.

First, for the target scheduling time interval (e.g., the k-th interval), the EV aggregator must determine whether to implement charging scheduling according to the existing plan. The judgment is based on three criteria: (1) whether the real-time clearing price curve has changed; (2) whether newly connected EVs have introduced additional electricity demand; (3) whether grid-connected EVs have failed to depart

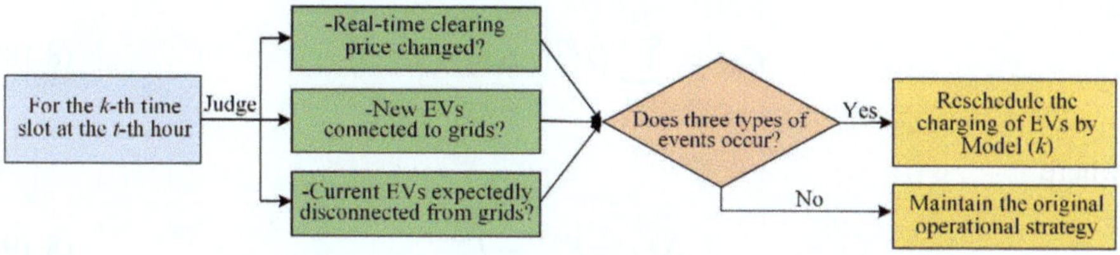

Fig. 8.4 EV real-time charging scheduling schematic diagram

at their scheduled time. If any of these three conditions occur, the EV aggregator shall reschedule the charging of EVs; otherwise, the original charging plan of EVs will be maintained.

For the real-time optimizer Model (k), its inputs include the data of grid-connected EVs, the real-time clearing price curve, and the day-ahead awarded power. The real-time clearing price is a real-time locational marginal price calculated by the power dispatch institution using security-constrained economic dispatch, aiming to minimize generation cost, based on the grid operating boundary conditions and ultra-short-term load forecasting information.

After the end of the k-th time interval, the charging data of electric vehicles in the system and the remaining awarded power of the current hour shall be updated, so as to proceed to the charging scheduling of electric vehicles in the next time interval.

Here, Model (k) is introduced to derive the optimal EV charging strategy for the k-th time slot at the t-th hour during charging rescheduling. Let S_k^1 represent the set of available EVs operating in the CMM during the k-th time slot, and S_k^2 represent the set of available EVs operating in the DIM during the k-th time slot. Let λ_i^{RT} denote the real-time clearing price released for the i-th hour. For the periods following the k-th time slot, the expected revenue that EV aggregators can generate from EVs operating in these two modes in real-time markets can be expressed as:

$$R_{RT} = \sum_{j=k}^{96}\left(\sum_{n\in S_k^1}\lambda_{n,j}^{Agr}\cdot E_{n,j}^{M_1}\right) + \sum_{j=k}^{96}\left((1+\theta_0)\lambda_h^{RT}\cdot\sum_{n\in S_i^2}E_{n,j}^{M_2}\right), \text{ where}$$
$$h = \text{int}((k-1)/4)+1 \tag{8.17}$$

where $\lambda_{n,j}^{Agr}$ denotes the agreement price that EV aggregators sell to the n-th EV in the set at the j-th time slot, $E_{n,j}^{M_1}$ denotes the charging quantity of the n-th EV in the set at the j-th time slot, λ_h^{RT} denotes the released real-time clearing price at the h-th hour, $E_{n,j}^{M_2}$ denotes the charging quantity of the n-th EV in the set at the j-th time slot.

Correspondingly, the expected costs that EV aggregators are required to pay for the periods after the k-th time slot at the t-th hour in real-time markets can be computed as follows:

$$C_{RT} = \sum_{i=t}^{24} \left(\lambda_i^{RT} \cdot \Delta E_i + C_i^{Pen}\right) \tag{8.18}$$

where

$$\Delta E_i = E_i^{RT} - E_i^{Day} \tag{8.19}$$

$$E_i^{RT} = \begin{cases} \sum_{j=4k-3}^{k-1} \delta_j + \sum_{n \in N_1} \sum_{j=k}^{4k} E_{n,j}^{M_1} + \sum_{n \in N_2} \sum_{j=k}^{4k} E_{n,j}^{M_2}, i = t \\ \sum_{n \in N_1} \sum_{j=4i-3}^{4i} E_{n,j}^{M_1} + \sum_{n \in N_2} \sum_{j=4i-3}^{4i} E_{n,j}^{M_2}, \quad i > t \end{cases} \tag{8.20}$$

$$C_i^{Pen} = \begin{cases} \left[E_i^{Day} - (1+\varepsilon_0)E_i^{RT}\right](\lambda_i^{RT} - \lambda_i^{Day}),\ E_i^{Day} > (1+\varepsilon_0)E_i^{RT} \ and \ \lambda_i^{RT} > \lambda_i^{Day} \\ \left[(1-\varepsilon_0)E_i^{RT} - E_i^{Day}\right](\lambda_i^{Day} - \lambda_i^{RT}),\ E_i^{Day} < (1-\varepsilon_0)E_i^{RT} \ and \ \lambda_i^{RT} < \lambda_i^{Day} \\ 0, \quad otherwise \end{cases} \tag{8.21}$$

where ΔE_i denotes the deviation between the day-ahead bidding energy and the consumed energy at the i-th hour, δ_j denotes the consumed energy at the j-th time slot, C_i^{Pen} denotes the penalty costs arising from the deviation at the i-th hour.

Hence, in Model (k), the objective function aimed at maximizing the profits of EV aggregators can be formulated as follows:

$$\begin{aligned} \max \sum_{j=k}^{96} \left(\sum_{n \in S_k^1} \lambda_{n,j}^{Agr} \cdot E_{n,j}^{M_1} \right) + \sum_{j=k}^{96} \left((1+\theta_0)\lambda_h^{RT} \cdot \sum_{n \in S_i^2} E_{n,j}^{M_2} \right) \\ - \sum_{i=t}^{24} \left(\lambda_i^{RT} \cdot \Delta E_i + C_i^{Pen} \right) \end{aligned} \tag{8.22}$$

subject to:

$$\sum_{j \in [k, \Gamma_n^d]} E_{n,j}^{M_1} \cdot \frac{\eta_n}{Bat_n} = SoC_n^d - SoC_n^c, \forall n \in S_k^1 \tag{8.23}$$

$$E_{n,j}^{M_1} \le P_n^{\text{Max}} \Delta t, \ \forall j \ge k, n \in S_k^1 \tag{8.24}$$

$$E_{n,j}^{M_1} \ge 0, \forall j \ge k, n \in S_k^1 \tag{8.25}$$

where Eqs. (8.23)–(8.25) specify that the charging requirements of EVs must be met before they disconnect from the grid, and the charging power of EVs should be constrained to the maximum capacity of the charger, consistent with the day-ahead bidding model. Additionally, the charging amount $E_{n,j}^{M_2}$ of EVs in the set N_2 is simulated.

The model developed above enables the derivation of the optimal EV charging strategy for the k-th time slot. In real-time energy management, if there are fluctuations in real-time clearing prices, or if any new EVs are connected to the system or any existing EVs are disconnected from the grid, the model will be activated with updated input parameters. Once EV aggregators acquire the optimal EV charging strategy, coordinated charging instructions will be transmitted to the charging piles for execution.

8.3 Data Preparation for Simulations

This section provides the method to generate both the EV charging data and the electricity market data.

EV Charging Data. In the day-ahead bidding model of an EV aggregator, it is necessary to model the randomness of EV cluster charging. In practice, it is difficult to accurately predict the individual charging demand of a single EV. However, for a specific EV cluster, the charging behavior usually follows a certain statistical distribution. To address the impact of EV charging uncertainty on the day-ahead bidding strategy of the EV aggregator, a specific probability distribution is adopted to simulate EV charging behavior, which serves as the input of the model.

For the EV slow-charging mode, the charging scenarios can be divided into two categories: residential charging and workplace charging. Data show that the proportions of EVs adopting residential charging and workplace charging are approximately 75% and 25%, respectively. EV charging behaviors (e.g., grid connection time, grid disconnection time) follow certain statistical distributions. Therefore, statistical distribution functions can be used to characterize EV charging behaviors. Table 8.2 presents the charging data generation methods for the two scenarios of residential charging and workplace charging.

In the residential charging scenario, EVs are usually charged at night. In this case, the grid connection time of EVs is mainly distributed around 20:00, and the grid

Table 8.2 Creation methods for EV-related data

Charge type	Proportion (%)	Plug-in time	Plug-out time
Overnight charging	75	$N(20, 2)$	$N(7.5, 1.5)$
Daytime charging	25	$N(8, 1.5)$	$N(17.5, 2)$
Charge type	**Plug-in SoC**	**Plug-out SoC**	**Charging level**
Overnight charging	$U(0.2, 0.5)$	Variable	Level 1: 16A/220 V
Daytime charging	$U(0.2, 0.5)$	Variable	Level 2: 32A/220 V

disconnection time is mainly distributed around 07:30. Thus, it is assumed that the grid connection time of EVs follows a normal distribution with a mean of 20:00 and a standard deviation of 2 h. In addition, the grid disconnection time of EVs is assumed to follow a normal distribution with a mean of 07:30 and a standard deviation of 1.5 h. For the EV charging scheduling problem, considering that the termination time of a scheduling cycle is 24:00, the charging termination time of EVs whose grid disconnection time exceeds 24:00 is set to 24:00, and their terminal SoC is set to 100% or the maximum achievable SoC within the grid connection period. Furthermore, the initial SoC of EVs when connected to the grid is assumed to follow a continuous uniform distribution between 0.2 and 0.5. The maximum operating power of the charger in the residential charging scenario is assumed to be 3.52 kW (Level 1: 16A/220 V).

In the workplace charging scenario, EVs are mostly charged during the daytime. In this case, the grid connection time of EVs is mainly distributed around 08:00, and the grid disconnection time is mainly distributed around 19:30. Here, the grid connection time of EVs is assumed to be characterized by a normal distribution with a mean of 08:00 and a standard deviation of 1.5 h, and the grid disconnection time is described by a normal distribution with a mean of 19:30 and a standard deviation of 2 h. In addition, the battery SoC of EVs is generally at a low level when connected to the grid, while EV users expect to charge the battery to a relatively high level. Therefore, the initial SoC of EVs is assumed to follow a uniform distribution between 0.2 and 0.5, and the terminal SoC is set to 100% or the maximum achievable SoC within the grid connection period. The maximum operating power of the charger in the workplace charging scenario is assumed to be 7.04 kW (Level 2: 32A/220V).

Electricity Market Data. In this work, the proposed solution is implemented within the context of the Guangdong electricity market to examine EV charging behaviors. As one of the most economically developed regions in China, Guangdong Province has introduced a series of robust policies to promote the adoption of EVs. While large-scale EV adoption will pose challenges to distribution networks, Guangdong, one of the first provinces in China to launch electricity market reforms, also offers a unique opportunity to study EV charging behaviors within a market-oriented framework.

In the day-ahead bidding model, EV aggregators must integrate day-ahead clearing prices with potential real-time clearing scenarios to formulate an optimal bidding strategy. It is important to note that both day-ahead and real-time clearing prices are unknown in advance. Generally, day-ahead clearing prices can be predicted with relative accuracy, as daily day-ahead price profiles exhibit certain regular patterns. However, real-time clearing prices are subject to extremely high uncertainty, as they are closely tied to the real-time supply and demand dynamics of the grid. To address such price uncertainties in real-time markets, several potential real-time clearing scenarios are considered.

Potential real-time clearing scenarios can be categorized into four types, i.e. low, medium, high and extra high, using a multi-class support vector machine to analyze historical market clearing data, as illustrated in Fig. 8.5. It can be observed that these four real-time clearing price curves follow the same trend of variation. In each

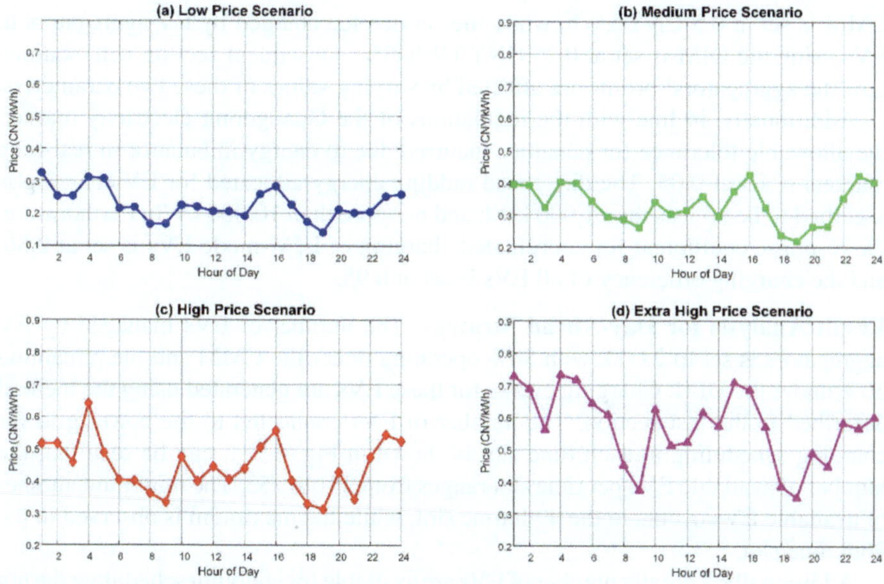

Fig. 8.5 Four types of potential real-time clearing scenarios

potential scenario, high prices typically occur during the periods 9:00–11:00, 15:00–17:00, and 20:00–23:00, while low prices appear at 13:00, 19:00, and 0:00–8:00. Specifically, in the extra high real-time clearing scenario, the lowest daily clearing price is 0.410 CNY/kWh, and the highest reaches 1.236 CNY/kWh. In contrast, in the low real-time clearing scenario, prices range from 0.268 CNY/kWh to 0.759 CNY/kWh. These four potential clearing scenarios are incorporated into the model through the use of probability functions.

From the perspective of EV aggregators, advanced prediction algorithms are required to forecast day-ahead clearing prices for bidding purposes; however, this falls outside the scope of this paper. Instead, the actual day-ahead market clearing data presented are utilized to validate the proposed method. Furthermore, actual real-time clearing prices are characterized by high uncertainty and volatility. In this work, real-time clearing prices are generated based on a normal distribution to enable real-time EV charging scheduling.

8.4 Case Study and Numerical Results

Simulation Setting. In the case study, detailed parameters required for the simulations are provided in Table 8.3. Contracts between EV aggregators and EV users exert a significant influence on the EV aggregators' profitability. Specifically, the agreement price at which EV aggregators sell electricity to EVs operating under the

CMM is set at 0.5 CNY/kWh, while the service fee charged by EV aggregators to EVs using the DIM is set at 0.25 CNY/kWh. The subsequent section will examine how the aggregators' profits are affected by varying values of these two parameters.

Additionally, in line with the regulations of the Guangdong electricity market, the allowable tolerance for penalties incurred due to energy imbalance in real-time markets is set at 0.05. The day-ahead bidding energy allocated for EV charging is specified to be no less than 1,000 kWh and no more than 10,000 kWh. Furthermore, the response coefficient for coordinated charging of DIM-mode EVs is set at 2.86, and the charging efficiency of all EVs is set at 0.95.

Result Analysis for Day-Ahead Strategy. The number of EVs managed by EV aggregators is set to 2,000, with 50% operating under the CMM and the remaining 50% under the DIM. Charging events for these EVs are generated using the method described in the last section. The number of EVs connected to the power grid for charging scheduling at each time slot is shown in Fig. 8.6. It can be seen that the number of available EVs per time slot ranges from 251 to 753. The minimum number of available EVs occurs in the 49th time slot, while the maximum is observed in the 96th time slot.

Additionally, a greater number of EVs are available for charging scheduling during two time windows: the 1st to 32nd time slots (i.e., 0:00–8:00) and the 80th–96th time slots (i.e., 20:00–24:00). At present, most EV users tend to charge their vehicles at night, so the simulated EV charging data align with real-world EV charging behaviors.

In the day-ahead time horizon, EV aggregators are required to bid for energy to meet EV charging needs. Simulation results indicate that the day-ahead bidding strategy of EV aggregators does not solely depend on day-ahead energy prices, as shown in Fig. 8.7. In fact, this strategy is jointly determined by both day-ahead and real-time markets. If the real-time clearing price is higher than the day-ahead clearing price, EV aggregators could profit by bidding for more energy in the day-ahead market in the absence of a penalty mechanism. However, given the penalty costs arising from energy imbalance, aggregators must adopt a suitable day-ahead bidding strategy to minimize operational costs. Overall, the price difference (PD) between

Table 8.3 Values of necessary parameters for simulations

Parameter	Value
$\lambda_{n,h}^{Agr}$ (CNY/kWh)	0.5
θ_0 (pu)	0.25
ε_0 (p.u.)	0.05
$E_{\min}^{Day}$ (kWh)	1000
$E_{\max}^{Day}$ (kwh)	10,000
$\varphi_{n,k}$ (p.u.)	2.86
η_n (p.u.)	0.95

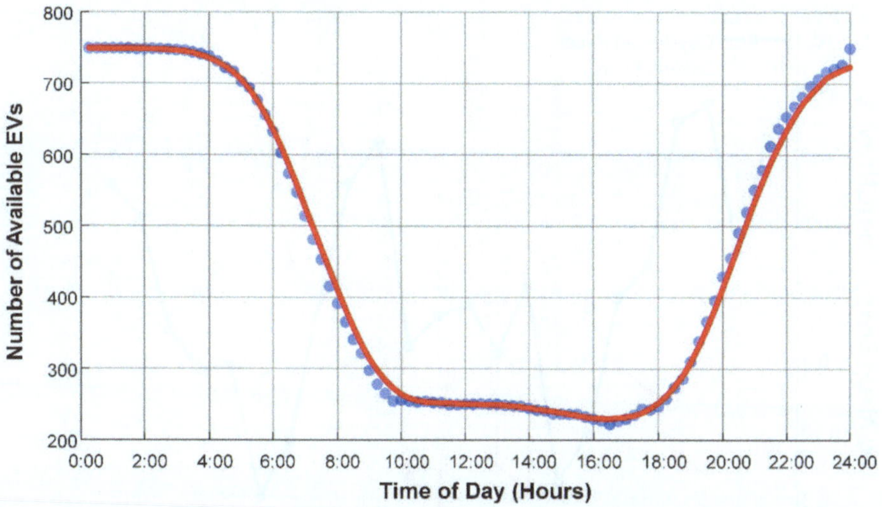

Fig. 8.6 Number of available EVs for charging scheduling at each time slot

real-time and day-ahead markets exerts a significant impact on the day-ahead bidding strategy of EV aggregators.

Figure 8.8 demonstrates the correlation between the day-ahead bidding strategy and the PD under the four potential real-time clearing scenarios. The PD is positive (greater than 0) across all potential scenarios during the 9th, 11th, 12th, 14th, 15th, 16th, 17th, 18th, and 19th hours. Nevertheless, EV aggregators only bid the maximum allowable energy during the 9th, 12th, 16th, 17th, and 19th hours. This is because, during these hours, aggregators can generate sufficient revenue by purchasing lower-cost electricity in the day-ahead market and reselling it in the real-time market, which offsets the penalty costs incurred due to energy imbalance.

When EV aggregators submit their day-ahead bidding strategies to the energy exchange center, their purchasing costs and expected penalty costs can be calculated, as illustrated in Fig. 8.9. It can be observed that the purchasing costs in day-ahead markets mostly range between 1,000 CNY and 2,000 CNY. Higher purchasing costs, exceeding 3,000 CNY, occur at the 9th, 12th, 16th, 17th, and 19th hours, with the highest cost (approximately 5,000 CNY) recorded at the 16th hour. These higher costs are attributed to both the large volume of purchased energy and the high energy prices during these periods. In contrast, lower purchasing costs, less than 1,000 CNY, are seen at the 10th, 11th, 13th, and 20th hours.

Regarding expected penalty costs, they remain below 500 CNY for most hours. However, they exceed 1,500 CNY at the 11th, 12th, 15th, 16th, and 17th hours. This is because the day-ahead bidding energy during these hours is extremely high, while the actual energy used for EV charging is very limited, leading to significant imbalance-related penalties.

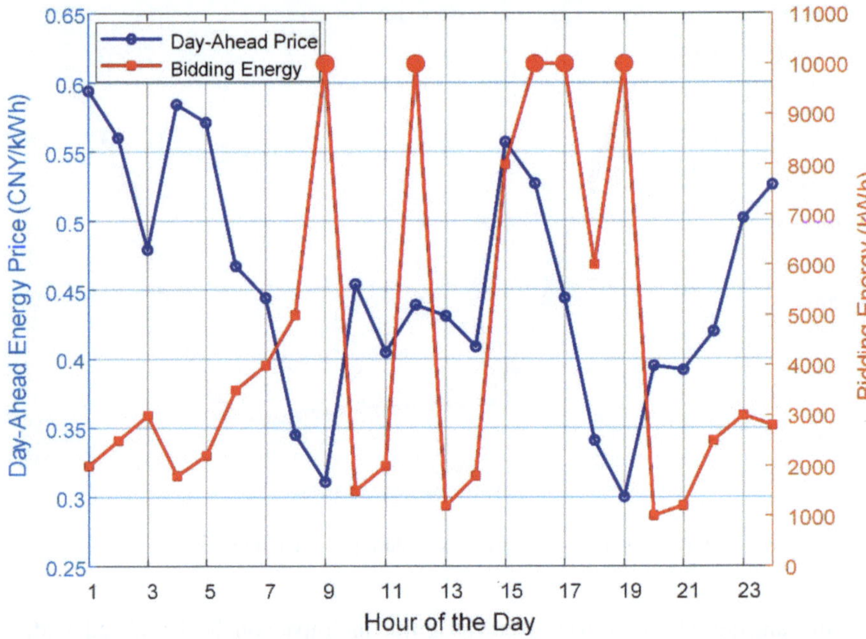

Fig. 8.7 Day-ahead bidding strategy of EV aggregators associated with day-ahead energy prices

Result Analysis for Real-Time Operation. In real-time operations, EV aggregators aim to adjust the charging power of available EVs to maximize profits, based on real-time market-clearing results and the day-ahead bidding strategy. Figure 8.10 compares the day-ahead bidding energy with the actual energy consumed by EVs. It shows that the energy consumed by EVs aligns with the day-ahead bidding energy during most hours of the day, indicating that EV charging tasks are scheduled as planned to achieve coordinated energy management of EVs. Notably, significant deviations between the day-ahead bidding energy and the energy consumed by EVs occur at the 9th, 12th, 16th, 17th, and 19th hours. This is because, although EV aggregators bid the maximum allowable energy at these hours, there are not enough available EVs to consume the bid energy. Overall, EV aggregators schedule EV charging to match the day-ahead bidding energy profiles, thereby reducing their penalty costs.

Figure 8.11 illustrates the relationship between the energy consumed by EVs, the price difference (PD), and the actual real-time clearing price. When PD is negative (PD < 0), the day-ahead price is higher than the real-time price, meaning EV aggregators purchased energy at a higher price in the day-ahead market. Consequently, during real-time operations, they strive to maximize energy consumption for EV charging to avoid incurring additional operational costs. Numerical results confirm that EV charging energy consumption is relatively high when PD < 0, such as at the 2nd, 3rd, and 22nd hours. In contrast, when PD is positive (PD > 0), the real-time

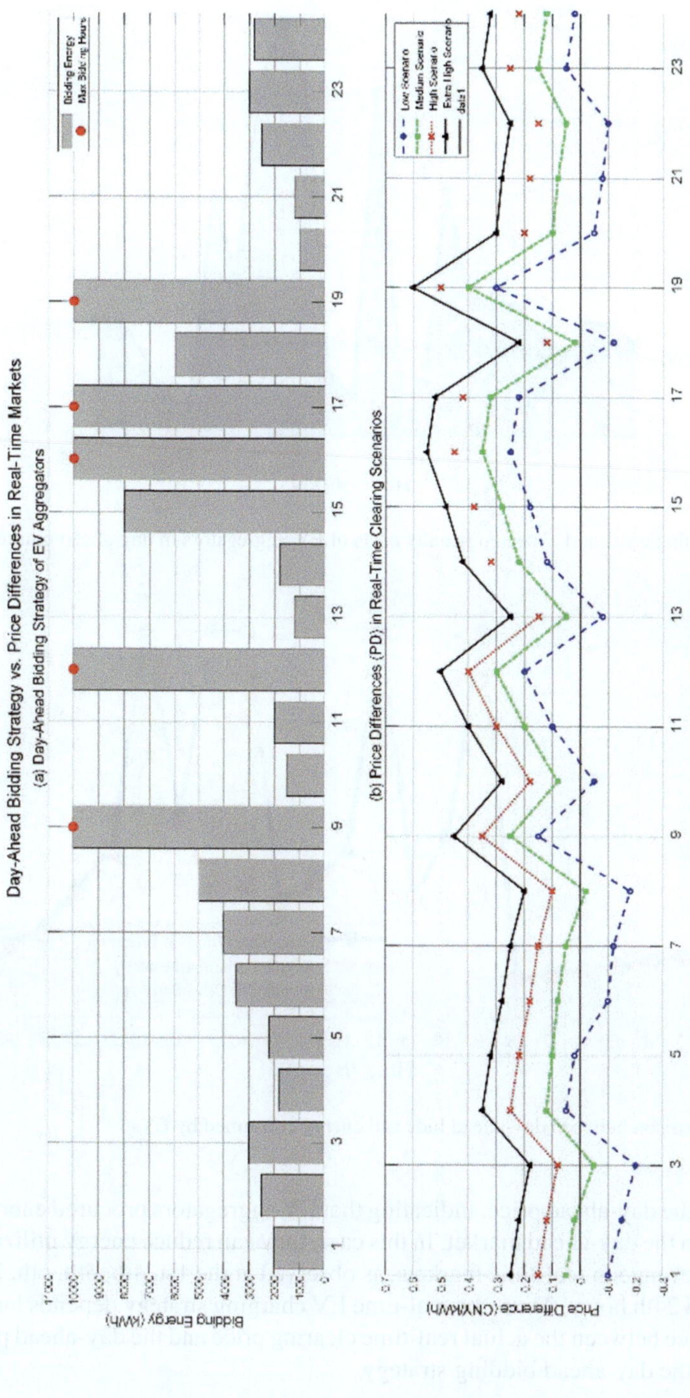

Fig. 8.8 Day-ahead bidding strategy of EV aggregators considering potential real-time clearing scenarios

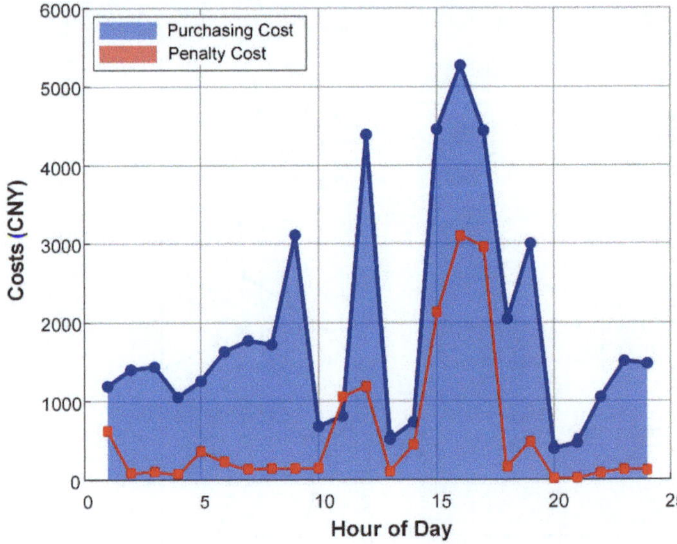

Fig. 8.9 Purchasing costs and expected penalty costs of EV aggregators in day-ahead markets

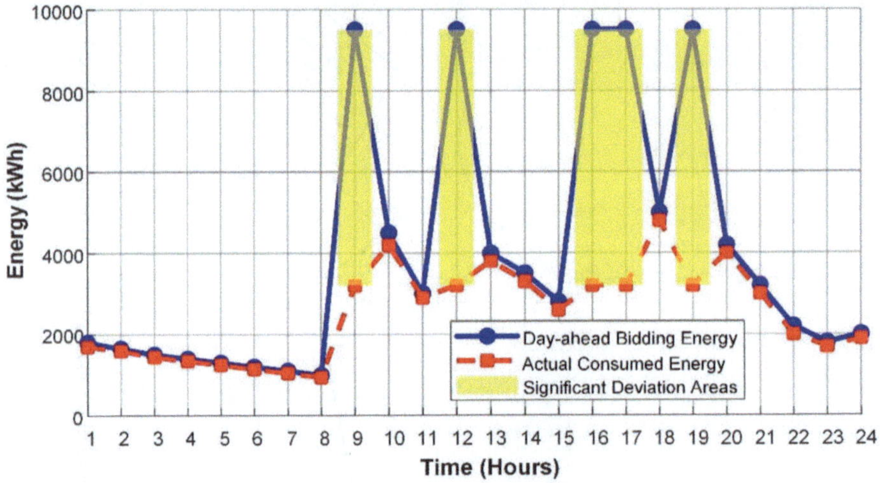

Fig. 8.10 Comparison between day-ahead bids and energy consumed by EVs

price exceeds the day-ahead price, indicating that EV aggregators procured energy at a lower price in the day-ahead market. In this case, they can reduce energy utilization to gain more revenue in real-time markets, as observed at the 1st, 4th, 5th, 6th, 20th, 21st, 23rd, and 24th hours. Thus, the real-time EV charging strategy depends largely on the difference between the actual real-time clearing price and the day-ahead price, in addition to the day-ahead bidding strategy.

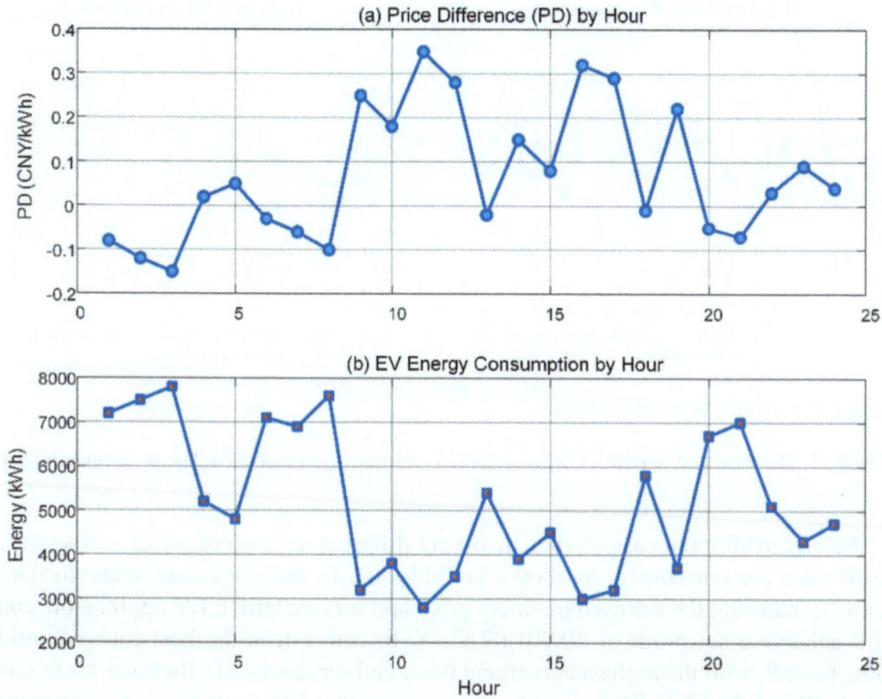

Fig. 8.11 Consumed energy by EVs associated with PD for actual real-time clearing results

A cost–benefit analysis of EV aggregators participating in real-time markets is presented in Fig. 8.12. EV aggregators generate revenue by selling energy to individual EVs for charging. Conversely, they are required to pay penalty costs for imbalanced energy in real-time markets and settle the imbalance energy.

Figure 8.12a illustrates the daily revenue of EV aggregators, showing that the overall variation trend of revenue from CMM-mode EVs is consistent with that from DIM-mode EVs. When the PD is negative, EV aggregators earn higher revenue from both CMM-mode and DIM-mode EVs. However, the revenue curve for DIM-mode EVs fluctuates significantly more than that for CMM-mode EVs. The costs incurred by EV aggregators are shown in Fig. 8.12b. Since the energy consumed by EVs aligns well with the day-ahead bidding strategy, penalty costs in real-time markets are extremely low, except for the periods when aggregators bid the maximum allowable energy. It is only during these hours that large deviations occur, leading to high penalty costs for aggregators. Nevertheless, the substantial surplus of energy purchased in the day-ahead market can be traded in real-time markets for secondary settlement. It is evident that during the 9th, 12th, 16th, 17th, and 19th hours, the revenue from trading the imbalanced energy exceeds the penalty costs associated with the imbalance quantity.

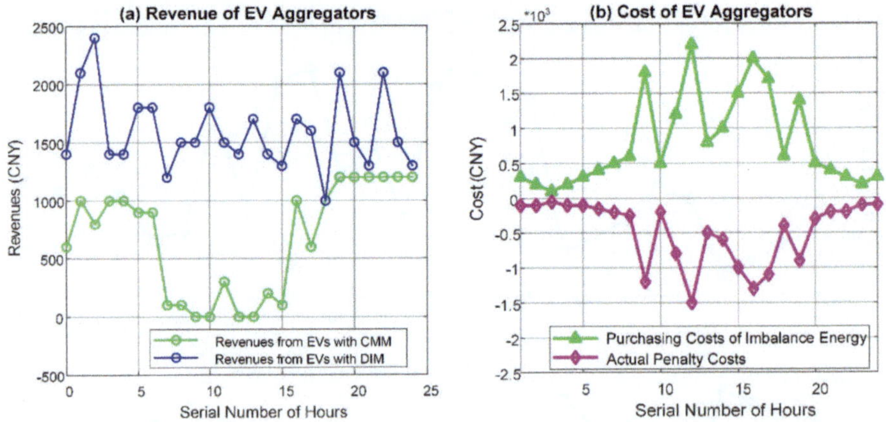

Fig. 8.12 Revenue and cost of EV aggregators in real-time markets over a day: **a** revenue; **b** cost

Finally, additional case studies involving different agreement prices and service tariff rates are considered, as shown in Table 8.4. In the worst-case scenario (i.e., Case 1, featuring the lowest agreement price and service tariff), EV aggregators can still achieve a net profit of 20,901.07 CNY. In contrast, in the best-case scenario (i.e., Case 9, with the highest agreement price and service tariff), their net profit can reach 30,385.22 CNY. These results demonstrate that EV aggregators can leverage the flexibility of EV charging to generate substantial profits under various conditions. This highlights significant business opportunities for EV aggregators to coordinate EV charging in energy markets while also helping EV users reduce their charging costs.

Comparison of Different Cases. The effectiveness of the proposed solution is further assessed under a scenario involving large-scale EV deployment. It is assumed that

Table 8.4 Profit analysis of EV aggregators for different agreement prices and service tariffing rates

Case	Agreement price for EVs with CMM (CNY/kWh)	Service tariffing for EVs with DIM (p.u.)	Net profit (CNY)
1	0.4	0.20	20,901.07
2	0.4	0.25	22,394.58
3	0.4	0.30	23,888.08
4	0.5	0.20	24,149.64
5	0.5	0.25	25,643.15
6	0.5	0.30	27,136.65
7	0.6	0.20	27,398.21
8	0.6	0.25	28,891.71
9	0.6	0.30	30,385.22

EV aggregators manage 4,000 EVs in energy markets, with 2,000 operating under the CMM and another 2,000 under the DIM. Additionally, a more generalized day-ahead price curve is considered, where the hourly day-ahead price is bounded below by the price in the low real-time clearing scenario and above by the price in the extra-high real-time clearing scenario. Unlike the previous case, EV aggregators do not bid the maximum allowable energy under these conditions.

In this work, potential real-time clearing scenarios are categorized into four types to enable EV aggregators to obtain an optimal bidding strategy efficiently. It should be noted that while considering all possible real-time clearing scenarios could yield a more optimal bidding solution for aggregators, this approach would require extensive computational time, particularly when aggregators manage large-scale EV fleets.

The model was run on a 64-bit personal computer equipped with an Intel Core i5-10,400 processor and 16 GB of RAM, using the Lingo 18.0 × 64 interface for solving. Figure 8.13 visually compares the computation time required to derive the bidding strategy when different numbers of real-time clearing scenarios are considered. For small EV fleets (e.g., 500 and 1,000 vehicles), the difference in computation time is negligible. However, for larger fleets (e.g., 1,000 and 1,500 vehicles), the model with 8 real-time clearing scenarios requires several times more computation time than the model with 4 scenarios. This confirms that the proposed strategy significantly reduces the time complexity of the bidding problem.

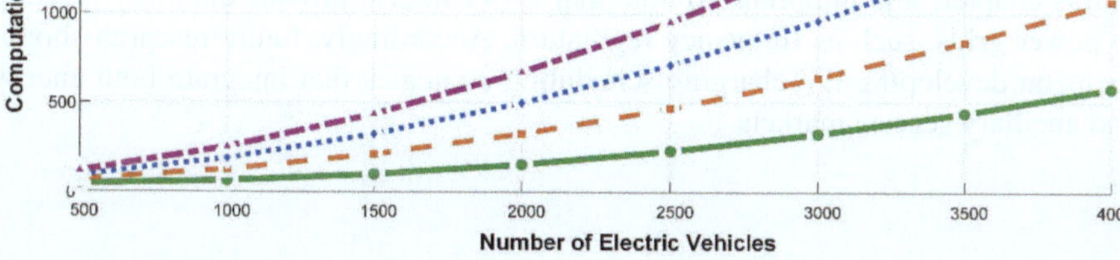

Fig. 8.13 Comparison of time consumed considering different numbers of real-time clearing scenarios

8.5 Summary

This chapter explores the optimal charging scheduling of EVs in a competitive electricity market from the perspective of EV aggregators. To this end, a two-phase energy management framework for EV charging is established in line with market regulations. In the day-ahead bidding phase, aggregators account for various uncertainties, including those associated with EV charging behaviors and market fluctuations, to formulate optimal bidding strategies. The real-time scheduling phase then involves dynamic adjustments to the charging plans of available EVs, ensuring alignment with the day-ahead bid energy profile and thus minimizing operational costs. Ultimately, the proposed approach enables the total EV charging load to match the pre-committed bidding energy.

Case study results yield three key insights: (1) EV aggregators' bidding decisions are closely related to EV grid-connection times and market-clearing prices. Numerical findings demonstrate that aggregators bid for more electricity at night than during the day, mainly due to higher nighttime EV charging demand driven by user preferences; (2) The day-ahead bidding strategy is largely shaped by the price differential between day-ahead and real-time clearing prices. Specifically, if the day-ahead price for a given hour is lower than the real-time clearing prices across all potential scenarios, aggregators tend to bid the maximum allowable energy in the day-ahead market (e.g., the 9th, 12th, 16th, 17th, and 19th hours in the case study); (3) Most notably, the proposed method allows aggregators to schedule EV charging loads to match the day-ahead bid energy, thereby cutting operational costs. Furthermore, simulation results indicate that the model incorporating four real-time clearing scenarios for optimal bidding is more efficient than those relying on a large set of historical real-time clearing data.

Beyond the coordinated EV charging strategies in the energy market discussed in this chapter, it is important to note that EVs can also provide ancillary services to power grids, such as frequency regulation. Accordingly, future research should focus on developing EV charging scheduling strategies that integrate both energy and ancillary service markets.

Chapter 9
PEV-to-Grid Operational Strategies in Active Power Distribution Systems

Abstract This chapter investigates game-theoretic operational strategies for plug-in electric vehicles (PEVs) providing vehicle-to-grid (V2G) services in active power distribution systems, considering dynamic pricing and the integration of distributed energy resources. Three distinct game models are developed and analyzed. First, a non-cooperative game model is established where multiple EV users independently optimize their charging/discheduling schedules to minimize individual costs, leading to a Nash equilibrium. Second, a cooperative game model is formulated where EVs form a coalition through an aggregator to minimize the total charging cost, demonstrating superior economic benefits compared to the non-cooperative approach. Third, a Stackelberg master–slave game is proposed modeling the hierarchical interaction between the grid operator (leader) utilizing energy storage for peak shaving and EV aggregators (followers). The operator sets dynamic prices to minimize peak-shaving costs, while aggregators optimize EV schedules accordingly. The model is solved using Karush–Kuhn–Tucker (KKT) conditions. Numerical simulations based on a parking lot scenario with 150 EVs validate the proposed strategies. Results show that cooperative gaming reduces total EV charging costs compared to the scenarios with uncontrolled V2G, while the Stackelberg equilibrium achieves a win–win outcome, reducing both EV charging costs and grid operator storage operating costs, effectively flattening the load curve and improving renewable energy utilization.

9.1 Introduction

The preceding chapter investigated optimal scheduling strategies for electric vehicle (EV) charging loads in active distribution networks. Based on the different requirements and characteristics of the two EV charging modes, a spatial scale scheduling approach was applied to fast-charging loads, while a temporal scale scheduling approach was applied to regular charging loads. A dynamic electricity pricing strategy was adopted to incentivize EVs to charge during off-peak hours and at locations with

Q. Yang et al., *Integration of Massive Electric Vehicles to Smart Grid*, Smart Energy Systems, https://doi.org/10.1007/978-981-92-0557-8_9

lower load, thereby optimizing grid operation and improving the utilization efficiency of distributed renewable energy sources.

In fact, electric vehicles are not merely simple electrical loads; they can also be viewed as controllable distributed energy storage and sources through their charging and discharging behaviors. This enables EVs to charge as storage units during grid off-peak periods, storing surplus electricity, and to feed power back to the grid as sources during peak load periods. Thus, their role transforms from "power demand" to "power generation", which is referred to as V2G (Vehicle to Grid). Reasonable scheduling strategies for EV V2G can not only mitigate the negative impacts of large-scale disordered EV charging on the power grid but also provide various ancillary services such as frequency regulation, peak shaving, and spinning reserve. This achieves multiple benefits, including significantly reducing user charging costs and grid operation costs, improving the grid load profile and power quality, and enhancing the utilization efficiency of distributed renewable energy sources. Therefore, how to formulate reasonable EV V2G strategies that satisfy EV charging demands while accounting for the uncertainty of distributed generation output, enabling EVs to better provide ancillary services to the grid, has become a highly important issue.

Furthermore, with the rapid development of energy storage technology, energy storage is playing an increasingly important role in power system ancillary services. Compared to traditional generating units, energy storage offers fast real-time response capabilities, lower investment and operating costs, and possesses the dual attributes of both source and load, allowing for more flexible frequency regulation and peak shaving. Consequently, this chapter will also consider utilizing energy storage for peak shaving services.

Additionally, the dynamic pricing strategy in the previous chapter did not account for the impact of EV charging load on electricity prices. With a significant increase in EV penetration, this approach could potentially turn the original grid load valley into a new peak for EV charging load, which is detrimental to grid stability and economic operation. Therefore, how to simultaneously obtain scheduling outcomes beneficial to both individual EV users and the grid operator, while considering the dynamic impact of EV charging load on electricity prices, becomes a very challenging problem.

In light of these three aspects, this chapter builds upon the previous one by considering the V2G mode where EVs act as sources discharging to the grid in the active distribution network. Based on game theory, it investigates the optimal decision-making problems for EV users and the grid operator under the influence of dynamic pricing, primarily including the following:

First, a charging and discharging model for multiple EVs based on non-cooperative game theory was established, along with a cooperative game-based charging and discharging model for multiple EVs considering an aggregator. The Nash equilibrium point for the non-cooperative game and the optimal strategy combination for the cooperative game were solved, and a comparative analysis of the total charging costs under several charging and discharging strategies was conducted.

Second, a Stackelberg game theory-based master–slave game model between the grid operator and EV users was established. This model utilizes both energy storage equipment and EV V2G behavior for peak shaving and valley filling. The KKT conditions were used for solving the model, demonstrating that this method reduces the charging cost for EV users while also lowering the operating cost of energy storage, thereby achieving a satisfactory win–win outcome for both parties.

9.2 Power System Auxiliary Services

Power system auxiliary services refer to a series of services provided by power generation enterprises, power grid operators, and even users in the power system, which are necessary to ensure the safe and stable operation of the power system and the compliance of power quality with requirements, mainly including peak shaving services, frequency modulation services, standby services, black start services, etc. Among them, peak shaving and frequency modulation services are mainly to ensuring power quality, while standby and black start services are mainly responsible for ensuring the safe operation of the system.

At present, the auxiliary services of China's power system are mainly required by grid operators to be provided by power generation enterprises as a condition for their access to the grid. Grid operators arrange and dispatch the auxiliary services provided by each power generation enterprise in a unified manner, and provide appropriate economic subsidies according to the actual implementation of each power generation enterprise. Although this approach is simple and effective, it has poor economic viability and may lead to duplication or overinvestment, resulting in wastage of resources. At the same time, China still uses traditional power sources such as hydroelectric units, coal-fired units, and gas-fired units to provide auxiliary services. These devices have slow climbing speeds, low real-time response capabilities, high investment and operating costs, and can also cause environmental pollution and other problems.

With the rapid development of energy storage technology, energy storage has played a good role in the ancillary services of the US electricity market, and has solved the problems that traditional power sources face in providing ancillary services. This provides us with good inspiration. Using energy storage as a new means to participate in the frequency regulation auxiliary service market and improve efficiency is an important direction for the future development of China's power auxiliary service market.

In addition, electric vehicles have also demonstrated significant advantages in providing auxiliary services to the power grid. Electric vehicles often stay at home for more than 12 h at night, while the time from empty to fully charged is generally 6–8 h. If electric vehicles can be reasonably scheduled to charge during the low peak periods of traditional loads in the power grid while ensuring the charging demand, and discharged as a power source during the peak periods of traditional loads in the power grid, it can not only alleviate the negative impact of disorderly charging of electric vehicles on the power grid, but also improve the peak valley difference of power grid loads, save the cost of power grid frequency regulation and peak shaving, and absorb more electricity generated by distributed power sources, thereby improving the utilization rate of renewable energy and fully leveraging the advantages of energy conservation and environmental protection of electric vehicles.

Compared to the conventional charging mode, which has a large margin in time scale, V2G is not suitable in the spatial scale because electric vehicles first need to meet the travel needs of users. Users are often unwilling to sacrifice their travel convenience and drive their electric vehicles to distant nodes for discharge. At the same time, fast charging of electric vehicles is not suitable for V2G, because when fast charging occurs, the battery level of electric vehicles is often very low and cannot be used as a power source to reverse transmit electrical energy to the grid. Therefore, this chapter only considers the conventional charging mode of electric vehicles providing auxiliary services to the grid through V2G on a time scale.

9.3 Game Theory

Game theory, also translated as game theory, is a mathematical theory and method that mainly studies how two or more participants make decisions that are most beneficial to themselves in competitive games when their interests are interrelated or even conflicting. In 1944, John von Neumann (1903–1957) and Oskar Morgenstern (1902–1977) coauthored the book "Game Theory and Economic Behavior", which first studied the problem of multiplayer games, marking the initial formation of game theory. Subsequently, American mathematician John Forbes Nash Jr. (1928–2015) systematically studied game theory in the early 1950s, proposed the concept of Nash equilibrium, and proved the existence of Nash equilibrium points, laying the foundation for modern game theory and earning him the Nobel Prize in Economics for his contributions. Since its establishment, game theory has continuously developed and improved, and now it has become one of the most crucial analytical and decision-making tools, widely applied in fields such as politics, economics, engineering, etc., and has had a profound impact on various disciplines.

A game model G typically has three basic elements: players, strategies, and payoffs. Participants refer to all decision-making entities involved in the game, and typically, the set of participants in a n human game model can be represented as $N = \{1, 2, \ldots, n\}$, and the strategy refers to the action plan that participants can take. If the vector composed of all strategies of the i th participant s is S_i, and S_i is not empty, then the strategy space $S = \{S_1, S_2, \ldots, S_n\}$ is the set of strategies that all participants can choose from. Payment can measure the payoff of participants at the end of a game, and the payoff of each participant is related to the strategies of all participants. Therefore, the payoff of each participant at the end of the game is a function of the set of strategies adopted by all participants in the game, commonly referred to as the payoff function. For participants i, the payoff function of the strategy combination $s = \{s_1, s_2, \ldots, s_n\}$ is $u_i(s_1, s_2, \ldots, s_n)$. Therefore, the entire game model can be represented using $G = \{N; S_1, \ldots, S_n; u_1, \ldots, u_n\}$.

There are many classification methods for game models. Generally speaking, according to different classification criteria, they can be divided into cooperative and non-cooperative games, zero-sum and non-zero-sum games, complete information and incomplete information games, peer-to-peer games, and master–slave games. In this section, we will mainly discuss three types of games: non-cooperative games, cooperative games, and master–slave games.

Non-cooperative game: A non-cooperative game is the most common type of game theory, in which all participants simultaneously decide their strategies and act according to their strategies. Each participant hopes to gain more benefits, and there is no common interest among the participants. Therefore, a non-cooperative game is named. The focus of non-cooperative games is on how participants can make reasonable predictions about the strategies of other participants and adopt the most advantageous strategy for themselves.

Firstly, in a non-cooperative game model $G = \{N; S_1, \ldots, S_n; u_1, \ldots, u_n\}$, the optimal strategy for the participants i is defined as the strategy that maximizes their payoff function while all other participants adopt the strategy, i.e., satisfies Eq. (9.1), where s_{-i} is the strategy for all participants except for the nth participant.

$$s_i^*(s_{-i}) = \arg\max u_i(s_i, s_{-i}) \tag{9.1}$$

When all participants choose their respective optimal strategies, that is, if each one s_i^* is the optimal strategy for s_{-i}^*, $i = 1, \ldots, N$, i.e. satisfies Eq. (9.2), then the strategy combination $s^* = (s_1^*, \ldots, s_i^*, \ldots, s_n^*)$ is called the Nash equilibrium of the game.

$$s_i^*(s_{-i}^*) = \arg\max u_i(s_i, s_{-i}^*) \tag{9.2}$$

It can be observed that when using the strategy combination of Nash equilibrium s^*, no participant can achieve greater returns by simply changing their own strategy, which can be expressed as:$s^* = (s_1^*, \ldots, s_i^*, \ldots, s_n^*)$ the strategy combination is a Nash equilibrium if and only if Eq. (9.3) is satisfied, where s_i' is the strategy adopted by the ith participant that is different from its optimal strategy s_i^*.

$$u_i(s_i^*, s_{-i}^*) \geq u_i(s_i', s_{-i}^*), \forall s_i \in S_i, i \in N, s_i' \neq s_i^* \tag{9.3}$$

From Eq. (9.3), it can be seen that when the game reaches the Nash equilibrium, the strategy is the best combination of strategies recognized by all participants, and no participant is willing to change their own strategy alone because no matter how they change their strategy, they will not gain more benefits.

At the same time, Nash also used the fixed point theorem of Jingfu Kakutani to prove that the Nash equilibrium is universally present in game models. That is, if the number of participants in the game $G = \{N; S_1, \ldots, S_n; u_1, \ldots, u_n\}$ is finite and all are finite sets, then there exists at least one Nash equilibrium in the game

G. Subsequently, other scholars have used various methods to prove that the Nash equilibrium is widely present in different game models.

Cooperative game: Nash used the existence of binding agreements in games to define non-cooperative games and cooperative games. Therefore, cooperative games refer to games where participants no longer communicate without strategic information, like non-cooperative games, but can cooperate and form alliances by reaching agreements. Therefore, they are also known as coalition games.

When certain participants unite, each participant can gain more benefits than in non-cooperative games, which is also a necessary condition for forming cooperative games. At this point, the alliance formed by the participants can be considered as a whole to participate in the game.

At present, research on cooperative games mainly focuses on how to establish alliances and how to allocate additional profit vectors in alliances, that is, payment transferable cooperative games. However, in this chapter, we will not consider how to allocate additional profit vectors, so we will not provide a more detailed introduction to the relevant theories of cooperative games.

Stackelberg game: The master–slave game is a special form of non-cooperative game, first proposed by German economist Heinrich Freiherr von Stackelberg in 1934. Therefore, the master–slave game is also known as the Stackelberg game. As the name suggests, a master–slave game consists of two layers: a leader and a follower. Unlike non-cooperative games where all participants make decisions simultaneously, in master–slave games, the leader first determines their own strategy, followed by the follower making the optimal strategy based on the leader's strategy, and finally the leader formulates their own optimal strategy based on the follower's decision.

There are two forms of Stackelberg games: "one-to-one", where there is one leader and one follower, and "one to many", where there is one leader and multiple followers.

In the case of only one leader and one follower, the leader first formulates a strategy s_l and publishes it to the followers, who then determine their optimal strategy s_l based on the strategy published by the leader, as shown in Eq. (9.4), where u_f is the follower's payment function.

$$s_f^*(s_l) = \arg\max u_f\left(s_l, s_f\right) \tag{9.4}$$

Finally, the leader determines their optimal strategy $s_f^*(s_l)$ based on the strategy adopted by the followers $s_l^*\left(s_f^*\right)$ and then announces it to them again. After multiple iterations as described above, both parties can eventually reach the Stackelberg Equilibrium (SE), at which point neither the leader nor the follower can unilaterally change their strategy to gain greater benefits.

In the case of a leader and multiple followers, the leader first formulates and announces the strategy s_l, and multiple followers can use non-cooperative or cooperative games to determine their respective optimal strategies $s_{f,i}^*(s_l)$. Finally, the leader comprehensively considers the strategies of each follower to determine their own optimal strategy $s_l^*\left(s_{f,i}^*\right)$, ultimately achieving the Stackelberg equilibrium.

Since leaders can prioritize the formulation of strategies and determine their optimal strategy based on the strategies of followers, the Stackelberg game is a master–slave game, in which the leader occupies a dominant position. Meanwhile, it should be noted that leaders have already fully understood the impact of the initial strategy on follower strategies and predicted the actions carried out by the followers.

In the actual electricity market, there is often a master–slave relationship between grid operators and users. Grid operators have a certain degree of dominance, can control the operation of a large number of power sources and loads, and also have pricing power. In addition, the users need to find the most favorable electricity consumption strategy under the operation and pricing strategy of the power grid operator. Therefore, the Stackelberg game model has been widely applied in the electricity market.

9.4 Non-cooperative Game Theory Based V2G Strategy for EVs

When considering the reverse impact of electric vehicle charging load on electricity prices, the original linear optimization problem becomes a game of multiple electric vehicles competing with each other to obtain maximum profits. Therefore, this chapter adopts three game theory models to study the V2G strategy of electric vehicles considering dynamic electricity prices.

To utilize the V2G behavior of electric vehicles as much as possible to provide peak shaving services, power grid operators will adopt a dynamic electricity pricing strategy and determine the electricity price as shown in Eq. (9.5), where $\rho(t)$ is the dynamic electricity price at that time;a The benchmark electricity price and electricity price multiplier determined by the power grid operator, and $a \geq 0, b \geq 0$;$P_{BD}(t)$, $P_{DG}(t)$ and $P_{EV}(t)$ respectively t represent the traditional load, distributed power output, and total charging load of electric vehicles at different times, and $P_{EV}(t) > 0$ represent that the electric vehicle is charging and discharging.

$$\rho(t) = a + b(P_{BD}(t) - P_{DG}(t) + P_{EV}(t)) \tag{9.5}$$

From Eq. (9.5), it can be seen that when a and b are given, the electricity price will be positively correlated with the total load of the nodes at that moment. The larger the node load, the higher the electricity price at that moment, and vice versa. Due to the desire of electric vehicle users to pursue lower charging costs, they choose to charge during low electricity price periods and discharge during high electricity price periods, thereby helping grid operators achieve peak shaving and valley filling effects.

It should be noted that the total charging cost of the first electric vehicle is not only related to its own charging and discharging strategy, but also to the strategies of other electric vehicles in the same period. When formulating the charging and discharging strategy, it is necessary to consider the game with other electric vehicles. This electricity pricing strategy can effectively prevent a large number of electric vehicles from choosing to charge at the same time, causing the period to shift from the original low load to a new high load.

If there is an n electric vehicle in a parking lot that can participate in V2G auxiliary services in the game, each vehicle can clarify its own start time, $t_{s.i}$ end time $t_{e,i}$, and charging needs $P_{ch,i}$, and hope to minimize the total charging cost by formulating a reasonable charging and discharging strategy while meeting its own charging needs, as shown in Eq. (9.6), where $x_{i,t}$ is the charging and discharging strategy of the electric vehicle at time t, representing the electric vehicle's choice to charge $x_{i,t} > 0$, the electric vehicle's choice to discharge $x_{i,t} < 0$, and the electric vehicle's choice not to perform charging and discharging behavior $x_{i,t} = 0;x_{i,t}$ The relationship between $P_{EV}(t)$ and satisfies Eq. (9.7), where $x_{-i,t}$ is the sum of the loads corresponding to the charging and discharging strategies of all electric vehicles except for the first i electric vehicle at t time.

$$\min u_i = \sum_{t=t_{s,i}}^{t_{e,i}} \rho_t \cdot x_{i,t} \tag{9.6}$$

$$P_{EV}(t) = \sum_{i=1}^{n} x_{i,t} = x_{i,t} + x_{-i,t} \tag{9.7}$$

By substituting the dynamic electricity price function shown in Eqs. (9.5) and (9.7) into Eq. (9.6), the payment function for the first electric vehicle can be obtained as shown in Eq. (9.8).

$$\min u_i = \sum_{t=t_{s,i}}^{t_{e,i}} \left[a + b\left(P_{BD,t} - P_{DG,t} + x_{i,t} + x_{-i,t}\right)\right] \cdot x_{i,t} \tag{9.8}$$

In the V2G process, the first i electric vehicle needs to meet the constraints of Eqs. (9.9)–(9.11).

$$\sum_{t=t_{s,i}}^{t_{e,i}} x_{i,t} = P_{ch,i} \tag{9.9}$$

$$P_{\min} \le SoC_{0,i} + \sum_{t=t_{s,i}}^{t} x_{i,t} \cdot \Delta t \le P_{\max} \tag{9.10}$$

$$-P_{nch} \le x_{i,t} \le P_{nch} \tag{9.11}$$

Equation (9.9) represents the total amount of electricity required to charge an electric vehicle when it leaves the parking lot after ending its V2G service, ensuring that the electric vehicle has sufficient electricity to meet its next travel needs at the end of the V2G service.

Equation (9.10) represents that at any given time t, the battery level of an electric vehicle must be between the set maximum and minimum levels $P_{\max}$, to prevent situations where the electric vehicle user suddenly needs to travel ahead of schedule and runs out of power $P_{\min}$, while also ensuring the battery's lifespan. Here, $SoC_{0,i}$ is the initial battery level of the electric vehicle; Δt is the length of a unit time interval.

Equation (9.11) represents the upper and lower limits of the charging and discharging power of electric vehicles, which cannot exceed the maximum rated charging and discharging power of electric vehicles P_{nch}. In this chapter, it is assumed that the charging and discharging power of electric vehicles can be continuously varied between $[-P_{nch}, P_{nch}]$.

In this game problem, each electric vehicle will minimize its total charging cost as shown in Eqs. (9.6)–(9.11), and the optimal strategy for the first i electric vehicle x_i^* can be represented by Eq. (9.12).

$$x_i^*(x_{-i}) = \arg\max u_i(x_i, x_{-i}) \tag{9.12}$$

Based on the game theory, it can be seen that the game problem exists in the Nash equilibrium, which satisfies Eq. (9.13), that is, when the game reaches the Nash equilibrium, each electric vehicle will be unable to further reduce its total charging cost by only changing its own charging and discharging strategy.

$$u_i(x_i^*, x_{-i}^*) \geq u_i(x_i', x_{-i}^*), \forall x_i \in S_i, i \in N, x_i' \neq x_i^* \tag{9.13}$$

There are various ways to solve the Nash equilibrium, e.g., the iterative method, the interval equation method, the Nikaido Isoda equation method, and the KKT method. In this chapter, the iterative method will be used to solve the problem. Each electric vehicle selects the most advantageous charging and discharging strategy based on other electric vehicle charging and discharging strategies, and continuously iterates and updates the new charging and discharging strategy until the difference between the two adjacent optimal strategies of each electric vehicle is less than the preset convergence factor r_{err}^0. This proves and obtains the Nash equilibrium of the non-cooperative game.

9.5 Cooperative Game Theory Based V2G Strategy for EVs

If all electric vehicles no longer engage in internal non-cooperative games, but form an alliance to jointly pursue the maximization of total revenue, then the problem becomes a cooperative game. In this game, it can be assumed that there exists an aggregator that can control the charging and discharging behavior of all electric vehicles, with the aim of minimizing the total charging cost of all electric vehicles.

It is still assumed that the power grid operator adopts a dynamic electricity price as shown in Eq. (9.5), and the aggregator of electric vehicles can arrange the charging and discharging behavior of all electric vehicles uniformly according to the situation of each electric vehicle, to minimize the total charging cost of all electric vehicles, as shown in Eq. (9.14).

$$\min u_{agg} = \sum_{i=1}^{n} \sum_{t=t_{s,i}}^{t_{e,i}} \rho_t \cdot x_{i,t} \tag{9.14}$$

Equation (9.14) still needs to satisfy the constraints shown in Eqs. (9.9)–(9.11).

In addition, to ensure that electric vehicles are willing to participate in cooperative games, it is necessary to ensure that the total charging cost of each electric vehicle in cooperative games is less than the cost at the Nash equilibrium in non-cooperative games, i.e., Eq. (9.15), where $c_{ne,i}$ is the total charging cost of the i th electric vehicle in non-cooperative games.

$$\sum_{t=t_{s,i}}^{t_{e,i}} \rho_t \cdot x_{i,t} > c_{ne,i}, i = 1, 2, \ldots, n \tag{9.15}$$

Because electric vehicle aggregators can uniformly control the charging and discharging behavior of all electric vehicles, the optimal charging and discharging strategy for each electric vehicle can be directly solved by the aggregators without going through the iterative process of game theory.

By substituting the dynamic electricity price function Eq. (9.5) into the payment function Eq. (9.15) of the electric vehicle aggregator, Eq. (9.16) can be obtained, where x_t is the total charging and discharging power of all electric vehicles at time.

$$\min u_{agg} = \sum_{t=1}^{24} \left[a + b\left(P_{BD,t} - P_{DG,t} + x_t\right)\right] \cdot x_t \tag{9.16}$$

At time t, for electric vehicle aggregators to obtain the minimum total charging cost, they need to solve for the derivative of the payment function u_{agg} with respect to the total charging and discharging power of electric vehicles x_t, as shown in Eq. (9.17), where $p_0 = P_{BD,t} - P_{DG,t}$ is the algebraic sum of all loads in the distribution network except for electric vehicle loads.

$$\frac{\partial u_{agg}}{\partial x_t} = 2bx_t + (a + bp_0) \tag{9.17}$$

Equation (9.18) can be obtained by considering the constraints (9.9)–(9.11) and (9.15), which represent the optimal charging and discharging power that the electric vehicle aggregator wishes to provide at the given time.

$$x_t^* = -\frac{a + bp_0}{2b} \tag{9.18}$$

In general, it can be seen from Eq. (9.18) that to achieve the minimum total charging cost, electric vehicle aggregators often need to choose charging and discharging power that is different from the same number, which improves the utilization rate of distributed power sources and achieves peak shaving and valley filling.

However, when there are constraints on the charging demand of electric vehicles, the total charging and discharging power of electric vehicle aggregators often cannot meet Eq. (9.18), so the optimization algorithms need to be adopted to solve for the optimal charging and discharging strategy.

9.6 Master–Slave Game Theory Based V2G Strategy for EVs

In this section, it is considered that power grid operators use energy storage devices to provide certain auxiliary services to the distribution network. By controlling the charging and discharging behavior of energy storage devices, they charge and store electricity during low load periods with low electricity prices, and discharge during peak load periods with high electricity prices, to achieve peak shaving and valley filling while minimizing the operating costs of peak shaving. Due to the incentive of dynamic electricity prices, both energy storage devices and electric vehicles hope to charge during low electricity price periods and discharge during high electricity price periods. Therefore, the energy storage devices of grid operators and electric vehicle users form a competitive relationship, and grid operators occupy a dominant position in the competition. Electric vehicle users often need to determine their optimal strategy based on the strategy formulated by grid operators, thus forming a Stackelberg master–slave game.

Assuming that the power grid operator has distributed power sources and large capacity energy storage devices for providing peak shaving and valley-filling auxiliary services at each distribution node i. Electric vehicles that perform regular charging at the same node will adopt a cooperative game approach, with a unified aggregator scheduling the charging and discharging behavior. The dynamic electricity price is still positively correlated with the net load of the node, as shown in Eq. (9.19), where $P_{ES}(t)$ is the charging and discharging power of the energy storage device. When $P_{ES}(t) > 0$, the energy storage is charged, and when $P_{ES}(t) < 0$, the energy storage is discharged.

$$\rho(t) = a + b(P_{BD}(t) - P_{DG}(t) + P_{EV}(t) + P_{ES}(t)) \tag{9.19}$$

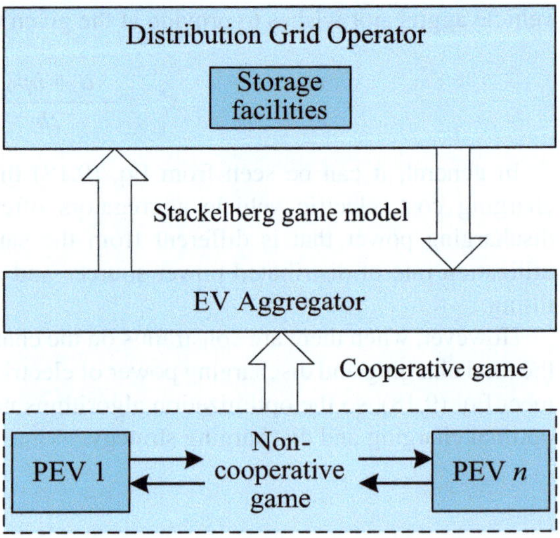

Fig. 9.1 Stackelberg gaming model of V2G

The Stackelberg master–slave game model is shown in Fig. 9.1, where the grid operator, as the leader, can prioritize the formulation of energy storage charging and discharging strategies. As followers, electric vehicle aggregators strive to minimize the total charging cost as much as possible under the strategies formulated by grid operators, ultimately achieving the Stackelberg equilibrium.

For power grid operators as leaders, the goal is to minimize peak shaving costs throughout the day, i.e. the payment function shown in Eq. (9.20), can be minimized.

$$\min_{P_{ES}(t)} U = \sum_{t=1}^{24} \rho(t) \cdot P_{ES}(t) \tag{9.20}$$

The constraints that Eq. (9.20) needs to satisfy are shown in Eqs. (9.21)–(9.23).

$$\sum_{t=1}^{24} P_{ES}(t) = 0 \tag{9.21}$$

$$P_{\min} \leq SoC_0 + \sum_{t=1}^{t} P_{ES}(t) \leq P_{\max} \tag{9.22}$$

$$-P_{ES}^{\max} \leq P_{ES}(t) \leq P_{ES}^{\max} \tag{9.23}$$

Equation (9.21) represents the algebraic sum of the amount of charge and discharge energy stored in one day that needs to be kept constant to ensure normal use the next day.

Equation (9.22) represents that the amount of energy stored at any time t needs to be between the preset minimum and maximum values to prevent overcharging and discharging of energy storage, where SoC_0 is the initial amount of energy stored; Δt Is the length of a unit time interval.

Equation (9.23) represents the maximum charging and discharging power range of energy storage, which means that energy storage can continuously vary within this range $\left[-P_{ES}^{\max}, P_{ES}^{\max}\right]$.

For electric vehicle aggregators as followers, their goal is to minimize the total charging cost of all participating electric vehicles. The payment function is the same as given in Eq. (9.14), and the constraints are also given in Eqs. (9.9)–(9.11) and (9.15).

According to the Stackelberg game model shown in Fig. 9.1, after estimating the basic load of the power grid and the output of distributed power sources for the day, the grid operator first formulates an energy storage charging and discharging strategy and notifies the electric vehicle aggregator. Subsequently, the electric vehicle aggregator determines the optimal charging and discharging strategy for each electric vehicle with the goal of minimizing Eq. (9.14) while satisfying various constraints, and reports it to the grid operator. Subsequently, the grid operator determines the optimal charging and discharging strategy for energy storage based on Eqs. (9.19)–(6.23). After multiple games, the Stackelberg equilibrium can ultimately be reached, which is the optimal energy storage strategy for grid operators and the optimal electric vehicle charging and discharging strategy for electric vehicle aggregators. Here, the KKT (Karush Kuhn Tucker) condition will be used to solve the V2G master–slave game model of the electric vehicle.

9.7 Experimental Assessment and Numerical Analysis

In this section, it is assumed that an electric vehicle parking lot is located in a commercial office building and is responsible for the V2G scheduling of electric vehicles at the distribution node during daytime hours. There is a total of 150 electric vehicles in the parking lot, and the charging power of each electric vehicle can vary within ±100 kW. The distribution of arrival and departure times conforms to the vehicle travel survey data. The distribution node is also connected to a distributed power source consisting of rooftop photovoltaics and small wind turbines. A day will be divided into 24 time intervals, and at the beginning of each interval, the charging and discharging strategy for that hour will be determined. The parameters in Eq. (9.5) are set to $a = 20$, $b = 0.1(\$)$.

Below, we will simulate and compare several charging and discharging strategies for electric vehicles.

Disorderly charging scenario. If all electric vehicles adopt a disorderly charging strategy, that is, they start charging immediately when they arrive at the parking lot, and remain parked after charging is completed without discharging. It should be highlighted that the charging load of electric vehicles not only fails to assist the distribution network in peak shaving and valley filling, but also increases the peak valley difference of the system, prolonging the duration of the system load peak, which is very detrimental to the safety and economic operation of the distribution network.

Disorderly charging and discharging scenario. Consider charging and discharging every electric vehicle, and choose to discharge at an appropriate time while meeting their own charging needs, in order to earn the maximum profit for themselves and minimize charging costs as much as possible. However, assuming that each electric vehicle has no information exchange with the other, that is, it acts completely according to its own optimal strategy without considering the strategies of other electric vehicles. Under this strategy, the total charge and discharge load of electric vehicles, as well as the total load curve of the distribution node, are shown in Fig. 9.2. It can be observed that all electric vehicles will choose to discharge during the peak period of the system's original load (approximately 9:00–13:00), while charging during the low period of the system's original load (approximately 14:00–18:00), thereby reducing the peak valley difference of distribution nodes to a certain extent, improving the efficiency of new energy utilization, and reducing the total charging cost of electric vehicles to $61.95.

However, it should be noted that, as shown in Fig. 9.2, 7:00–9:00 and 14:00–18:00 have become the new peak load periods for distribution nodes. If more electric vehicles adopt this strategy for charging and discharging, the load during these two time periods will further increase, causing the charging cost of electric vehicles to rise instead of decrease, and also making load fluctuations more severe, which will pose potential risks to the safe and stable operation of the distribution network. Therefore, it is necessary to coordinate and schedule the charging and discharging behavior of each electric vehicle.

Non-cooperative game theory based scheduling strategy. If all electric vehicles adopt the V2G strategy based on a non-cooperative game, the Nash equilibrium of the

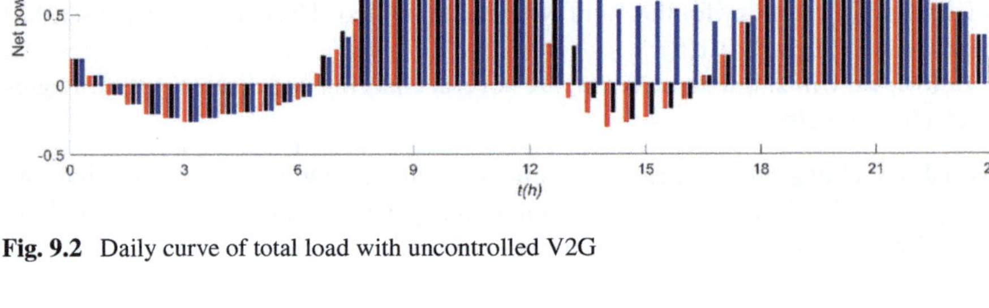

Fig. 9.2 Daily curve of total load with uncontrolled V2G

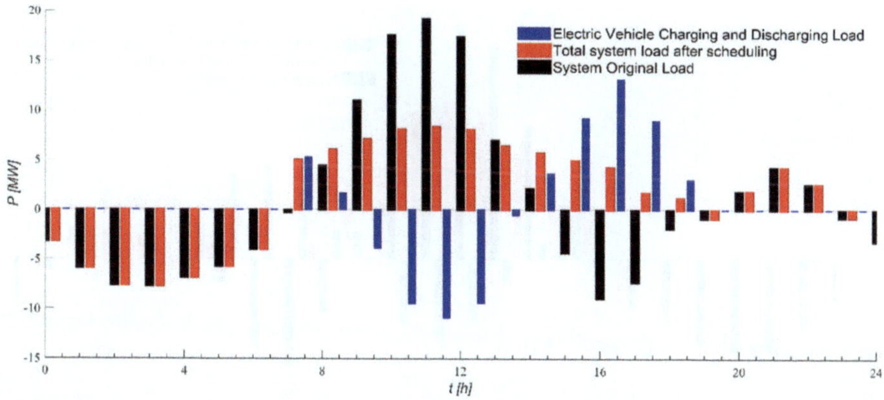

Fig. 9.3 Daily curve of total load with non-cooperative gaming V2G

game can be reached after multiple iterations. At this point, the total load of electric vehicle charging and discharging, as well as the total load curve of the distribution node, are shown in Fig. 9.3. The fluctuation of the distribution network load has been significantly improved, increasing the utilization efficiency of new energy, and the total charging cost of all electric vehicles has been reduced to \$12.53.

(1) **Cooperative Game Based Scheduling Strategy**

As mentioned earlier, cooperative games generally yield more benefits than non-cooperative games. If all electric vehicles form an alliance and the aggregator of each electric vehicle uniformly schedules its charging and discharging behavior, the total charging and discharging load of electric vehicles and the total load curve of the distribution node are shown in Fig. 9.4. At this time, the total charging cost of all electric vehicles is only \$1.64. It can be observed that most of the time, electric vehicle aggregators will determine the total charging and discharging power of electric vehicles using Eq. (9.18). Comparing Figs. 9.3 and 9.4, it can also be found that although non-cooperative games can better improve the load fluctuations of the distribution network during the game process, their cost is 7.64 times that of cooperative games. Electric vehicle clusters can significantly reduce their total charging cost through cooperative games.

(2) **Stackelberg Master–slave Game Based Scheduling Strategy**

Assuming the maximum charging and discharging power of the energy storage equipment of the power grid operator is 10 MW, when both parties reach the Stackelberg equilibrium, the charging and discharging load curves of the electric vehicle cluster and energy storage equipment are shown in Fig. 9.5. At this time, the energy storage equipment of the power grid operator can earn \$27.75 in a day, while the total charging cost of electric vehicles is \$13.91. It can be observed that after the energy storage equipment of the power grid operator is connected to the peak shaving auxiliary

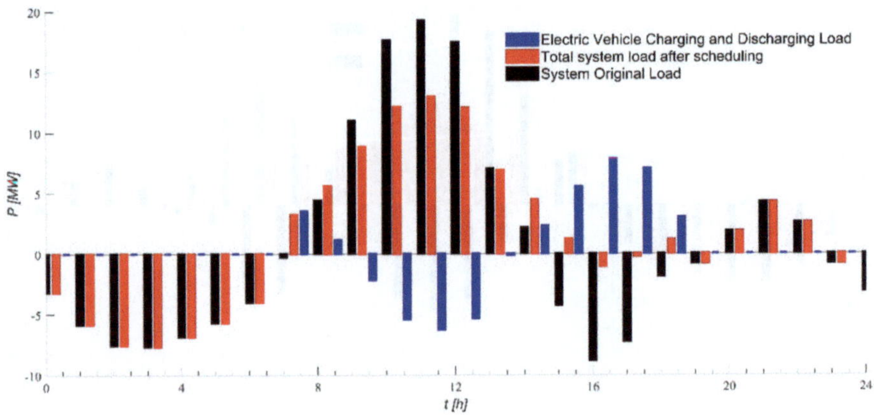

Fig. 9.4 Daily curve of total load with cooperative gaming V2G (minimal costs)

service, the charging cost of the electric vehicle cluster has increased, but it is still much lower than the cost of disorderly charging and discharging.

At the same time, it should be noted that due to electric vehicles leaving the parking lot of the commercial office building after work in the evening, the power grid operator cannot use the V2G behavior of electric vehicles to peak and valley load the distribution network at night. After considering the energy storage equipment of the power grid operator, a certain amount of peak shaving auxiliary services can be provided throughout the day to better alleviate fluctuations in node load.

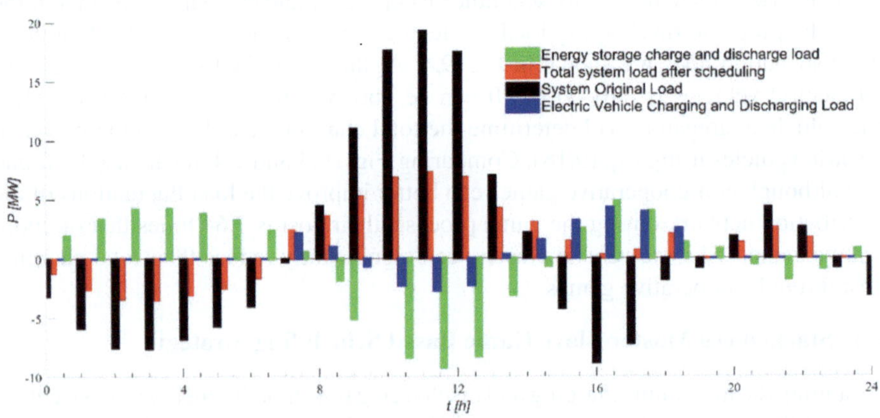

Fig. 9.5 Daily curve of load demand with Stackelberg gaming

9.8 Summary

V2G technology is an important way to explore the potential advantages of electric vehicles in auxiliary services, that is, to control the conversion of electric vehicles between "load" and "source", to provide various auxiliary services such as frequency regulation, peak shaving, and rotating backup for the power grid, greatly reducing the charging cost of users and the operating cost of the power grid, improving the load curve and power quality of the power grid, and coordinating and complementing with distributed renewable power sources to improve the utilization efficiency of clean energy. In order to guide electric vehicle users to participate in auxiliary services, certain price incentives are needed.

In view of this, this chapter studies the V2G scheduling problem of electric vehicles in active distribution networks, considering dynamic electricity prices, and based on game theory, investigates the optimal charging and discharging decision-making problem between electric vehicle users and grid operators under the influence of dynamic electricity prices. Firstly, non-cooperative game models and cooperative game models for electric vehicle charging and discharging were established, and their Nash equilibrium points were solved. The results were compared and analyzed. At the same time, the advantages of energy storage devices providing auxiliary services were also considered, and a master–slave game model between grid operators and electric vehicle users was established. The importance of the V2G behavior of electric vehicles in peak shaving of distribution networks was also demonstrated.

Chapter 10
Modeling Method for Power-Traffic Coupled Networks Considering Bounded Rationality of Users

Abstract This chapter proposes two distinct modeling frameworks for power-traffic coupled networks that incorporate the bounded rationality of users to better reflect realistic travel and charging behaviors. First, a bounded rationality decision model is developed using a tolerance parameter to quantify the degree of rationality, contrasting it with the fully rational model. Based on this, a multi-period static modeling method is introduced, where traffic flow is characterized as a bounded rationality static user equilibrium without dynamic propagation. Subsequently, a dynamic modeling method is proposed, modeling traffic flow as a bounded rationality dynamic user equilibrium with full propagation characteristics. Within the dynamic framework, an improved point queue model based on difference equations is designed to accurately capture the entry, queuing, charging, and departure processes of electric vehicles at fast-charging stations, enabling precise tracking of the transformation from traffic flow to charging load. In both approaches, the distribution network is represented by a multi-period optimal power flow model, with nodal marginal prices serving as charging prices. The interaction between traffic flow and power flow is characterized via a fixed-point model, defining the equilibrium state of the coupled network. Iterative solution algorithms are developed accordingly. Case studies validate the effectiveness of both methods: the multi-period static model offers high computational efficiency for identifying critical periods and infrastructure, while the dynamic model provides finer temporal resolution and reveals microscopic coupling mechanisms, offering robust theoretical support for the flexible scheduling of fast-charging infrastructure.

10.1 Introduction

With the increasing popularity of electric vehicles (EVs) and the continuous growth of charging power, random charging behaviors of EV users have become more frequent, highlighting the dynamic operational characteristics of power-traffic coupled networks. Conventional single-period static modeling methods are only applicable to steady-state analysis within a single time slice, which makes them

Q. Yang et al., *Integration of Massive Electric Vehicles to Smart Grid*, Smart Energy Systems, https://doi.org/10.1007/978-981-92-0557-8_10

inadequate for capturing the spatiotemporal evolution of traffic flow and power flow over short periods. In contrast, multi-period static modeling and dynamic modeling methods can effectively characterize the spatiotemporal distribution features of traffic flows and charging loads, thereby providing critical support for the flexible scheduling of fast charging facilities in power-traffic coupled systems.

In such a context, the fast charging station, as the core hub of the power-traffic coupled network, plays a decisive role in the effectiveness of the overall dynamic modeling. Therefore, one of the key challenges in modeling such networks lies in the design of a fast-charging station model that can accurately characterize the real-world charging processes.

In addition, the travel and charging behaviors of EVs can be significantly influenced by their decision rationality. Most existing studies assumed that users are fully rational. However, EV user behavior generally exhibits bounded rationality in practice. Firstly, the EV users cannot accurately predict congestion and queuing conditions in the power-traffic coupled network, and hence they tend to misestimate time costs, service delays, and charging costs. In addition, users may not be sensitive to small cost differences, and near-optimal solutions may also be selected instead of the true optimum. Although users subjectively pursue the lowest travel and charging costs, their final decisions often result in suboptimal outcomes. To more accurately depict real-world EV user behaviors and enhance the effectiveness and accuracy of spatio-temporal charging demand estimation, it is necessary to consider bounded rationality in the dynamic modeling of power-traffic coupled networks.

To assess the impact of bounded rationality on travel and charging behaviors and to reveal the complex coupling mechanisms between transportation traffic flow and power flow under dynamic operation, this chapter proposes a modeling method for power-traffic coupled networks that incorporates the bounded rationality of users. Specifically, this chapter first develops a bounded rationality decision model by introducing a user tolerance parameter to quantify the degree of rationality and compares it with the fully rational decision model. Further, two types of modeling methods for power-traffic coupled networks are proposed: a multi-period static modeling method and a dynamic modeling method. In the multi-period static modeling method, traffic flow is described as a multi-period bounded rationality static user equilibrium without dynamic propagation characteristics, while in the dynamic modeling method, traffic flow is modeled as a bounded rationality dynamic user equilibrium with dynamic propagation characteristics. It is considered that these two methods can be suitable for different application scenarios.

Furthermore, within the dynamic modeling framework, this chapter proposes an improved point queue model based on a system of difference equations. This model is used to accurately describe the arrival, queuing, charging, and departure processes of EV traffic flow within fast charging stations, enabling the tracking of the transformation from traffic flow to charging load. The mathematical properties, physical interpretation, and consistency of this improved point queue model with real-world traffic dynamics are also discussed.

In both modeling approaches, the distribution network's power flow is represented using a multi-period optimal power flow model. The nodal marginal electricity price is taken as the charging price at the station, which increases with the charging load. A fixed-point model is then used to characterize the interaction between traffic flow and power flow until equilibrium is reached. This equilibrium state is defined as the network equilibrium of the power-traffic coupled network. An iterative solution algorithm is subsequently designed to solve this equilibrium.

Finally, the effectiveness of the proposed modeling method is validated through two case studies. This chapter aims to provide a novel theoretical framework for the dynamic modeling of power-traffic coupled networks and to offer theoretical support for the optimal scheduling of fast charging infrastructure.

The structure of this chapter is organized as follows: Sect. 10.2 introduces the concept and definition of user decision rationality and analyzes the differences between full rationality and bounded rationality. Section 10.3 presents the multi-period static modeling method for power-traffic coupled networks based on bounded rationality static user equilibrium and the corresponding solution approach. Section 10.4 proposes a dynamic modeling method for power-traffic coupled networks based on bounded rationality, dynamic user equilibrium and the corresponding algorithms. Section 10.5 verifies the proposed modeling and solution methods through two case studies. Section 10.6 concludes this chapter.

10.2 User Decision Rationality Models

In power-traffic coupled networks, user decision rationality is reflected in their travel and charging choices (only for EV users). This behavior is of critical importance for analyzing the spatiotemporal distribution of traffic flow and charging demand. This section introduces the decision models of fully rational and bounded rational users and analyzes their differences.

Absolute Rational Decision-Making Model for Users. The absolute rational decision-making model assumes that electric vehicle (EV) users and internal combustion engine vehicle (ICEV) users are fully rational, selecting only the travel plan with the lowest overall cost. Taking EV users as an example, a travel plan e includes elements such as departure time, route p, and fast-charging station f, while ICEV users' travel plans do not include the option of charging stations.

Let the set of all feasible travel plans for EV users be $\Omega = e_1, e_2, ..., e_n$, where the travel cost of each plan $e_i \in \Omega$ is denoted as $f(e_i)$. Absolute rational users will choose the travel plan with the minimum travel cost e_*, where the minimum cost c_* satisfies:

$$c_* = f(e_*) = \min_{e_i \in \Omega} f(e_i) \tag{10.1}$$

The set of optimal travel plans chosen by EV users is defined as:

$$\Omega_* = \{e_i | f(e_i) = c_*, e_i \in \Omega\} \tag{10.2}$$

Similarly, the minimum travel cost and optimal travel plan set for ICEV users can be defined. The above modeling assumptions are based on two important premises:

First, users are assumed to accurately perceive the complete and real-time evolving information within the power-traffic coupled network, including congestion, queuing, and electricity pricing trends, thus supporting their ability to identify the global optimal travel plan. That is, users are presumed to have zero estimation deviation in the travel cost function $f(e_i)$.

Second, users are considered to have absolute sensitivity to cost differences between different travel plans and will continue to adjust their choice until the travel cost reaches the global minimum.

Bounded Rational Decision-Making Model for Users. In practical scenarios, due to limitations in information channels (e.g., navigation software or historical experience), and the existence of cognitive bias and delays, users often cannot obtain complete information on the power-traffic coupled network. As a result, they are unable to accurately estimate the true travel cost $f(e_i)$ of a given travel plan and may instead base their decisions on a perceived travel cost $\hat{f}(e_i)$ derived from limited information. Moreover, users tend to exhibit insensitivity within a certain tolerance range to cost differences between different travel plans, and the plans with slightly higher costs than the optimal one may still be acceptable to boundedly rational users.

In summary, users' travel and charging decisions often show bounded rationality. To better model users' bounded rational behavior in traffic flow modeling, this study introduces a user tolerance parameter $\varepsilon \geq 0$ to quantify the degree of bounded rationality. This parameter defines users' acceptable deviation range between the perceived cost of a suboptimal plan and the actual minimum cost plan, and simultaneously reflects the maximum error between the estimated perceived cost and the true travel cost.

Let the perceived travel cost function be denoted by $\hat{f}(e_i)$, and let the error between the perceived cost and true cost $f(e_i)$ be σ_i, where $\sigma_i \in [-\varepsilon/2, \varepsilon/2]$. Then, we have:

$$\hat{f}(e_i) = f(e_i) + \sigma_i, \forall e_i \in \Omega \tag{10.3}$$

When the perceived cost of a suboptimal travel plan is not significantly greater than the perceived cost of the optimal plan, that travel plan is still acceptable to boundedly rational users. That is:

$$\hat{f}(e_i) - \hat{f}(e_*) \leq 0, \exists e_i \in \Omega \tag{10.4}$$

Which is equivalent to:

$$\begin{aligned} &\hat{f}(e_i) - \hat{f}(e_*) = f(e_i) - f(e_*) + \sigma_i - \sigma_* \leq 0, \exists e_i \in \Omega \\ &\Rightarrow f(e_i) - f(e_*) \leq \sigma_* - \sigma_i, \exists e_i \in \Omega \end{aligned} \tag{10.5}$$

Since the upper bound of $\sigma_i - \sigma_*$ is ε, the above inequality holds under the condition:

$$f(e_i) - f(e_*) \leq \varepsilon \tag{10.6}$$

And since $c_* = f(e_*)$ is the minimum travel cost, the above expression can be rewritten as:

$$f(e_i) \in [c_*, c_* + \varepsilon] \tag{10.7}$$

The above condition is defined as the ε-bounded rationality suboptimal solution condition, and the set of travel options satisfying this condition is denoted as ε-bounded rationality suboptimal solution set:

$$\Omega_*^\varepsilon = \{e_i | f(e_i) \in [c_*, c_* + \varepsilon], e_i \in \Omega\} \tag{10.8}$$

The set of travel options acceptable to boundedly rational users includes the optimal travel solutions as well as certain suboptimal ones whose costs are close to the minimum, i.e. $\Omega_* \subset \Omega_*^\varepsilon \subset \Omega$. As the user's tolerance ε increases, the number of acceptable travel solutions $|\Omega_*^\varepsilon|$ will also increase, leading to more diversity in traffic flow distribution. Meanwhile, users' sensitivity to cost differences will decrease accordingly, indicating that the tolerance ε reflects the degree of rationality. When the tolerance equals 0, users are considered fully rational; when $\varepsilon > 0$, users exhibit bounded rationality, and their rationality level declines as ε increases. The fully rational model can be viewed as a special case of the bounded rationality model when $\varepsilon = 0$, while the bounded rationality decision model extends the fully rational model under more general circumstances.

10.3 Modeling Method for Multi-period Static Coupling of Electric-Traffic Networks Considering User Bounded Rationality

Based on the bounded rationality decision model, this section further introduces a multi-period static modeling method for electric-traffic coupled networks with the consideration of user bounded rationality. This includes a multi-period bounded rational static user equilibrium traffic flow model in the transportation network, considering the dynamic propagation of traffic flow, a multi-period optimal power flow model in the distribution network, the coupled network equilibrium model, as well as the corresponding solutions formed by integrating the two subsystems.

Multi-Period Bounded Rational User Equilibrium Traffic Network Model.

(1) Basic Assumptions.

The multi-period static traffic network model proposed in this section is based on the following assumptions:

(a) The traffic network is represented as a directed graph $G^{\mathrm{TN}} = (V^{\mathrm{TN}}, L^{\mathrm{TN}})$, where $V^{\mathrm{TN}} = \{v\}$ is the set of nodes and $L^{\mathrm{TN}} = \{l\}$ is the set of edges (or links). The node set includes the set of origin nodes $O = \{o\}$, destination nodes $D = \{d\}$, and fast-charging station nodes $F = \{f\}$. The set of origin–destination pairs is denoted by od, with each od corresponding to an OD pair (Origin–Destination). The edge set includes both regular road segments L^{road} and fast-charging road segments L^{FCS}. Fast-charging stations are treated as special nodes in the traffic network, each of which only connects one inbound road and one outbound road. The inbound road represents the entrance to the charging station, and the outbound road represents the exit. All other roads in the traffic network are considered regular roads.

(b) The travel paths of electric vehicles and fuel vehicles are both defined as a set of paths connecting OD pairs. Let the path sets of electric vehicles and fuel vehicles be denoted by $P_{od}^{\mathrm{EV}} = \{p\}$ and $P_{od}^{\mathrm{GV}} = \{p\}$, respectively. Each electric vehicle path $p \in P_{od}^{\mathrm{EV}}$ consists of regular road segments and one fast-charging road segment (corresponding to a fast-charging station $\upsilon(p)$), whereas each fuel vehicle path $p \in P_{od}^{\mathrm{GV}}$ consists only of regular road segments. Therefore, $P_{od}^{\mathrm{EV}} \cap P_{od}^{\mathrm{GV}} = \emptyset$. Since each electric vehicle path corresponds to a unique fast-charging station, the electric vehicle user's choice of fast-charging station can be transformed into the selection of a travel path. If no charging is required, the electric vehicle does not need to visit a fast-charging road segment, and its path selection is equivalent to that of a fuel vehicle. In this case, it can be regarded as a fuel vehicle.

(c) The travel demand of electric vehicle (EV) and gasoline vehicle (GV) users is defined based on different travel requirements (including departure time, origin, and destination). The number of EV users departing from o and arriving at d during the time period t is denoted as $Q_{od}^{\mathrm{EV}}(t)$. Similarly, the travel demand of GV users is denoted as $Q_{od}^{\mathrm{GV}}(t)$. The set of periods is defined as $T = 0, \Delta t, 2\Delta t, ..., T$, where Δt is the time interval and T is the upper time limit.

(d) The EV and GV traffic flow at each period is assumed to be in a steady state, meaning that path traffic flows are uniformly distributed, without considering delays due to traffic wave propagation, congestion, or other dynamic features. The traffic volume assigned to each path can be used to represent user flow distribution and is referred to as path traffic flow.

(e) Users' travel strategies include path selection. For EV users in $O_{od}^{\mathrm{EV}}(t)$, the number of users selecting path $p \in P_{od}^{\mathrm{EV}}$ is defined as the path traffic flow $f_p^{\mathrm{EV}}(t)$, which represents the number of EV users traveling from origin o to destination d via path p at time t. Similarly, the path traffic flow for GV users is denoted as $f_p^{\mathrm{GV}}(t)$.

(2) Path Traffic Flow and Path Travel Cost.

To satisfy the travel demand of EV and GV users, the corresponding path traffic flows must satisfy the flow conservation conditions, which state that the total path traffic flow between an OD pair must equal the travel demand for that OD pair during each period:

$$\sum_{p\in P_{od}^{\mathrm{EV}}} f_p^{\mathrm{EV}}(t) = Q_{od}^{\mathrm{EV}}(t), \forall od \in OD, t \in \mathcal{T} \tag{10.9}$$

$$\sum_{p\in P_{od}^{\mathrm{GV}}} f_p^{\mathrm{GV}}(t) = Q_{od}^{\mathrm{GV}}(t), \forall od \in OD, t \in \mathcal{T} \tag{10.10}$$

Since path traffic flow represents a steady-state distribution, namely, that the traffic at the origin and destination nodes of a path occurs within the same period, there is no need to explicitly consider the travel time conservation in the multi-period static traffic network model.

The travel cost corresponding to the path traffic flow is primarily determined by the users' departure time. Let the arrival time at the destination node (i.e., the path traversal time) of the EV path flow $f_p^{\mathrm{EV}}(t)$ and GV path flow $f_p^{\mathrm{GV}}(t)$ be denoted as $\xi_p^{\mathrm{EV}}(t)$ and $\xi_p^{\mathrm{GV}}(t)$, respectively. Then, the path travel cost is defined as follows:

$$\begin{aligned}\phi_p^{\mathrm{EV}}(t) &= \varphi_p^{\mathrm{EV}}(t) + \theta_p^{\mathrm{EV}}(t)\\ &= \gamma^{\mathrm{EV,t}}\Big[\xi_p^{\mathrm{EV}}(t) - t\Big] + \lambda_{\upsilon(p)}^{\mathrm{ch}}(t)P_{\mathrm{ch}}\tau^{\mathrm{ch}}, \forall p \in P^{\mathrm{EV}}, t \in \mathcal{T}\end{aligned} \tag{10.11}$$

$$\phi_p^{\mathrm{GV}}(t) = \varphi_p^{\mathrm{GV}}(t) = \gamma^{\mathrm{GV,t}}\Big[\xi_p^{\mathrm{GV}}(t) - t\Big], \forall p \in P^{\mathrm{GV}}, t \in \mathcal{T} \tag{10.12}$$

Here, $\phi_p^{\mathrm{EV}}(t)$ represents the total travel cost at time t for EV users choosing path p (corresponding to path flow $f_p^{\mathrm{EV}}(t)$), which includes the time cost $\varphi_p^{\mathrm{EV}}(t)$ and the charging cost $\theta_p^{\mathrm{EV}}(t)$; $\gamma^{\mathrm{EV},t}$ and $\gamma^{\mathrm{GV},t}$ represent the time cost coefficients for EV and GV users at time t, respectively; $\lambda_{\upsilon(p)}^{\mathrm{ch}}(t)$ is the nodal locational marginal price of the fast charging station $c(p)$ passed by path p at time t, which also serves as the charging electricity price; P_{ch} is the charging power at fast charging station $\upsilon(p)$; $\phi_p^{\mathrm{GV}}(t)$ represents the travel cost at time t for GV users choosing path p (corresponding to path flow $f_p^{\mathrm{GV}}(t)$), which includes only the time cost $\varphi_p^{\mathrm{GV}}(t)$.

In the multi-period static traffic network, for a certain path segment, the travel time cost is distributed over the links along the path, and the total link flow over a time period is positively correlated with the path traffic flow (including both EV and GV). Therefore, it is necessary to define the mapping between path traffic flow and link flow, as well as the mapping between path traffic flow and time-indexed link flow.

The relationship between link flow and the sum of path flows passing through that link is expressed in (2.13) and (2.14), where the total path flow passing through a certain link is the sum of all path flows over periods that traverse this link.

$$\Lambda_p^l = \begin{cases} 0, \ l \in p \\ 1, \ l \notin p \end{cases}, \forall l \in L^{\mathrm{TN}}, p \in P \tag{10.13}$$

Fig. 10.1 An illustrative example of a multi-period static traffic network model

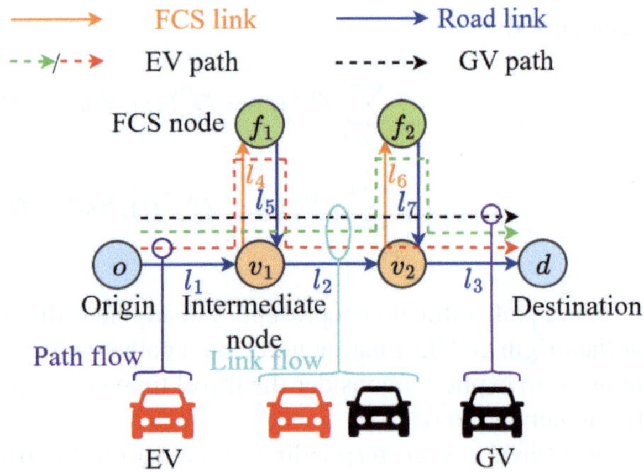

$$x_l(t) = \sum_{p \in P^{\text{EV}}} \Lambda_p^l f_p^{\text{EV}}(t) + \sum_{p \in P^{\text{GV}}} \Lambda_p^l f_p^{\text{GV}}(t), \forall l \in L^{\text{TN}}, t \in \mathcal{T} \tag{10.14}$$

In the above, Λ_p^l is a 0–1 parameter indicating whether link l belongs to path p; $x_l(t)$ denotes the traffic flow on link l at time t.

The multi-period static traffic network model considered in this section is illustrated by the example in Fig. 10.1, which consists of 6 nodes (1 origin node, 2 intermediate nodes, 2 fast charging station nodes, and 1 destination node), 7 links (2 fast charging station links, 5 regular links, shown as solid lines), and 3 paths (2 EV paths and 1 GV path, shown as dashed lines). Among them, each EV path includes one fast charging station, whereas the GV path does not include any fast charging stations. In addition, Fig. 10.1 also illustrates the inclusion relationship between path traffic flows and link traffic flows.

The functional relationship between travel time delay and link flow is given in (2.15)–(2.17), where the delay on regular links and fast charging station links is modeled using the BPR model and M/M/c/K queuing theory, respectively.

$$\varsigma_l(t) = \begin{cases} \text{RL}[x_l(t)], & l \in L^{\text{road}} \\ \text{FL}[x_l(t)], & l \in L^{\text{FCS}} \end{cases}, \forall t \in \mathcal{T} \tag{10.15}$$

$$\text{RL}_{\text{BPR}}[x_l(t)] = \tau_l^{\text{free}} + 0.15\tau_l^{\text{free}} \left[\frac{x_l(t)}{C_l} \right]^4, \forall l \in L^{\text{road}}, t \in \mathcal{T} \tag{10.16}$$

$$\text{FL}_{\text{MMcK}}[x_l(t)] = \tau_l^{\text{free}} + \tau_l^{\text{q}} \left[\frac{x_l(t)}{N_l} \right]^3 + \tau^{\text{ch}}, \forall l \in L^{\text{FCS}}, t \in \mathcal{T} \tag{10.17}$$

In the equations above, $\zeta_l(t)$ denotes the travel time of link l at time t; $RL(\cdot)$ and $FL(\cdot)$ represent the travel time functions of regular road segments and fast charging station segments, respectively, both of which are functions of the traffic flow $x_l(t)$

on link l at time t. (10.16) corresponds to the BPR model for regular roads; τ_l^{free} is the free-flow travel time when there is no congestion on link l; C_l is the capacity of link l. (10.17) is based on the $M/M/c/K$ queuing model for fast charging stations; τ_q is the service time parameter; N_j is the number of chargers at the charging station j; c^{ch} is the charging time.

Furthermore, the relationship between the route travel time and the travel times of the constituent links is shown in (10.18), where the route travel time equals the departure time plus the sum of travel times across all links on the route. The total route travel time must not exceed the time step length Δt of the modeling period; otherwise, it violates the constraint of (10.19), which would prevent the vehicle from reaching the destination by the end of the time step.

Equation (10.19) ensures the rationality of the multi-period static transportation network model but also implies that the time resolution cannot be arbitrarily small, which constrains the time granularity of the model.

$$\xi_p(t) = t + \sum_{l \in p} \varsigma_l(t), \forall p \in P, t \in \mathcal{T} \tag{10.18}$$

$$\sum_{l \in p} \varsigma_l(t) \le \Delta t, \forall p \in P, t \in \mathcal{T} \tag{10.19}$$

From the above analysis, it can be seen that due to overlapping links among different routes, the route travel cost is not only a function of the corresponding route traffic flow but is also affected by the traffic flows of other routes sharing the same links. Regular road segments and fast charging segments will simultaneously compete for limited capacity resources, so the mutual interference will affect both the travel time and the corresponding route cost.

(3) Multi-Period Static User Equilibrium Model with Bounded Rationality.

For any intended route traffic flow, the corresponding route travel cost can be calculated using the above methods. In conjunction with the bounded rational decision model in Sect. 10.2.2, the travel cost associated with the route flow must satisfy the following multi-period user equilibrium condition under bounded rationality:

$$\forall od \in OD, p \in P_{od}^{\text{EV}}, t \in \mathcal{T} \Rightarrow \begin{cases} f_p^{\text{EV}}(t) > 0 \Rightarrow \phi_p^{\text{EV}}(t) \in \left[\mu_{od}^{\text{EV}}(t), \mu_{od}^{\text{EV}}(t) + \varepsilon_{od}^{\text{EV}}\right] \\ \phi_p^{\text{EV}}(t) > \mu_{od}^{\text{EV}}(t) + \varepsilon_{od}^{\text{EV}} \Rightarrow f_p^{\text{EV}}(t) = 0 \end{cases} \tag{10.20}$$

$$\forall od \in OD, p \in P_{od}^{\text{GV}}, t \in \mathcal{T} \Rightarrow \begin{cases} f_p^{\text{GV}}(t) > 0 \Rightarrow \phi_p^{\text{GV}}(t) \in \left[\mu_{od}^{\text{GV}}(t), \mu_{od}^{\text{GV}}(t) + \varepsilon_{od}^{\text{GV}}\right] \\ \phi_p^{\text{GV}}(t) > \mu_{od}^{\text{GV}}(t) + \varepsilon_{od}^{\text{GV}} \Rightarrow f_p^{\text{GV}}(t) = 0 \end{cases} \tag{10.21}$$

In the above, ε_{od}^{EV} and ε_{od}^{GV} represent the user tolerance levels for electric vehicles and gasoline vehicles, respectively; $\mu_{od}^{EV}(t)$ and $\mu_{od}^{GV}(t)$ denote the minimum travel cost of electric and gasoline vehicle users at OD pair od and time t, expressed as:

$$\mu_{od}^{\mathrm{EV}}(t) = \min_{p \in P_{od}^{\mathrm{EV}}} \phi_p^{\mathrm{EV}}(t), \forall od \in OD, t \in \mathcal{T} \tag{10.22}$$

$$\mu_{od}^{\mathrm{GV}}(t) = \min_{p \in P_{od}^{\mathrm{GV}}} \phi_p^{\mathrm{GV}}(t), \forall od \in OD, t \in \mathcal{T} \tag{10.23}$$

In the multi-period static user equilibrium with bounded rationality, if a route is selected by users and has a non-zero flow, then the corresponding travel cost must meet the bounded rational acceptable cost condition; otherwise, if the cost does not satisfy the bounded rational condition, the flow on that route must be zero, meaning it is not selected by any users.

Furthermore, the multi-period bounded rational static user equilibrium conditions can be equivalently reformulated into the following nonlinear complementary constraints:

$$0 \le f_p^{\mathrm{EV}}(t) \perp \phi_p^{\mathrm{EV}}(t) - \pi_{od}^{\mathrm{EV}}(t) + \sigma_p^{\mathrm{EV}}(t) \ge 0, \forall od \in OD, p \in P_{od}^{\mathrm{EV}}, t \in \mathcal{T} \tag{10.24}$$

$$0 \le f_p^{\mathrm{GV}}(t) \perp \phi_p^{\mathrm{GV}}(t) - \pi_{od}^{\mathrm{GV}}(t) + \sigma_p^{\mathrm{GV}}(t) \ge 0, \forall od \in OD, p \in P_{od}^{\mathrm{GV}}, t \in \mathcal{T} \tag{10.25}$$

$$0 \le \sigma_p^{\mathrm{EV}}(t) \le \varepsilon_{od}^{\mathrm{EV}}, \forall od \in OD, p \in P_{od}^{\mathrm{EV}}, t \in \mathcal{T} \tag{10.26}$$

$$0 \le \sigma_p^{\mathrm{GV}}(t) \le \varepsilon_{od}^{\mathrm{GV}}, \forall od \in OD, p \in P_{od}^{\mathrm{GV}}, t \in \mathcal{T} \tag{10.27}$$

In the above, $0 \le x \perp y \ge 0$ indicates $x \ge 0, y \ge 0$, and either x or y must be zero. $\pi_{od}^{EV}(t)$ and $\pi_{od}^{GV}(t)$ are continuous auxiliary variables.

According to the above analysis, when the user tolerance level ε equals 0, the above multi-period static user equilibrium model with bounded rationality degenerates into a multi-period static user equilibrium model under absolute rationality. To illustrate the difference in traffic flow distribution between the two, this section constructs an example of traffic flow distribution that satisfies both the multi-period absolute rational and bounded rational user equilibrium (for a certain period), as shown in Fig. 10.2. Under absolute rationality, users only choose the optimal route (route 4), while under bounded rationality ($\varepsilon = 1.5$ \$), users can also accept suboptimal routes (route 3), as long as the difference between the cost of the selected route and the minimum travel cost does not exceed the user's tolerance ε. Therefore, the traffic flow under bounded rationality is more diversified.

To illustrate the difference in route traffic flow between absolute and bounded rationality, Fig. 10.2 assumes the travel cost of each route remains unchanged. However, since the route travel cost is a function of traffic flow, when the traffic flow changes over time, the corresponding route cost usually also changes.

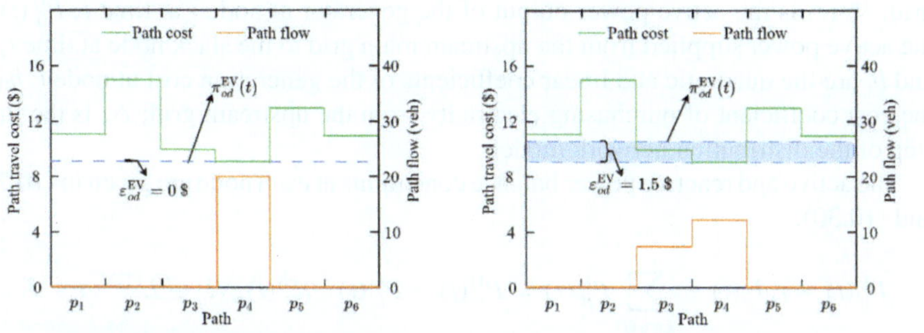

(a) Static user equilibrium model with absolute rationality

(b) Static user equilibrium model with bounded rationality

Fig. 10.2 Distribution comparison of traffic flow under absolute rational and bounded rational static user equilibrium models

Distribution Network Model Based on Multi-period Optimal Power Flow. The distribution network typically has a radial topology, as shown in Fig. 10.3. Therefore, the Branch Flow model is adopted to model the power flow distribution. Assume that the distribution network is represented as a connected graph $G^{PDN} = (V^{PDN}, L^{PDN})$, where $V^{PDN} = V^{PDN+} \cup \{0\}$ and L^{PDN} are the sets of nodes and lines, respectively. Node 0 is the slack node connected to the upstream main grid, and the nodes corresponding to fast-charging stations form a subset $V^{FCS} \subseteq V^{PDN+}$ of the user-side distribution network nodes.

To ensure economic efficiency, the operation of the distribution network satisfies the optimal power flow model, with the objective function given in (10.28):

$$\min O^{PDN} = \sum_{t\in T} \sum_{v_i \in V^{PDN+}} \left\{ a_i \left[P_i^G(t) \right]^2 + b_i P_i^G(t) \right\} \Delta t + \sum_{t\in T} b_0 P_0^G(t) \Delta t \quad (10.28)$$

Here, o^{PDN} is the objective function of the distribution network operation, including the generation cost and the cost of electricity purchased from the upstream main

Fig. 10.3 Schematic example of a radial distribution network

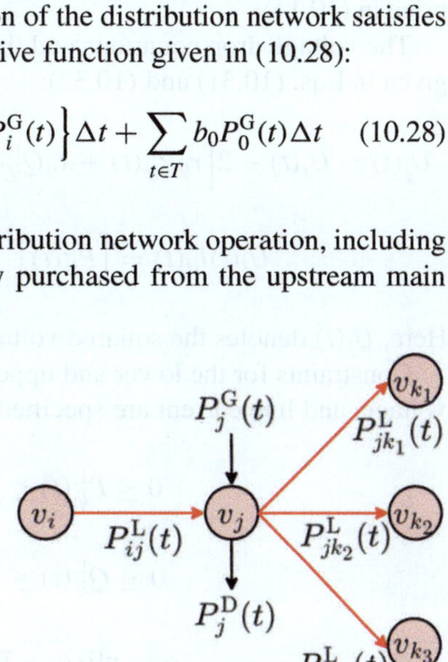

grid; $P_i^G(t)$ is the active power output of the generator at node i at time t; $P_0^G(t)$ is the active power supplied from the upstream main grid to the slack node at time t; a_i and b_i are the quadratic and linear coefficients of the generation cost at node i; b_0 is the cost coefficient of purchasing electricity from the upstream grid; Δt is the time step of the distribution network model.

The active and reactive power balance constraints at each node are given in (10.29) and (10.30):

$$P_{ij}^{\mathrm{L}}(t) - r_{ij}I_{ij}(t) - \sum_{k\in\pi(j)} P_{jk}^{\mathrm{L}}(t) = P_j^{\mathrm{D}}(t) - P_j^{\mathrm{G}}(t) : \lambda_j^{\mathrm{ch}}(t), \forall l_{ij} \in L^{\mathrm{PDN}}, t \in \mathcal{T} \tag{10.29}$$

$$Q_{ij}^{\mathrm{L}}(t) - x_{ij}I_{ij}(t) - \sum_{k\in\pi(j)} Q_{jk}^{\mathrm{L}}(t) = Q_j^{\mathrm{D}}(t) - Q_j^{\mathrm{G}}(t), \forall l_{ij} \in L^{\mathrm{PDN}}, t \in \mathcal{T} \tag{10.30}$$

In these equations, $P_{ij}^L(t)$ and $Q_{ij}^L(t)$ denote the active and reactive power transmitted over line l_{ij} at time t; r_{ij} and x_{ij} are the resistance and reactance of line l_{ij}; $I_{ij}(t)$ represents the squared magnitude of the current over line l_{ij}; $P_j^L(t)$ and $Q_j^L(t)$ are the active and reactive power transmitted to the downstream nodes v_j at time t; $\pi(j)$ denotes the set of downstream nodes of node v_j; $P_j^D(t)$ and $Q_j^D(t)$ are the active and reactive power loads at node v_j at time t; $P_j^G(t)$ is the reactive power generated by the generator at node v_j at time t; $\lambda_j^{\mathrm{ch}}(t)$ is the marginal price at node v_j at time t, corresponding to the dual variable of constraint (10.29), which typically acts as the electricity selling price in the distribution network or as the fast-charging electricity price in (10.11).

The voltage drop equations and the relationship between current and voltage are given in Eqs. (10.31) and (10.32):

$$U_j(t) = U_i(t) - 2\left[r_{ij}P_{ij}^{\mathrm{L}}(t) + x_{ij}Q_{ij}^{\mathrm{L}}(t)\right] + \left[r_{ij}^2 + x_{ij}^2\right]I_{ij}, \forall l_{ij} \in L^{\mathrm{PDN}}, t \in \mathcal{T} \tag{10.31}$$

$$U_i(t)I_{ij}(t) = \left[P_{ij}^{\mathrm{L}}(t)\right]^2 + \left[Q_{ij}^{\mathrm{L}}(t)\right]^2, \forall l_{ij} \in L^{\mathrm{PDN}}, t \in \mathcal{T} \tag{10.32}$$

Here, $U_j(t)$ denotes the squared voltage magnitude at node v_j at time t.

Constraints for the lower and upper bounds of line power, generator power, node voltage, and line current are specified in Eq. (10.33):

$$0 \le P_{ij}^{\mathrm{L}}(t) \le \overline{P}_{ij}^{\mathrm{L}}, \forall l_{ij} \in L^{\mathrm{PDN}}, t \in \mathcal{T} \tag{10.33}$$

$$0 \le Q_{ij}^{\mathrm{L}}(t) \le \overline{Q}_{ij}^{\mathrm{L}}, \forall l_{ij} \in L^{\mathrm{PDN}}, t \in \mathcal{T} \tag{10.34}$$

$$0 \le P_i^{\mathrm{G}}(t) \le \overline{P}_i^{\mathrm{G}}, \forall v_i \in V^{\mathrm{PDN}+}, t \in \mathcal{T} \tag{10.35}$$

$$0 \leq Q_i^{\mathrm{G}}(t) \leq \overline{Q}_i^{\mathrm{G}}, \forall v_i \in V^{\mathrm{PDN}+}, t \in \mathcal{T} \tag{10.36}$$

$$0 \leq P_0^{\mathrm{G}}(t) \leq \overline{P}_0^{\mathrm{G}}, \forall t \in \mathcal{T} \tag{10.37}$$

$$\underline{U}^2 \leq U_i(t) \leq \overline{U}^2, \forall v_i \in V^{\mathrm{PDN}+}, t \in \mathcal{T} \tag{10.38}$$

$$0 \leq I_{ij}(t) \leq \overline{I}_{ij}^2, \forall l_{ij} \in L^{\mathrm{PDN}}, t \in \mathcal{T} \tag{10.39}$$

To describe the coupling relationship between the composition of node-side loads and the charging loads brought by transportation power flows, Eqs. (10.40) and (10.41) are given as follows:

$$P_i^{\mathrm{D}}(t) = P_i^{\mathrm{con}}(t) + P_i^{\mathrm{ch}}(t), \forall v_i \in V^{\mathrm{PDN}+}, t \in \mathcal{T} \tag{10.40}$$

$$P_i^{\mathrm{ch}}(t) = \begin{cases} x_{\kappa(i)}(t) P_{\mathrm{ch}}, & v_i \in V^{\mathrm{FCS}} \\ 0, & v_i \notin V^{\mathrm{FCS}} \end{cases}, \forall t \in \mathcal{T} \tag{10.41}$$

Here, $P_i^{\mathrm{con}}(t)$ and $P_i^{\mathrm{ch}}(t)$ represent the regular active power load and the charging load at node v_i at time t; $\kappa(i)$ denotes the road segment corresponding to the fast-charging station connected to node v_i.

Multi-period Static Equilibrium of Power-Traffic Coupled Network Considering Bounded Rationality of Users. According to (10.11), the electricity price at fast-charging stations affects the temporal distribution of electric vehicle traffic. On the other hand, Eq. (10.41) indicates that electric vehicle traffic also affects the temporal distribution of power flow at fast-charging stations, thereby influencing charging electricity prices and forming feedback. Therefore, the power-traffic coupled network exhibits mutual influence between power flow and traffic flow, eventually reaching an equilibrium state. This equilibrium is defined as the multi-period static equilibrium of the power-traffic coupled network, as shown in Fig. 10.4.

Although fuel vehicles do not directly affect the temporal distribution of power flow, the coexistence of electric and fuel vehicles on ordinary roads will still compete for the limited traffic capacity. This could lead to longer travel times and reduced charging efficiency, and hence, the fuel vehicle traffic and electric vehicle traffic are considered to jointly participate in this coupling.

The set of nodal marginal prices is defined as $\lambda^{\mathrm{ch}} = \{\lambda_i^{\mathrm{ch}}(t) | v_i \in V^{\mathrm{FCS}}, t \in \mathcal{T}\}$; the traffic flow set is defined as $f = \{f_p(t) | p \in P, t \in \mathcal{T}\}$, where nodal marginal price includes information about power flow, and λ^{ch} and f are dual variables.

The interaction between power flow and traffic flow can be represented as:

$$\lambda^{\mathrm{ch}} = f^{\mathrm{OPF}}(f) \tag{10.42}$$

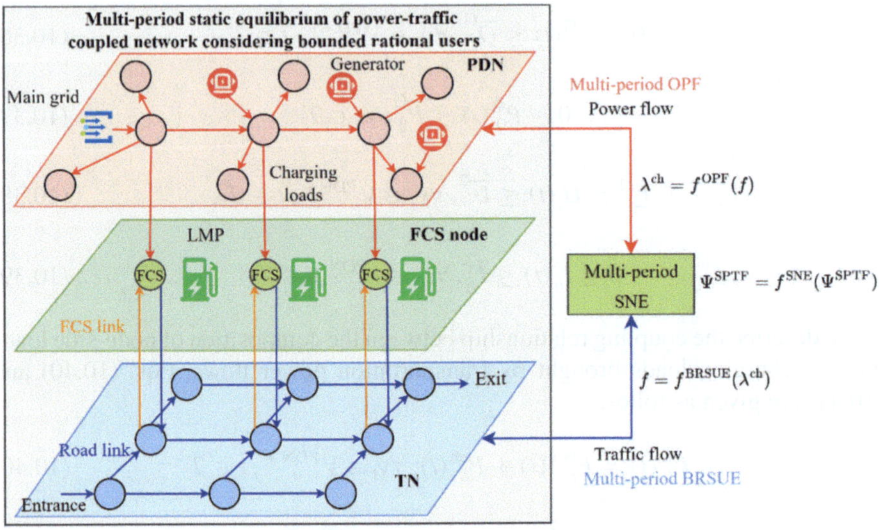

Fig. 10.4 Multi-period static equilibrium of power-traffic coupled network considering bounded rational users

$$f = f^{\mathrm{BRSUE}}(\lambda^{\mathrm{ch}}) \tag{10.43}$$

In this equation, $f^{\mathrm{OPF}}(\cdot)$ represents the response function of nodal marginal electricity prices to traffic flow distribution under the multi-period optimal power flow model, where the relationship between traffic flow and charging load can be directly obtained from (10.41); $f^{\mathrm{BRSUE}}(\cdot)$ represents the response function of traffic flow to nodal electricity prices under the multi-period bounded rational static user equilibrium model.

To further define the multi-period static equilibrium of the power-traffic coupled network, let the multi-period static electricity-traffic flow $\psi^{\mathrm{SPTF}} = \{\lambda^{\mathrm{ch}}, f\}$, then the multi-period static equilibrium of the power-traffic coupled network can be modeled as the fixed-point problem solution:

$$\psi^{\mathrm{SPTF}} = f^{\mathrm{SNE}}(\psi^{\mathrm{SPTF}}) \tag{10.44}$$

where $f^{\mathrm{SNE}}(\cdot)$ denotes the response function of the multi-period static electricity-traffic flow under the user equilibrium of the power-traffic coupled network.

Solution Method. The above fixed-point problem can be solved through a nested iteration algorithm to obtain the time-varying distributions of nodal electricity prices and traffic flows.

In Step 2, when solving the multi-period optimal power flow model, the non-convex constraint (10.32) can be relaxed by the following second-order cone constraint:

$$U_i(t)I_{ij}(t) \geq \left[P_{ij}^{\mathrm{L}}(t)\right]^2 + \left[Q_{ij}^{\mathrm{L}}(t)\right]^2, \forall l_{ij} \in L^{\mathrm{PDN}}, t \in \mathcal{T} \tag{10.45}$$

At this point, the optimal power flow problem is transformed into a second-order cone optimization problem, where the complementary variable of constraint (10.29) can be solved directly through a solver.

In Step 3, the multi-period bounded rational static user equilibrium model contains nonlinear constraints (10.16), (10.17), (10.24), (10.25), where the nonlinear constraints (10.16) and (10.17) can be handled through piecewise linearization methods. The nonlinear complementarity constraints (10.24) and (10.25) can be linearized by introducing 0–1 variables.

Algorithm 1: Single-layer Fixed-Point Iterative Algorithm for Multi-Period Static Model of Power-Traffic Coupled Network

Step 1	Set the maximum number of iterations K, convergence thresholds $\overline{\delta}^{\lambda}$, $\overline{\delta}^{f}$, and initialize the iteration counter $k = 0$, initial nodal marginal electricity price $\lambda^{\mathrm{ch},0}$, and traffic flow f^0
Step 2	According to the traffic flow in the k-th iteration, directly obtain the charging load of each fast-charging station node in the distribution network: $P^{\mathrm{ch}} = \{P_i^{\mathrm{ch}}(t) \mid \forall v_i \in V^{\mathrm{FCS}}, t \in \mathcal{T}\}$ Solve the multi-period optimal power flow model to update the nodal marginal electricity price in the $(k+1)$-th iteration, that is: $\lambda^{\mathrm{ch},k+1} = f^{\mathrm{OPF}}(f^k)$
Step 3	Based on the nodal marginal electricity price from the k-th iteration, solve the multi-period bounded rational static user equilibrium model to update the traffic flow in the $(k+1)$-th iteration, that is: $f^{k+1} = f^{\mathrm{BRSUE}}(\lambda^{\mathrm{ch},k})$
Step 4	Update the convergence errors: $\delta_{k+1}^{\lambda} = \frac{\lVert \lambda^{\mathrm{ch},k+1} - \lambda^{\mathrm{ch},k} \rVert_2}{\lVert \lambda^{\mathrm{ch},k} \rVert_2}$, $\delta_{k+1}^{f} = \frac{\lVert f^{k+1} - f^k \rVert_2}{\lVert f^k \rVert_2}$。 If $\delta_{k+1}^{\lambda} \leq \overline{\delta}^{\lambda}$ and $\delta_{k+1}^{f} \leq \overline{\delta}^{f}$, or $k \geq K$, then output the multi-period static electricity-traffic flow: $\psi^{\mathrm{SPTF}} = \{\lambda^{\mathrm{ch},k+1}, f^{k+1}\}$.Otherwise, let $k = k + 1$, and return to Step 2

$$0 \leq f_p^{\mathrm{EV}}(t) \leq M\left[1 - z_p^{\mathrm{EV}}(t)\right], \forall od \in OD, p \in P_{od}^{\mathrm{EV}}, t \in \mathcal{T} \tag{10.46}$$

$$0 \leq \phi_p^{\mathrm{EV}}(t) - \pi_{od}^{\mathrm{EV}}(t) + \sigma_p^{\mathrm{EV}}(t) \leq M z_p^{\mathrm{EV}}(t), \forall od \in OD, p \in P_{od}^{\mathrm{EV}}, t \in \mathcal{T} \tag{10.47}$$

$$0 \leq f_p^{\mathrm{GV}}(t) \leq M\left[1 - z_p^{\mathrm{GV}}(t)\right], \forall od \in OD, p \in P_{od}^{\mathrm{GV}}, t \in \mathcal{T} \tag{10.48}$$

$$0 \leq \phi_p^{\mathrm{GV}}(t) - \pi_{od}^{\mathrm{GV}}(t) + \sigma_p^{\mathrm{GV}}(t) \leq M z_p^{\mathrm{GV}}(t), \forall od \in OD, p \in P_{od}^{\mathrm{GV}}, t \in \mathcal{T} \tag{10.49}$$

where M is a sufficiently large positive constant; $z_p^{\mathrm{EV}}(t)$ and $z_p^{\mathrm{GV}}(t)$ are 0–1 auxiliary variables.

In Step 4, $\|x\|_2$ denotes the 2-norm of vector x.

Additionally, since the bounded rational static user equilibrium at each period is independently solvable, the equilibrium for each period can be computed in parallel to improve computational efficiency.

10.4 Modeling Method for Dynamic Power-Traffic Coupled Networks Considering User Bounded Rationality

In light of the dynamic propagation characteristics of traffic flow, this section introduces a modeling method for dynamic power-traffic coupled networks considering user bounded rationality. It includes the bounded rational dynamic user equilibrium traffic flow model within the transportation network, the multi-period optimal power flow model within the distribution network, and the dynamic equilibrium model of the power-traffic coupled network formed by the integration of both subsystems, along with the corresponding solution approaches.

Dynamic Traffic Network Model Based on Bounded Rational Dynamic User Equilibrium and Improved Node Queues.

(1) Basic Assumptions.

The dynamic traffic network model proposed in this section is based on the following assumptions:

a. Same as Assumption (1) in Sect. 10.3.1.
b. Same as Assumption (2) in Sect. 10.3.1.
c. Electric vehicle (EV) users and gasoline vehicle (GV) users have different travel demands (origin, destination, and desired arrival time), and are divided into different user groups accordingly. For EV users departing from origin o to destination d and desiring to arrive on time $q \in \widehat{T}$, the total travel demand is denoted by $Q_{od,q}^{\mathrm{EV}}$; similarly, the total travel demand for gasoline vehicle users is denoted by $Q_{od,q}^{\mathrm{GV}}$. Here, $\widehat{T}$ denotes the set of user-desired arrival times.
d. Considering the dynamic propagation of traffic flow, the flow is viewed as granular. At a certain time, users depart from the origin and move along a specific route toward the destination. During this process, they may experience transitions and stops (due to queueing delays at road segments, charging queue delays at EV charging stations, etc.), resulting in state switches. Since traffic flow across a path is not evenly distributed and exhibits divergence, it is not feasible to assign aggregate variables to represent path flow. Therefore, the model must construct independent node queues along the path to represent the flow dynamics, including inflow, queuing, and charging (only for charging station segments) for each segment.

e. After traffic enters a road segment, vehicles first travel at a predetermined speed. Upon reaching the exit end of the segment, they are appended to the end of the queue in sequence. The queue follows the First-In-First-Out (FIFO) principle. Since the vehicles in the queue are treated as discrete units, the queue is modeled as a point queue, and the flow in the point queue is referred to as queuing traffic. For a regular road segment, if the queue length S is less than or equal to the capacity of the exit C_l, then all vehicles in the queue are allowed to proceed to the next segment at the current time, and the queue is cleared. Otherwise, only C_l vehicles at the head of the queue are allowed to pass, while the remaining $S - C_l$ vehicles continue to wait for the next time step. For a fast-charging station segment, let the total number of chargers be N_l, and the number of chargers in use at time t be $n_l(t)$, corresponding to charging traffic. If the queue length S is less than or equal to the number of idle chargers, i.e., $N_l - n_l(t)$, then all queued EVs at the current time are allowed to start charging, the queue is cleared, and the number of chargers in use is updated to O. Otherwise, only the first $N_l - n_l(t)$ EVs are allowed to start charging, while the remaining $S - N_l + n_l(t)$ EVs continue to wait in line, and the number of idle chargers is updated accordingly. Once EVs finish charging, they leave the fast-charging station segment, enter the next road segment, and the chargers they occupied are automatically released and marked as idle.
f. The user's travel plan includes the choice of departure time and travel route. For electric vehicles in $Q_{od,\varrho}^{\text{EV}}$: For each user, the total number of EV users who choose departure time $t \in \mathcal{T}$ and travel route $p \in P_{od}^{\text{EV}}$ is defined as the path departure flow $h_{p,\varrho}^{\text{EV}}(t)$, representing the number of EV users departing from origin o at time t along route p to destination d. Similarly, the path departure flow $h_{p,\varrho}^{\text{GV}}(t)$ can be defined for gasoline vehicle users. The path departure flow only indicates the input traffic at the source of the route, i.e., the input traffic at the entrance of the first segment of the route. This is the source of all traffic (including inflow, queuing, charging, and flowing traffic) in the transportation network.

The comparison between multi-period static network path traffic and dynamic network path departure flow is illustrated in Fig. 10.5. In the multi-period static transportation network model, each time step's path traffic is assumed to be stable and evenly distributed. In contrast, in the dynamic transportation network model, only the input traffic at the route source is described as the path departure flow.

(2) Path Departure Flow and Path Travel Cost.

In the dynamic traffic network model, users have desired arrival times and can flexibly choose their departure times and travel routes. Compared to Sect. 10.3.1, the departure time in the traffic flow is also a decision variable for users. Therefore, this study adopts a path departure time-based user travel choice model. To satisfy the corresponding travel demand, the path departure flow must meet the following requirements:

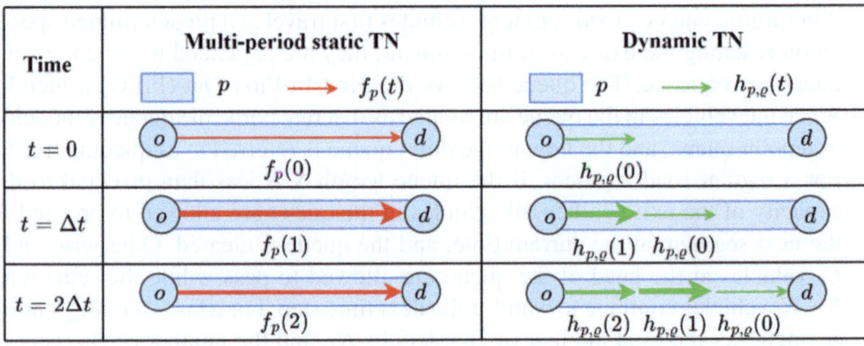

Fig. 10.5 Comparison between multi-period static traffic network path flows and dynamic traffic network path departure flows

$$\sum_{p \in P_{od}^{\mathrm{EV}}} \sum_{t \in \mathcal{T}} h_{p,\varrho}^{\mathrm{EV}}(t) = Q_{od,\varrho}^{\mathrm{EV}}, \forall od \in OD, \varrho \in \hat{\mathcal{T}} \tag{10.50}$$

$$\sum_{p \in P_{od}^{\mathrm{GV}}} \sum_{t \in \mathcal{T}} h_{p,\varrho}^{\mathrm{GV}}(t) = Q_{od,\varrho}^{\mathrm{GV}}, \forall od \in OD, \varrho \in \hat{\mathcal{T}} \tag{10.51}$$

Here, $h_{p,\varrho}^{\mathrm{EV}}(t)$ and $h_{p,\varrho}^{\mathrm{GV}}(t)$ represent the path departure flows of electric vehicles and gasoline vehicles in $Q_{od,\varrho}^{\mathrm{EV}}$ and $Q_{od,\varrho}^{\mathrm{GV}}$, respectively, i.e., the traffic flow of electric/gasoline vehicle users choosing to depart along path p at time t to arrive at destination d from origin o within the expected arrival time ϱ.

The corresponding travel costs for the path departure flow are defined in (10.52) to (10.59), where (10.52) and (10.53) describe the travel costs for electric and gasoline vehicle users in the respective path departure flows $h_{p,\varrho}^{\mathrm{EV}}(t)$ and $h_{p,\varrho}^{\mathrm{GV}}(t)$, who choose to depart from o to d at time t via path p.

$$\phi_{p,\varrho}^{\mathrm{EV}}(t) = \varphi_p^{\mathrm{EV}}(t) + \rho_{p,\varrho}^{\mathrm{EV}}(t) + \theta_p^{\mathrm{EV}}(t), \forall od \in OD, p \in P_{od}^{\mathrm{EV}}, \varrho \in \hat{\mathcal{T}}, t \in \mathcal{T} \tag{10.52}$$

$$\phi_{p,\varrho}^{\mathrm{GV}}(t) = \varphi_p^{\mathrm{GV}}(t) + \rho_{p,\varrho}^{\mathrm{GV}}(t), \forall od \in OD, p \in P_{od}^{\mathrm{GV}}, \varrho \in \hat{\mathcal{T}}, t \in \mathcal{T} \tag{10.53}$$

The travel cost $\phi_{p,\varrho}^{\mathrm{EV}}(t)$ for electric vehicle users on path p consists of the travel time cost $\varphi_p^{\mathrm{EV}}(t)$ and the time deviation penalty cost $\rho_{p,\varrho}^{\mathrm{EV}}(t)$, while the travel cost $\varphi_{p,e}^{\mathrm{GV}}(t)$ for gasoline vehicle users includes the travel time cost $\varphi_p^{\mathrm{GV}}(t)$ and the time deviation penalty cost $\rho_{p,\varrho}^{\mathrm{GV}}(t)$.

The travel time cost for both electric and gasoline vehicle users is defined in (10.54) and (10.55), which is only related to the travel time and independent of the desired arrival time window ϱ.

$$\varphi_p^{\mathrm{EV}}(t) = \gamma^{\mathrm{EV,t}}\left[\xi_p^{\mathrm{EV}}(t) - t\right], \forall od \in OD, p \in P_{od}^{\mathrm{EV}}, t \in \mathcal{T} \tag{10.54}$$

$$\varphi_p^{\mathrm{GV}}(t) = \gamma^{\mathrm{GV,t}}\left[\xi_p^{\mathrm{GV}}(t) - t\right], \forall od \in OD, p \in P_{od}^{\mathrm{GV}}, t \in \mathcal{T} \tag{10.55}$$

Here, $\xi_p^{\mathrm{EV}}(t)$ and $\xi_p^{\mathrm{GV}}(t)$ denote the travel time functions for electric and gasoline vehicles along path p corresponding to the departure flow $h_{p,\varrho}^{\mathrm{EV}}(t)$ and $h_{p,\varrho}^{\mathrm{GV}}(t)$, respectively.

The time deviation penalty costs for both electric and gasoline vehicle users are defined in (10.56) and (10.57). For a user with a desired arrival window e, whose desired arrival time window is $[o - \Delta\tau, o + \Delta\tau]$, if the actual arrival time is earlier than $o - \Delta\tau$ or later than $o + \Delta\tau$, a corresponding penalty is incurred.

$$\rho_{p,\varrho}^{\mathrm{EV}}(t) = \gamma^{\mathrm{EV,e}} \min\left[0, \xi_p^{\mathrm{EV}}(t) - \varrho + \Delta\tau\right] + \gamma^{\mathrm{EV,l}} \max\left[0, \xi_p^{\mathrm{EV}}(t) - \varrho - \Delta\tau\right],$$
$$\forall od \in OD, p \in P_{od}^{\mathrm{EV}}, \varrho \in \hat{\mathcal{T}}, t \in \mathcal{T} \tag{10.56}$$

$$\rho_{p,\varrho}^{\mathrm{GV}}(t) = \gamma^{\mathrm{GV,e}} \min\left[0, \xi_p^{\mathrm{GV}}(t) - \varrho + \Delta\tau\right] + \gamma^{\mathrm{GV,l}} \max\left[0, \xi_p^{\mathrm{GV}}(t) - \varrho - \Delta\tau\right],$$
$$\forall od \in OD, p \in P_{od}^{\mathrm{GV}}, \varrho \in \hat{\mathcal{T}}, t \in \mathcal{T} \tag{10.57}$$

Here, $\Delta\tau$ denotes the half-width of the arrival time window; $\gamma^{\mathrm{EV},e}$ and $\gamma^{\mathrm{EV},l}$ are the early and late penalty coefficients for electric vehicles, respectively; $\gamma^{\mathrm{GV},e}$ and $\gamma^{\mathrm{GV},l}$ are the corresponding coefficients for gasoline vehicles.

The charging cost for electric vehicle users is given by (10.58):

$$\theta_p^{\mathrm{EV}}(t) = \sum_{t' \in T_p^{\mathrm{ch}}(t)} \lambda_{\upsilon(p)}^{\mathrm{ch}}(t') P_{\mathrm{ch}} \Delta t, \forall od \in OD, p \in P_{od}^{\mathrm{EV}}, t \in \mathcal{T} \tag{10.58}$$

$$T_p^{\mathrm{ch}}(t) = \left(\xi_p^{\mathrm{EV}*}(t) - \tau^{\mathrm{ch}}, \xi_p^{\mathrm{EV}*}(t)\right], \forall od \in OD, p \in P_{od}^{\mathrm{EV}}, t \in \mathcal{T} \tag{10.59}$$

Here, $T_p^{\mathrm{ch}}(t)$ represents the charging duration, and $\xi_p^{\mathrm{EV}*}(t)$ is the charging completion time function of electric vehicles corresponding to the departure flow $h_{p,e}^{\mathrm{EV}}(t)$.

(3) Path and Segment Travel Time Model.

To compute the travel cost associated with path departure flow, it is necessary to further clarify the path travel time functions $\xi_p^{\mathrm{EV}}(t)$ and $\xi_p^{\mathrm{GV}}(t)$ for electric and gasoline vehicles, as well as the charging completion time function $\xi_p^{\mathrm{EV}*}(t)$. However, unlike the static traffic flow model, the path travel time is no longer a linear sum of segment travel times $S_l(t)$ under a dynamic traffic flow framework. This is because each segment's travel time function is related to the entry time into that segment. Since the departure time of upstream segments (i.e., the time of inflow into the next segment) is determined by the departure time from the previous segment, the path travel time should be calculated sequentially by chaining the segments in turn until the final segment's exit time is obtained. Similarly, the charging completion time of electric vehicles should be determined by chaining the segments from the departure point to the fast-charging station, resulting in the charging completion time function.

To this end, we first define the segment travel time function as in (10.60):

$$\xi_l(t) = t + \varsigma_l(t), \forall l \in L^{\mathrm{TN}}, t \in \mathcal{T} \tag{10.60}$$

Here, $\xi_l(t)$ denotes the time when traffic enters segment l at time t.

Then, the path travel time functions for electric and gasoline vehicles, as well as the charging completion time function, can be defined via segment chaining, as shown in (10.61)–(2.63):

$$\xi_p^{\mathrm{EV}}(t) = \xi_{l_1} \circ \xi_{l_2} \circ \cdots \circ \xi_{l_{\upsilon(p)}} \circ \cdots \circ \xi_{l_{|p|}}(t), \forall od \in OD, p \in P_{od}^{\mathrm{EV}}, t \in \mathcal{T} \tag{10.61}$$

$$\xi_p^{\mathrm{GV}}(t) = \xi_{l_1} \circ \xi_{l_2} \circ \cdots \circ \xi_{l_{|p|}}(t), \forall od \in OD, p \in P_{od}^{\mathrm{GV}}, t \in \mathcal{T} \tag{10.62}$$

$$\xi_p^{\mathrm{EV}*}(t) = \xi_{l_1} \circ \xi_{l_2} \circ \cdots \circ \xi_{l_{\upsilon(p)}}(t), \forall od \in OD, p \in P_{od}^{\mathrm{EV}}, t \in \mathcal{T} \tag{10.63}$$

In the electric vehicle path $p = l_1 \to l_2 \to \cdots \to l_{\upsilon(p)} \to \cdots \to l_{|p|}$, $l_{\upsilon(p)}$ denotes the fast-charging segment is through; while for gasoline vehicle path $p = l_1 \to l_2 \to \cdots \to l_{|p|}$, there is no fast-charging segment included. The symbol $\circ$ represents function nesting, e.g., $g \circ f(x) = g[f(x)]$.

An example of the chaining computation for the path travel time function is illustrated in Fig. 10.6, where the exit time of the upstream segment equals the entry time of the downstream segment.

On this basis, it is necessary to further define the relationship between the segment travel time function $S_l(t)$ and traffic flow. Unlike the segment travel time function model in the multi-period static model (see (10.15)), in the dynamic model, the traffic flow entering a segment does not exit instantly but goes through a dynamic propagation process, including free movement, queuing due to congestion (i.e., with preceding inflow traffic), original parking or charging, and subsequent departure. Therefore, the segment travel time is not only related to the current entry flow of that segment but also to the preceding inflow queue of the segment. Based on this, the segment travel time function model is defined as follows.

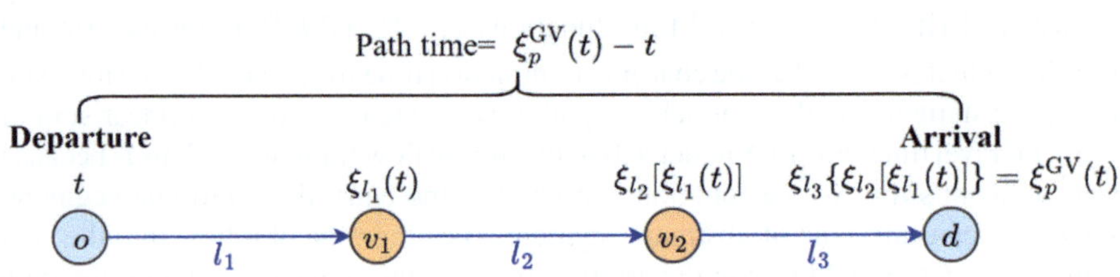

Fig. 10.6 Example of the chaining computation of path travel time function

$$\varsigma_l(t) = \begin{cases} \mathrm{RL}(\boldsymbol{Q}_{l,t}^{\mathrm{in}}), & l \in L^{\mathrm{road}} \\ \mathrm{FL}(\boldsymbol{Q}_{l,t}^{\mathrm{in}}), & l \in L^{\mathrm{FCS}} \end{cases}, \forall t \in \mathcal{T} \tag{10.64}$$

$$\boldsymbol{Q}_{l,t}^{\mathrm{in}} = \left\{Q_l^{\mathrm{in}}(t') | t' \in [0, t]\right\}, \forall l \in L^{\mathrm{TN}}, t \in \mathcal{T} \tag{10.65}$$

Let $\boldsymbol{Q}_{l,t}^{\mathrm{in}}$ denote the inflow queue of the segment l at time t.

The segment travel time function $S_l(t)$ in a dynamic traffic network is important for the evaluation of traffic travel cost. It affects users' time costs, penalties for schedule deviations, and charging costs. It also serves as the basis for constructing the dynamic propagation model of regular segments and fast-charging station segments (including entry, queuing, charging, and departure). However, due to the absence of the entry flow queue assumption in upstream segments, (10.16) and (10.17) in the static model cannot directly calculate segment travel time based on upstream inflow. Therefore, it is necessary to analyze the dynamic propagation process in detail to determine the segment-level outflow time and travel delay.

(4) FIFO and Flow Conservation Conditions in the Point Queue Model.

To establish the segment travel time function model, this section designs a regular point queue model based on a differential system and modifies it for fast-charging station segments to characterize the dynamic propagation features of both types. Before introducing the model equations for each segment type, the FIFO and flow conservation conditions applicable to both regular segments and fast-charging station segments are first presented:

(1) Outflow $Q_l^{\mathrm{out}}(t_2)$ of a downstream segment must not occur earlier than the inflow $Q_l^{\mathrm{in}}(t_1)$ of the upstream segment. In other words, the segment travel time function $\xi_l(t)$ must be non-decreasing, i.e.:

$$\xi_l(t_1) \le \xi_l(t_2), \forall l \in L^{\mathrm{TN}}, 0 \le t_1 \le t_2 \le T \tag{10.66}$$

(2) For the inflow of a segment (which may consist of multiple path subflows within $Q_l^{\mathrm{in}}(t)$), the corresponding outflow $Q_l^{\mathrm{out}}(t)$ must maintain the same path ratio as the inflow at time $Q_l^{\mathrm{in}}(t)$, ensuring flow conservation. Next, the path-based inflow and outflow queues $Q_l^{\mathrm{in}}(t)$ and $Q_l^{\mathrm{out}}(t)$ are defined.

$$Q_l^{\mathrm{in}}(t) = \begin{cases} \sum\limits_{p \in P} \Lambda_p^l Q_{l,p}^{\mathrm{in}}(t), & t \ge 0 \\ 0, & t < 0 \end{cases}, \forall l \in L^{\mathrm{TN}} \tag{10.67}$$

$$Q_l^{\mathrm{out}}(t) = \begin{cases} \sum\limits_{p \in P} \Lambda_p^l Q_{l,p}^{\mathrm{out}}(t), & t \ge 0 \\ 0, & t < 0 \end{cases}, \forall l \in L^{\mathrm{TN}} \tag{10.68}$$

Let $Q_{l,p}^{\mathrm{in}}(t)$ and $Q_{l,p}^{\mathrm{out}}(t)$ denote the share of path p in $Q_l^{\mathrm{in}}(t)$ and $Q_l^{\mathrm{out}}(t)$, respectively, i.e., the path-specific subflows. Thus, the FIFO condition (2) can be expressed by (10.69), which means: if at time t the inflow to segment l is $Q_l^{\text{in}}(t)$, then the corresponding outflow at time $\xi_l(t)$, i.e., $Q_l^{\text{out}}[\xi_l(t)]$, will retain the same path share as the inflow at time t. When $Q_l^{\text{in}}(t) = 0$, all subflows are also 0.

$$Q_{l,p}^{\text{out}}[\xi_l(t)] = \begin{cases} \frac{Q_{l,p}^{\text{in}}(t)}{Q_l^{\text{in}}(t)} Q_l^{\text{out}}[\xi_l(t)], & Q_l^{\text{in}}(t) \neq 0 \\ 0, & Q_l^{\text{in}}(t) = 0 \end{cases}, \forall l \in L^{\text{TN}}, t \in \mathcal{T} \quad (10.69)$$

The handover and transfer of subflows across adjacent segments correspond to path transitions at the segment interfaces. Therefore, based on the path flow model, the inflow model at each segment interface is further defined as follows:

$$Q_{l,p}^{\text{in}}(t) = \begin{cases} \sum\limits_{\varrho \in T} h_{p,\varrho}^{\text{EV}}(t), \ l = p_1, p \in P_{od}^{\text{EV}} \\ \sum\limits_{\varrho \in T} h_{p,\varrho}^{\text{GV}}(t), \ l = p_1, p \in P_{od}^{\text{GV}} \\ Q_{m,p}^{\text{out}}(t), \quad l \neq p_1, \{m \to l\} \subseteq p \in P_{od} \end{cases}, \forall od \in OD, t \in \mathcal{T} \quad (10.70)$$

For a path p with two adjacent segments $m \to l$, the upstream outflow $Q_m^{\text{out}}(t)$ of segment m corresponding to path p equals the downstream inflow $Q_l^{\text{in}}(t)$ of segment l regarding path p, ensuring flow continuity for $Q_{l,p}^{\text{in}}(t)$ at the path level.

It is important to point out that when path p has no upstream segment, the first segment l starts from the network origin node, and its subflow is directly equal to the path departure flow (as shown in Fig. 10.7), so the path departure flow can be regarded as the inflow from an external network.

The upstream subflow includes all path-specific subflows originating from outside the network at this interface, maintaining flow conservation and reflecting the inflow to the path's first segment in the network.

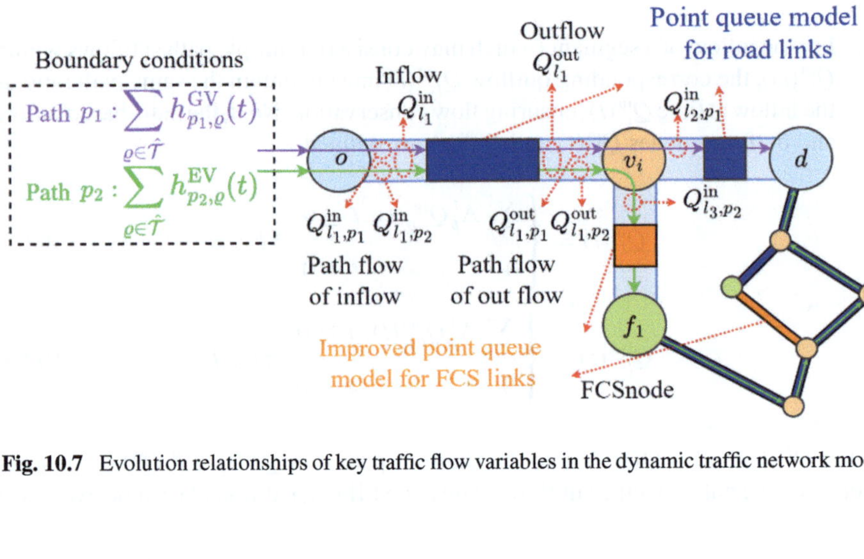

Fig. 10.7 Evolution relationships of key traffic flow variables in the dynamic traffic network model

Once the path departure flow enters the transportation network, it is only distinguished by the departure time t and the path p, not by the expected arrival time θ. Therefore, a single path can include path departure flows $h_{p,\varrho}^{\mathrm{EV}}(t)$ or $h_{p,\varrho}^{\mathrm{GV}}(t)$ with different expected arrival times θ, which can be aggregated as a single subflow of electric (or gasoline) vehicles. This is because, in the dynamic traffic network, the path travel time function $\xi_p^{\mathrm{EV}}(t)$ and the charging completion time $\xi_p^{\mathrm{EV}*}(t)$ are only related to the selected departure time t and path p, and are independent of the expected arrival time θ. Therefore, to calculate the travel cost $\phi_{p,\varrho}^{\mathrm{EV}}(t)$ of the path departure flow $h_{p,\varrho}^{\mathrm{EV}}(t)$, one only needs to compute the travel time $\xi_p^{\mathrm{EV}}(t)$ and charging completion time $\xi_p^{\mathrm{EV}*}(t)$ corresponding to the departure flow on path p at time t. Then, the time cost $\phi_p^{\mathrm{EV}}(t)$ and charging cost $\theta_p^{\mathrm{EV}}(t)$ can be derived from (10.54) to (10.58), while the schedule deviation penalty cost due to expected arrival time θ and path travel time $\xi_p^{\mathrm{EV}}(t)$ is computed using (10.56). In this case, the expected arrival time θ is implicitly encoded in $h_{p,\varrho}^{\mathrm{EV}}(t)$ and does not require being modeled along path p. That is, multiple expected arrival times θ can share (reuse) the same path p. This treatment significantly reduces the computational overhead introduced by enumerating combinations of expected arrival time and path departure flow in dynamic traffic network models.

The evolution relationships among the path departure flow, link inflow traffic, and its path split, as well as the link outflow traffic and its path split, are illustrated in Fig. 10.7. Among them, the path departure flow serves as the input boundary condition for the dynamic traffic network model, while the link inflow and outflow traffic act as the input and output quantities, respectively, for the link-based point queue model.

The above first-in-first-out (FIFO) condition only characterizes the nature of flow splitting and merging at the entrance and exit of a link. The detailed dynamic propagation of traffic flow within the link (including movement, queuing, charging, etc.) will be described by the following point queue model.

(5) Regular Point Queue Model for Normal Road Segments.

For regular road segments, the regular point queue model mainly depicts the flow propagation relationship among electric vehicles (EVs) and gasoline vehicles (GVs) entering the traffic flow, queuing traffic flow, and exiting the traffic flow.

Let (10.71) define the queue length of a normal road segment at time t. In this expression, $S_l(t)$ represents the queue length of link l at time t, and τ_i^{free} denotes the free-flow travel time on link l.

When $t \leq \tau_i^{\mathrm{free}}$, even the earliest inflow traffic $Q_l^{\mathrm{in}}(0)$ has not yet reached the link tail, the queue is empty.

When $t > \tau_i^{\mathrm{free}}$, the current queue length $S_l(t)$ equals the previous queue length $S_l(t-\Delta t)$ plus the new inflow traffic added to the queue tail at time $t-\tau_i^{\mathrm{free}}-\Delta t$, and then subtracting the outflow traffic departing from the queue head at time $t-\Delta t$.

Since the traffic inflow $Q_l^{\mathrm{in}}(t)$ has experienced the free-flow travel time and thus reaches the end of the link (i.e., the queue tail), the new queued inflow equals the inflow entering the link at $t-\tau_i^{\mathrm{free}}-\Delta t$.

$$S_l(t) = \begin{cases} 0, & t \in [0, \tau_l^{\text{free}}] \\ S_l(t - \Delta t) + Q_l^{\text{in}}(t - \tau_l^{\text{free}} - \Delta t) - Q_l^{\text{out}}(t - \Delta t), & t \in (\tau_l^{\text{free}}, T] \end{cases},$$
$$\forall l \in L^{\text{road}} \tag{10.71}$$

Equation (10.72) defines the traffic outflow at the tail of a normal road segment at time t. When $t < \tau_i^{\text{free}}$, even the earliest inflow $Q_l^{\text{in}}(0)$ has not yet reached the tail of link l, so the outflow $Q_l^{\text{out}}(t)$ is zero. When $t \geq \tau_i^{\text{free}}$, the inflow $Q_l^{\text{in}}(t - \tau_i^{\text{free}})$ has just arrived at the link tail and joined the queue. At this time, the number of queued vehicles is $S_l(t) + Q_l^{\text{in}}(t - \tau_i^{\text{free}})$, and the actual traffic outflow from the road segment equals the minimum between the number of queued vehicles and the maximum discharge capacity C_l at the exit.

$$Q_l^{\text{out}}(t) = \begin{cases} 0, & t \in [0, \tau_l^{\text{free}}) \\ \min\left[S_l(t) + Q_l^{\text{in}}(t - \tau_i^{\text{free}}), C_l\right], & t \in [\tau_l^{\text{free}}, T] \end{cases}, \forall l \in L^{\text{road}} \tag{10.72}$$

It is worth noting that $S_l(t)$ represents the queue length at the beginning of the current time step (i.e., the queue length at the start of t), while $Q_l^{\text{out}}(t)$ represents the outflow traffic at the end of the current time step. Therefore, the actual conditions at time $t = \tau_i^{\text{free}}$ are defined as follows:

$$S_l(\tau_l^{\text{free}}) = 0, \forall l \in L^{\text{road}} \tag{10.73}$$

$$Q_l^{\text{out}}(\tau_l^{\text{free}}) = \min\left[S_l(\tau_l^{\text{free}}) + Q_l^{\text{in}}(0), C_l\right] = \min\left[Q_l^{\text{in}}(0), C_l\right], \forall l \in L^{\text{road}} \tag{10.74}$$

The dynamic propagation process of dynamic traffic flow in the regular point queue model is illustrated in Fig. 10.8. In this example, the maximum number of vehicles that can move freely per unit time is 1. At $t = 0$, two EVs (EV1 and EV2) and one gasoline vehicle (GV1) enter the regular road segment. After a delay of Δt due to queuing, EV1 leaves the road segment at $t = \Delta t$, with a traffic flow of 1. EV2 and GV1 are still queued. At $t = 2\Delta t$ and $t = 3\Delta t$, they leave the road segment in order. During this process, the queue length gradually reduces to zero.

(6) Improved Point Queue Model for Fast Charging Station Segments.

The conventional point queue model describes the dynamic propagation of traffic flows along regular road segments, but it fails to represent segments that include fast charging station processes. Therefore, based on the standard point queue model, this section proposes an improved model to characterize the flow transmission relationships among electric vehicle (EV) inflows, queued flows, pre-charging flows, actual charging flows, and outflows at fast charging station segments.

Equation (10.75) defines the queue length of a fast charging segment at time t. When $t \leq \tau_i^{\text{free}}$, the earliest traffic flow entering segment l, i.e., $Q_l^{\text{in}}(0)$, has not yet joined the queue, so the queue length is zero. When $t > \tau_i^{\text{free}}$, unlike the regular segment queue model in (10.71), the queued EVs at the head of the fast charging

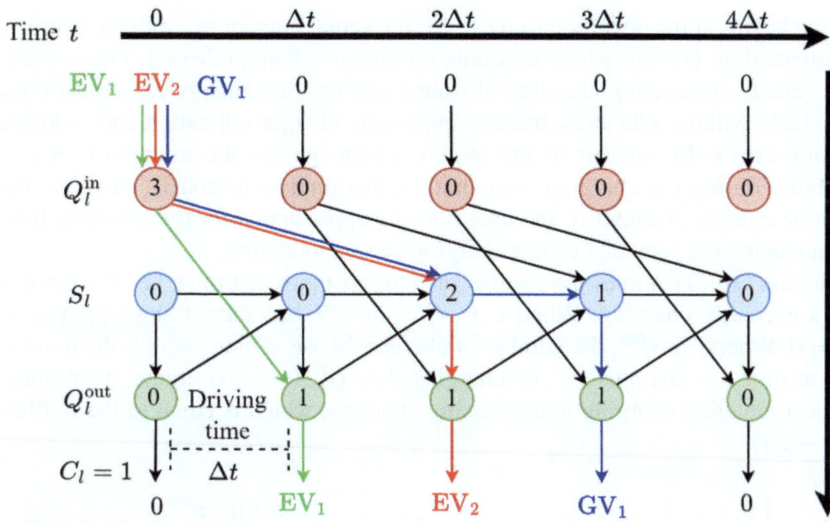

Fig. 10.8 Evolution of dynamic traffic flow in the regular point queue model

queue do not directly pass through the segment, but rather begin charging upon entering the station. Therefore, the queue update rule is that the current queue length $S_l(t)$ equals the previous time step's queue length $S_l(t-\Delta t)$ plus the newly added inflow $Q_l^{in}(t-\tau_i^{free}-\Delta t)$ at the queue tail, minus the pre-charging flow $Q_l^{pre}(t-\Delta t)$ from the queue head at the previous time step.

$$S_l(t) = \begin{cases} 0, & t \in [0, \tau_l^{free}] \\ S_l(t-\Delta t) + Q_l^{in}(t-\tau_l^{free}-\Delta t) - Q_l^{pre}(t-\Delta t), & t \in (\tau_l^{free}, T] \end{cases}, \quad \forall l \in L^{FCS} \tag{10.75}$$

(10.76) defines the flow permitted to begin charging at time t. When $t \le \tau_i^{free}$, the inflow $Q_l^{in}(0)$ has not yet entered a state where charging is allowed, so the charging flow is zero. When $t \ge \tau_i^{free}$, the traffic that entered the queue at $t-\tau_i^{free}$ reaches the charging station and joins the end of the queue. The demand for charging is given by $S_l(t) + Q_l^{in}(t-\tau_i^{free})$, while the number of available chargers equals the total number of chargers N_l minus the number of chargers currently in use $n_l(t)$. Hence, the actual permitted charging flow is the minimum of demand and availability.

$$Q_l^{pre}(t) = \begin{cases} 0, & t \in [0, \tau_l^{free}) \\ \min\left[S_l(t) + Q_l^{in}(t-\tau_l^{free}), N_l - n_l(t) + Q_l^{out}(t)\right], & t \in [\tau_l^{free}, T] \end{cases}, \quad \forall l \in L^{FCS} \tag{10.76}$$

It should be noted that $Q_l^{pre}(t)$ represents the traffic flow entering the pre-charging state at time t (i.e., leaving the queue and waiting for charger availability). Since charging begins at the next time step $t + \Delta t$, the actual number of available chargers is augmented by $Q_l^{\text{out}}(t)$, which accounts for chargers being released. This ensures that a vehicle completing charging at time t can be immediately replaced by the next vehicle without idle slots, thereby improving charger utilization and reducing modeling errors. In contrast to the model, which ignores the impact of charger availability during pre-charging computation, this improved model first calculates remaining charger availability and uses it as the upper bound for pre-charging flow, thus enhancing the accuracy of fast charging station modeling.

Equation (10.77) defines the number of chargers in use at time t, i.e., the number of EVs currently charging. When $t \leq \tau_i^{\text{free}}$, no EV has started charging yet, so $n_l(t) = 0$. When $t > \tau_i^{\text{free}}$, the number of chargers in use at time t equals the number in use at time $t - \Delta t$, plus the pre-charging flow $Q_l^{\text{pre}}(t - \Delta t)$, minus the number of EVs completing charging and releasing chargers, which is equal to the outflow $Q_l^{\text{out}}(t - \Delta t)$.

$$n_l(t) = \begin{cases} 0, & t \in [0, \tau_l^{\text{free}}] \\ n_l(t - \Delta t) + Q_l^{\text{pre}}(t - \Delta t) - Q_l^{\text{out}}(t - \Delta t), & t \in (\tau_l^{\text{free}}, T] \end{cases}, \forall l \in L^{\text{FCS}} \tag{10.77}$$

Equation (10.78) defines the traffic flow completing charging at time t. When $t < \tau_i^{\text{free}} + \tau^{\text{ch}}$, the earliest inflow $Q_l^{\text{in}}(0)$ to segment l has not yet completed charging, so the completed charging flow is zero. When $t \geq \tau_i^{\text{free}} + \tau^{\text{ch}}$, the completed charging flow at time t is given by the pre-charging flow $Q_l^{\text{pre}}(t - \tau^{\text{ch}})$ that entered the pre-charging state in the previous period, specifically in the interval $[t - \tau^{\text{ch}} + \Delta t, t]$.

$$Q_l^{\text{out}}(t) = \begin{cases} 0, & t \in [0, \tau_l^{\text{free}} + \tau^{\text{ch}}) \\ Q_l^{\text{pre}}(t - \tau^{\text{ch}}), & t \in [\tau_l^{\text{free}} + \tau^{\text{ch}}, T] \end{cases}, \forall l \in L^{\text{FCS}} \tag{10.78}$$

Similarly, in the improved point queue model for fast charging station segments, $S_l(t)$ represents the queue length at the beginning of period t, i.e., the queue length at the end of the previous period $t - \Delta t$. $Q_l^{\text{begin}}(t)$ is the pre-charging traffic flow at the end of period t. $n_l(t)$ is the number of charging piles in use at the beginning of period t, which is the same as the number in use at the end of period $t - \Delta t$. $Q_l^{\text{out}}(t)$ is the traffic flow completing charging at the end of period t.

Therefore, the boundary conditions for the four variables at $t = \tau_i^{\text{free}}$ are defined as follows:

$$S_l(\tau_l^{\text{free}}) = 0, \forall l \in L^{\text{FCS}} \tag{10.79}$$

$$\begin{aligned} Q_l^{\text{pre}}(\tau_l^{\text{free}}) &= \min\left[S_l(\tau_l^{\text{free}}) + Q_l^{\text{in}}(0), N_l - n_l(\tau_l^{\text{free}}) + Q_l^{\text{out}}(\tau_l^{\text{free}})\right], \forall l \in L^{\text{FCS}} \\ &= \min\left[Q_l^{\text{in}}(0), N_l\right] \end{aligned} \tag{10.80}$$

$$n_l(\tau_l^{\text{free}}) = 0, \forall l \in L^{\text{FCS}} \tag{10.81}$$

$$Q_l^{\text{out}}(\tau_l^{\text{free}}) = 0, \forall l \in L^{\text{FCS}} \tag{10.82}$$

In addition, the boundary condition for $Q_l^{\text{out}}(t)$ at $t = \tau_i^{\text{free}} + \tau^{\text{ch}}$ is defined as:

$$Q_l^{\text{out}}(\tau_l^{\text{free}} + \tau^{\text{ch}}) = Q_l^{\text{pre}}(\tau_l^{\text{free}}) = \min\left[Q_l^{\text{in}}(0), N_l\right] \tag{10.83}$$

The dynamic propagation process of traffic flow in the improved point queue model is illustrated in Fig. 10.9. In this example, the free travel time on the fast charging station segment is Δt, the charging duration is $2\Delta t$, and the number of charging piles is 1. At $t = 0$, two electric vehicles (EV1 and EV2) enter the fast charging station segment. After traveling for a free travel time Δt, both arrive at the end of the segment at $t = \Delta t$ (i.e., they enter the fast charging station). Since only one charging pile is available, EV1 enters the pre-charging state, i.e., it is allowed to enter the charging preparation queue and will start charging in the next period. Meanwhile, EV2 waits at the edge of the queue, forming a queued traffic flow.

At $t = 3\Delta t$, EV1 completes charging and leaves the charging station. The charging pile is released in the next period, so EV2 enters the pre-charging state. At $t = 4\Delta t$, EV2 starts charging, and the charging pile becomes occupied again. At $t = 5\Delta t$, EV2 completes charging and departs from the fast charging station, and the charging pile is released again.

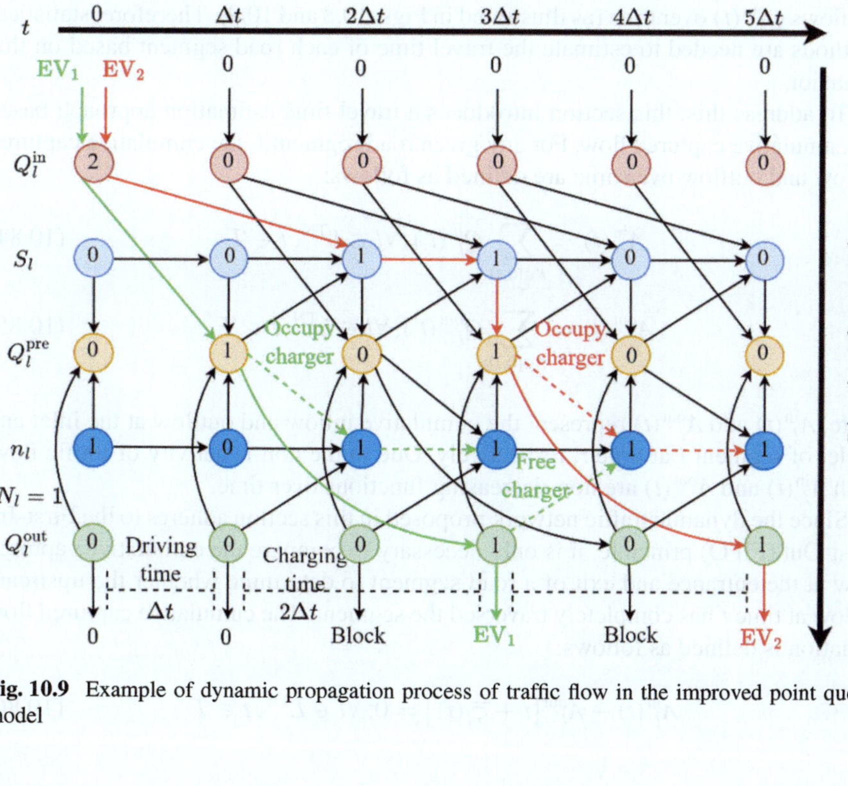

Fig. 10.9 Example of dynamic propagation process of traffic flow in the improved point queue model

If at any time all the charging piles are occupied, the fast charging segment enters a fully saturated state. That is, even if there is queued traffic waiting to proceed, the outflow of traffic is still zero, as observed at $t = 2\Delta t$ and $t = 4\Delta t$. In contrast, this fully blocked state does not occur in the conventional point queue model illustrated in Fig. 10.8, where the outflow can still be greater than zero as long as there is waiting traffic. This example further highlights the difference between the conventional and the improved point queue models.

(7) Point Queue Model-Based Travel Time Estimation Method Considering Cumulative Captured Flow

The aforementioned point queue model establishes a different system to represent the dynamics of mixed traffic on ordinary road segments, including entry, movement, queuing, and departure operations, as well as the processes of electric vehicles entering, moving, queuing, charging, and departing on fast charging station segments. Next, to determine the travel time of traffic flow along road segments, it is necessary to trace the relationship between inflow $Q_i^{\text{in}}(t)$ and outflow $Q_i^{\text{out}}(t)$ of each road segment. Through analysis of the point queue models defined in (10.71) and (10.75), it can be concluded that due to capacity limitations on ordinary road segments and at charging piles of fast charging stations, once a large volume of traffic flows into a segment, queues will form. In such cases, the traffic entering the tail of the queue at a given moment cannot immediately exit at the same moment. That is, traffic in $Q_i^{\text{in}}(t)$ may be distributed across multiple adjacent segments in the form of a sequence of outflows $Q_i^{\text{out}}(t)$ over time (as illustrated in Figs. 10.8 and 10.9). Therefore, statistical methods are needed to estimate the travel time of each road segment based on this behavior.

To address this, this section introduces a travel time estimation approach based on cumulative captured flow. For any given road segment l, the cumulative captured inflow and outflow over time are defined as follows:

$$A_l^{\text{in}}(t) = \sum_{t' \in [0,t]} Q_l^{\text{in}}(t'), \forall l \in L^{\text{TN}}, t \in \mathcal{T} \tag{10.84}$$

$$A_l^{\text{out}}(t) = \sum_{t' \in [0,t]} Q_l^{\text{out}}(t'), \forall l \in L^{\text{TN}}, t \in \mathcal{T} \tag{10.85}$$

Here, $A_l^{\text{in}}(t)$ and $A_l^{\text{out}}(t)$ represent the cumulative inflow and outflow at the inlet and outlet of segment l at time t, respectively. Due to the non-negativity of traffic flow, both $A_l^{\text{in}}(t)$ and $A_l^{\text{out}}(t)$ are non-decreasing functions over time.

Since the dynamic traffic network proposed in this section adheres to the First-In-First-Out (FIFO) principle, it is only necessary to compare the cumulative captured flow at the entrance and exit of a road segment to determine whether the upstream inflow at time t has completely traversed the segment. The cumulative captured flow equation is defined as follows:

$$A_l^{\text{in}}(t) - A_l^{\text{out}}\left[t + \overline{\varsigma}_l(t)\right] = 0, \forall l \in L^{\text{TN}}, t \in \mathcal{T} \tag{10.86}$$

Here, $\overline{\varsigma}_l(t)$ represents the upper bound estimate of the travel time for traffic flow $Q_i^{\text{in}}(t)$ to exit the road segment.

Due to the monotonicity of $A_l^{\text{in}}(t)$ and $A_l^{\text{out}}(t)$, the above (10.93) may have multiple solutions for a given $A_l^{\text{in}}(t)$. For ordinary road segments, the minimum $\overline{\varsigma}_l(t)$ satisfying $\overline{\varsigma}_l(t) \geq \tau_l^{\text{free}}$ is selected as the solution to (10.93); for fast charging station road segments, the minimum $\overline{\varsigma}_l(t)$ satisfying $\overline{\varsigma}_l(t) \geq \tau_l^{\text{free}} + \tau^{\text{ch}}$ is chosen. The purpose of introducing the lower bound on $\overline{\varsigma}_l(t)$ is to ensure that the estimated travel time for any segment is no less than its minimum travel time, thereby ensuring the rationality of the estimation.

If (10.93) holds, it can be inferred that the inflow $Q_l^{\text{in}}(t)$ at time t has completely exited the road segment by time $t + \overline{\varsigma}_l(t)$. However, as previously described, $Q_l^{\text{in}}(t)$ may be distributed across multiple adjacent $Q_l^{\text{out}}(t)$ values over time. Equation (10.93) only confirms that the last portion of $Q_l^{\text{in}}(t)$ exits the segment at time $t+\overline{\varsigma}_l(t)$, meaning that the final portion of $Q_l^{\text{in}}(t)$ is included within $Q_l^{\text{out}}\left[t + \overline{\varsigma}_l(t)\right]$. In contrast, the first portion of $Q_l^{\text{in}}(t)$ likely exited before time $t+\overline{\varsigma}_l(t)$. Therefore, $\overline{\varsigma}_l(t)$ is only an upper bound estimate of the travel time for traffic flow exiting the road segment.

On the other hand, based on the First-In-First-Out (FIFO) principle, it is known that the traffic flows $Q_l^{\text{in}}(t - \Delta t)$ and $Q_l^{\text{in}}(t)$ enter the road segment sequentially and will also exit the segment in order. Therefore, the upper bound of the time when $Q_l^{\text{in}}(t - \Delta t)$ exits the road segment (i.e., the time when the last portion of the traffic exits) is also the lower bound of the time when $Q_l^{\text{in}}(t)$ begins to exit the road segment. Based on this observation, the lower bound of the travel time for traffic flow $Q_l^{\text{in}}(t)$ to exit the road segment can be estimated, and the corresponding cumulative captured flow equation is defined as:

$$A_l^{\text{in}}(t - \Delta t) - A_l^{\text{out}}\left[t + \underline{\varsigma}_l(t)\right] = 0, \forall l \in L^{\text{TN}}, t \in \mathcal{T} \tag{10.87}$$

where $\underline{\varsigma}_l(t)$ denotes the lower bound estimate of the travel time $Q_l^{\text{in}}(t)$ to exit the road segment.

Rearranging (10.93) and (10.94), we obtain:

$$\begin{aligned} Q_l^{\text{in}}(t) &= Q_l^{\text{out}}\left[t + \underline{\varsigma}_l(t) + \Delta t\right] + Q_l^{\text{out}}\left[t + \underline{\varsigma}_l(t) + 2\Delta t\right] + \cdots + Q_l^{\text{out}}\left[t + \overline{\varsigma}_l(t)\right] \\ &= \sum_{\varsigma \in \left[\underline{\varsigma}_l(t)+\Delta t, \overline{\varsigma}_l(t)\right]} Q_l^{\text{out}}(t + \varsigma), \forall l \in L^{\text{TN}}, t \in \mathcal{T} \end{aligned} \tag{10.88}$$

The traffic flow in $Q_l^{\text{in}}(t)$ may be distributed across multiple neighboring $Q_l^{\text{in}}(t)$ values, corresponding to the sequence in which it exits the road segment. Since $\overline{\varsigma}_l(t)$ and $\underline{\varsigma}_l(t)$ are both estimates with an inherent margin of error, to unbiasedly estimate the travel time of $Q_l^{\text{in}}(t)$ on the road segment, this section selects the midpoint flow in $Q_l^{\text{in}}(t)$ as the reference, treating its travel time as $\hat{\xi}_i(t)$, i.e. the unbiased estimate of the travel time for the overall flow. This also enables the derivation of the unbiased cumulative captured flow equation.

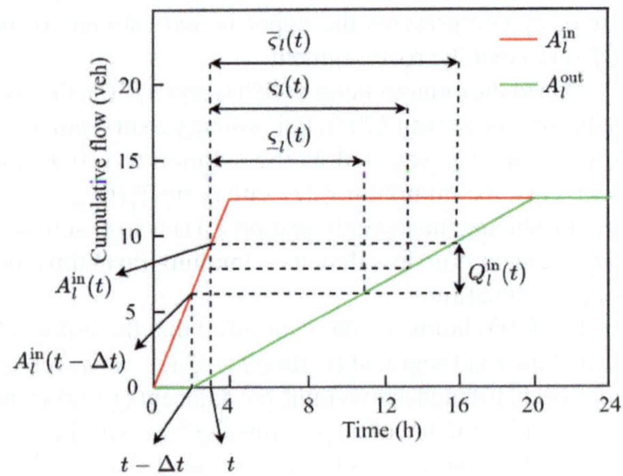

Fig. 10.10 Illustration of upper bound, lower bound, and unbiased estimates of road segment travel time based on cumulative captured flow

$$
\begin{aligned}
&A_l^{\text{in}}(t-\Delta t)+\frac{Q_l^{\text{in}}(t)}{2}-A_l^{\text{out}}\left[t+\varsigma_l(t)\right] \\
&=\frac{A_l^{\text{in}}(t-\Delta t)+A_l^{\text{in}}(t)}{2}-A_l^{\text{out}}\left[t+\varsigma_l(t)\right]=0, \forall l \in L^{\text{TN}}, t \in \mathcal{T}
\end{aligned}
\tag{10.89}
$$

The upper bound, lower bound and unbiased estimates of the travel time of traffic flow on the road segment based on the cumulative captured flow are illustrated in Fig. 10.10. As shown, the unbiased estimate of the segment travel time lies between the upper and lower bounds.

By substituting the function $\varsigma_l(t)$ into (10.60), the corresponding segment-level travel time function $\xi_l(t)$ can be obtained. Subsequently, by substituting it into the function sets defined in (10.61) through (10.63), the path travel time function $\xi_p^{\text{EV}}(t)$ for electric vehicles and $\xi_p^{\text{GV}}(t)$ for gasoline vehicles can be derived. Furthermore, the corresponding path departure times $h_{p,\varrho}^{\text{EV}}(t)$ and $h_{p,\varrho}^{\text{GV}}(t)$, as well as the resulting path travel costs $\phi_{p,\varrho}^{\text{EV}}(t)$ and $\phi_{p,\varrho}^{\text{GV}}(t)$, can finally be calculated by returning to (10.52) to (10.59).

It is worth noting that the path travel cost is not merely a function of the corresponding path departure flow. For electric vehicle users, the path travel cost $\phi_{p,\varrho}^{\text{EV}}(t)$ is not only influenced by the departure flow $h_{p,\varrho}^{\text{EV}}(t)$ along the path but also by the queuing and discharging of traffic. In particular, the departure flow ahead of a certain position may potentially block the later flow, thereby impacting the cost of downstream travel. Consequently, the path travel cost $\phi_{p,\varrho}^{\text{EV}}(t)$ is also influenced by the prior departure sequence $h_{p,\varrho}^{\text{EV}}(t)$, where $t' \in [0, t)$. In addition, due to potential overlaps among different paths, the cost $\phi_{p,\varrho}^{\text{EV}}(t)$ is further influenced by the sequence of all flows on the shared road segments. Similarly, the gasoline vehicle path travel cost is also affected by these flow interactions.

(8) Bounded Rational Dynamic User Equilibrium Model.

For a given path departure flow, by employing the aforementioned differential–algebraic system's point queue model and the cumulative flow-based travel time estimation method, the spatiotemporal propagation process of traffic flow across the entire road network and across all periods can be derived. Eventually, the path travel cost corresponding to each departure flow can be computed. This derivation process defines the dynamic network loading.

The input to the dynamic network loading model is the set of path departure flows $\mathcal{H}$, which contains all departure information of both electric and gasoline vehicle users. It is defined as follows:

$$\begin{aligned}\mathcal{H} = &\left\{h_{p,\varrho}^{\mathrm{EV}}(t) \geq 0 | od \in OD, p \in P_{od}^{\mathrm{EV}}, \varrho \in \hat{\mathcal{T}}, t \in \mathcal{T}\right\} \\ &\cup\left\{h_{p,\varrho}^{\mathrm{GV}}(t) \geq 0 | od \in OD, p \in P_{od}^{\mathrm{GV}}, \varrho \in \hat{\mathcal{T}}, t \in \mathcal{T}\right\}\end{aligned} \tag{10.90}$$

For a given set of path departure flows $\mathcal{H}$ and time-varying electricity price λ^{ch}, the corresponding set of path travel costs Φ calculated through dynamic network loading satisfies:

$$\begin{aligned}\Phi(\mathcal{H}, \lambda^{\mathrm{ch}}) = &\left\{\phi_{p,\varrho}^{\mathrm{EV}}(t) | od \in OD, p \in P_{od}^{\mathrm{EV}}, \varrho \in \hat{\mathcal{T}}, t \in \mathcal{T}\right\} \\ &\cup\left\{\phi_{p,\varrho}^{\mathrm{GV}}(t) | od \in OD, p \in P_{od}^{\mathrm{GV}}, \varrho \in \hat{\mathcal{T}}, t \in \mathcal{T}\right\}\end{aligned} \tag{10.91}$$

Equations (10.20), (10.21), (10.99), and (10.100) respectively define the bounded rational dynamic user equilibrium conditions for electric and gasoline vehicle traffic.

$$\begin{aligned}&\forall od \in \mathrm{OD}, p \in P_{od}^{\mathrm{EV}}, \varrho \in \widehat{\mathcal{T}}, t \in \mathcal{T} \\ &\Rightarrow \begin{cases} h_{p,\varrho}^{\mathrm{EV}}(t) > 0 \Rightarrow \phi_{p,\varrho}^{\mathrm{EV}}(t) \in \left[\mu_{od,\varrho}^{\mathrm{EV}}, \mu_{od,\varrho}^{\mathrm{EV}} + \varepsilon_{od}^{\mathrm{EV}}\right] \\ \phi_{p,\varrho}^{\mathrm{EV}}(t) > \mu_{od,\varrho}^{\mathrm{EV}} + \varepsilon_{od}^{\mathrm{EV}} \Rightarrow h_{p,\varrho}^{\mathrm{EV}}(t) = 0 \end{cases}\end{aligned} \tag{10.92}$$

$$\begin{aligned}&\forall od \in \mathrm{OD}, p \in P_{od}^{\mathrm{GV}}, \varrho \in \hat{\mathcal{T}}, t \in \mathcal{T} \\ &\Rightarrow \begin{cases} h_{p,\varrho}^{\mathrm{GV}}(t) > 0 \Rightarrow \phi_{p,\varrho}^{\mathrm{GV}}(t) \in [\mu_{od,\varrho}^{\mathrm{GV}}, \mu_{od,\varrho}^{\mathrm{GV}} + \varepsilon_{od}^{\mathrm{GV}}] \\ \phi_{p,\varrho}^{\mathrm{GV}}(t) > \mu_{od,\varrho}^{\mathrm{GV}} + \varepsilon_{od}^{\mathrm{GV}} \Rightarrow h_{p,\varrho}^{\mathrm{GV}}(t) = 0 \end{cases}\end{aligned} \tag{10.93}$$

The variables $\mu_{od,\varrho}^{EV}$ and $\mu_{od,\varrho}^{GV}$ denote the minimum path travel cost for electric vehicle and gasoline vehicle users under OD pair (o, d) and departure time ϱ, respectively. Specifically:

$$\mu_{od,\varrho}^{EV} = \min_{p \in P_{od}^{EV}} \min_{t \in \mathcal{T}} \phi_{p,\varrho}^{EV}(t), \forall od \in OD, \varrho \in \widehat{\mathcal{T}} \tag{10.94}$$

$$\mu_{od,\varrho}^{\mathrm{GV}} = \min_{p \in P_{od}^{\mathrm{GV}}} \min_{t \in \mathcal{T}} \phi_{p,\varrho}^{\mathrm{GV}}(t), \forall od \in OD, \varrho \in \hat{\mathcal{T}} \tag{10.95}$$

Under the bounded rational dynamic user equilibrium (BR-DUE), for electric vehicle users belonging to demand $Q_{od,\varrho}^{EV}$, if the departure flow $h_{p,\varrho}^{EV}(t) > 0$, then the corresponding travel cost $\phi_{p,\varrho}^{EV}(t)$ must satisfy the bounded rationality condition within a tolerance ε. Conversely, if the travel cost $\phi_{p,\varrho}^{EV}(t)$ does not satisfy the condition, then $h_{p,\varrho}^{EV}(t) = 0$ for that path and time. For gasoline vehicle users, the same condition applies and is omitted here for brevity.

To further illustrate the differences between absolute rationality and bounded rationality-based dynamic user equilibrium models, this section presents a traffic distribution comparison under both frameworks, as shown in Fig. 10.11. In particular, users under absolute rationality only select the most optimal path (Path 1) and departure time. In contrast, users under bounded rationality with a tolerance level of $\varepsilon = 2.5\$$ are allowed to accept and choose suboptimal paths (e.g., Path 1 and Path 2) and departure times, as long as the resulting travel cost is within the ε tolerance threshold above the minimal cost.

Power Grid Allocation Model Based on Multi-Period Optimal Flow Distribution

Since the system adjustment process of the power grid is significantly slower than the propagation of traffic flow in the fast-charging station traffic network, this section adopts the multi-period optimal flow distribution model proposed in Sect. 10.3.2. (10.41) is revised as follows:

$$P_i^{ch}(t) = \begin{cases} n_{\kappa(i)}(t)P_{ch}, & v_i \in V^{FCS} \\ 0, & v_i \notin V^{FCS} \end{cases}, \forall t \in \mathcal{T} \tag{10.96}$$

In the above equation, the charging power flow $n_{\kappa(i)}(t)$ can be obtained via the dynamic network loading process.

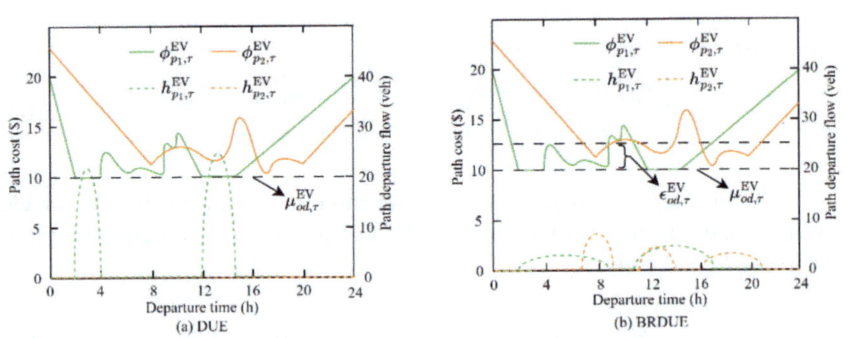

(a) Absolute Rational Dynamic User Equilibrium Model (b) Bounded Rational Dynamic User Equilibrium Model

Fig. 10.11 Traffic flow distribution comparison under absolute rational and bounded rational dynamic user equilibrium models

Since the time step of the dynamic traffic network model is smaller than that of the static traffic network model, the corresponding charging load resolution is higher than that in (10.41). Moreover, according to Theorem 2.1, the charging power flow $n_{\kappa(i)}(t)$ in the dynamic traffic network model is subject to strict lower and upper bounds and is constrained by the number of available charging piles $N_{\kappa(i)}$. Therefore, (10.103) ensures that the charging load does not exceed the maximum number of charging piles in the fast-charging station, i.e.

$$0 \leq P_i^{\text{ch}}(t) \leq N_{\kappa(i)} P_{\text{ch}}, \forall v_i \in V^{\text{FCS}}, t \in \mathcal{T} \tag{10.97}$$

In contrast, the static traffic network model in (10.41) does not impose such constraints on the number of charging piles for the charging load.

Dynamic Network Equilibrium of Power-Traffic Coupled Networks Considering Bounded User Rationality. In a dynamic transportation network, the marginal electricity price at fast-charging station nodes directly affects the dynamic distribution of electric vehicle (EV) and internal combustion engine vehicle (ICEV) traffic flows. Compared with multi-period static traffic networks, the traffic flow distribution in dynamic models is more intricate, as it not only reflects the propagation dynamics of traffic flow but also incorporates user behavior characteristics such as flexible departure time selection. This leads to a more diverse distribution of traffic flows and better captures the realistic operational states of transportation systems.

On the other hand, the distribution of EV traffic flows implicitly reflects users' selection of charging stations, i.e., the spatiotemporal distribution of charging demand. This distribution alters the active load at the power nodes connected to fast-charging stations, thereby influencing the active power balance constraints in the power system. As a result, the corresponding dual variables (i.e., nodal marginal electricity prices) will change, ultimately increasing the charging costs at fast-charging stations.

Furthermore, similar to the analysis in Sect. 10.3.3, interactions also exist between EV and ICEV traffic flows within the dynamic model. Therefore, the entire traffic flow must be treated as an integrated whole and modeled jointly as part of the coupled system. Under this framework, the mutual influence and feedback between traffic flow and power flow lead to a state in which neither transportation nor power networks induces further changes in each other. This steady condition defines the dynamic network equilibrium of the power-traffic coupled system, as illustrated in Fig. 10.12.

Unlike static traffic flow models, the distribution of dynamic traffic flows on each path is no longer uniform. Therefore, the variable used to characterize users' route choices is no longer the path traffic flow, but rather the path departure flow. The path departure flow $h_{p,\rho}^r(t)$ describes the dynamic departure behavior of traffic over time and is a crucial variable for modeling the temporal evolution of the network.

For a given path departure flow, the entire traffic flow evolution in the network, as well as its spatiotemporal distribution, can be simulated using the dynamic traffic network model. This, in turn, allows for the estimation of the dynamic changes in charging load, forming the basis for analyzing the dynamic network equilibrium of the power-traffic coupled network.

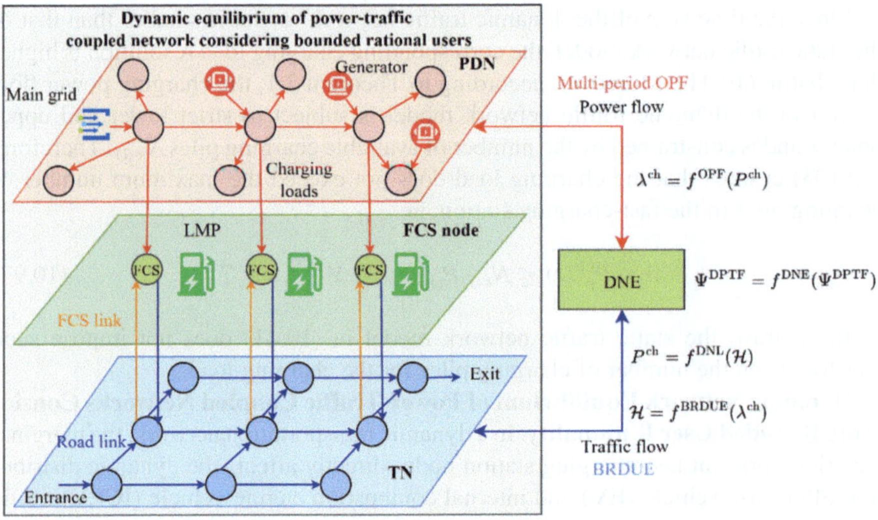

Fig. 10.12 Dynamic equilibrium of power-traffic coupled network considering bounded rational users

The dynamic network equilibrium can be expressed as:

$$\lambda^{ch} = f^{OPF}(P^{ch}) \tag{10.98}$$

$$\mathcal{H} = f^{BRDUE}(\lambda^{ch}) \tag{10.99}$$

$$P^{ch} = f^{DNL}(\mathcal{H}) \tag{10.100}$$

where $f^{OPF}(\cdot)$ represents the response function of the optimal power flow model that maps the charging load P^{ch} of fast-charging stations to the nodal marginal electricity price λ^{ch};$f^{BRDUE}(\cdot)$ represents the response function of the bounded rational dynamic user equilibrium model that maps λ^{ch} to the path departure flow set $\mathcal{H}$ under a given marginal charging price;$f^{DNL}(\cdot)$ represents the dynamic network loading function that maps the path departure flows $\mathcal{H}$ to the charging load P^{ch} of fast-charging stations;P^{ch} denotes the spatiotemporal distribution of charging load at fast-charging stations.

Likewise, the dynamic power-traffic flow in the coupled network can be defined as $\psi^{DPTF} = [\lambda^{ch}, \mathcal{H}, P^{ch}]$. The equilibrium condition of the power-traffic coupled dynamic network can be formulated as the fixed-point solution to the following problem:

$$\psi^{DPTF} = f^{DNE}(\psi^{DPTF}) \tag{10.101}$$

where $f^{\text{DNE}}(\cdot)$ denotes the self-mapping function of the dynamic power-traffic flow under network equilibrium.

Solution Method. The above fixed-point problem can be solved using a two-level fixed-point iterative algorithm to obtain the equilibrium values of nodal marginal electricity price, path departure flow, and spatial–temporal distribution of charging load, as follows:

Algorithm 2: Two-level fixed-point iterative algorithm	
Step 1	Set the maximum number of iterations K, and the tolerance thresholds $\overline{\delta}^{\lambda}, \overline{\delta}^{\mathcal{H}}, \overline{\delta}^{P}$ Initialize iteration index $k = 0$, initial marginal electricity price $\lambda^{\text{ch},0}$, initial path departure flow $\mathcal{H}^0$, and initial charging load $P^{\text{ch},0}$
Step 2	According to the marginal electricity price in the k-th iteration, solve the multi-period optimal power flow model to update the marginal price in the $(k+1)$-th iteration: $\lambda^{\text{ch},k+1} = f^{\text{OPF}}(P^{\text{ch},k})$
Step 3	According to the marginal price $\lambda^{\text{ch},k}$, solve the boundedly rational dynamic user equilibrium model to obtain the updated path departure flow: $\mathcal{H}^{k+1} = f^{\text{BRDUE}}(\lambda^{\text{ch},k})$
Step 4	According to the path departure flow $\mathcal{H}^{k+1}$, execute dynamic network loading to update the charging load: $P^{\text{ch},k+1} = f^{\text{DNL}}(\mathcal{H}^k)$
Step 5	Update the convergence errors: $\delta_{k+1}^{\lambda} = \frac{\Vert\lambda^{\text{ch},k+1} - \lambda^{\text{ch},k}\Vert_2}{\Vert\lambda^{\text{ch},k}\Vert_2}, \quad \delta_{k+1}^{\mathcal{H}} = \frac{\Vert\mathcal{H}^{k+1} - \mathcal{H}^k\Vert_2}{\Vert\mathcal{H}^k\Vert_2}, \quad \delta_{k+1}^{P} = \frac{\Vert P^{\text{ch},k+1} - P^{\text{ch},k}\Vert_2}{\Vert P^{\text{ch},k}\Vert_2}$ 。 If $\delta_{k+1}^{\lambda} \le \overline{\delta}^{\lambda}, \delta_{k+1}^{\mathcal{H}} \le \overline{\delta}^{\mathcal{H}}, \delta_{k+1}^{P} \le \overline{\delta}^{P}$ are all satisfied or $k \ge K$, output the fixed-point solution $\psi^{\text{DPTF}} = \{\lambda^{\text{ch},k+1}, \mathcal{H}^{k+1}, P^{\text{ch},k+1}\}$; Otherwise, let \$k = k + 1\$ and return to Step 2

Step 3: In the bounded rational dynamic user equilibrium model, the decision variables are significantly more complex than those in the multi-period static user equilibrium model. The problem involves numerous nonlinear constraints, such as the nested multilevel functional constraints in (10.61)–(2.63), the piecewise constraints in (10.69), and the nonlinear complementary constraints in (10.99) and (10.100), which together constitute an NP-hard problem. Linearization of these constraints would introduce a large number of 0–1 variables, severely reducing the efficiency of mathematical optimization methods and making it infeasible to obtain a solution within a reasonable amount of time. Therefore, for bounded rational dynamic user equilibrium problems, the fixed-point iterative method is typically used. The bounded rational dynamic user equilibrium conditions defined in (10.99) and (10.100) can be equivalently formulated as the following fixed-point projection problem:

$$\mathcal{H} = P_{\Pi}\left[\mathcal{H} - \alpha\tilde{\Phi}(\mathcal{H}, \lambda^{\text{ch}})\right] \tag{10.102}$$

Here, Π denotes the feasible set of path departure flows that satisfy the demand constraints defined in (10.50) and (10.51); α is a given positive constant; $P_{\Pi}(x)$

denotes the projection of vector x onto the space Π; and $\tilde{\Phi}(\mathcal{H}, \lambda^{ch})$ is the modified path travel cost under bounded rationality, satisfying:

$$\tilde{\Phi}(\mathcal{H}, \gamma^{ch}) = \left\{\tilde{\phi}_{p,\varrho}^{EV}(t)|od \in OD, p \in P_{od}^{EV}, \varrho \in \hat{\mathcal{T}}, t \in \mathcal{T}\right\} \cup \left\{\tilde{\phi}_{p,\varrho}^{GV}(t)|od \in OD, p \in P_{od}^{GV}, \varrho \in \hat{\mathcal{T}}, t \in \mathcal{T}\right\} \quad (10.103)$$

$$\tilde{\phi}_{p,\varrho}^{EV}(t) = \max\left[\phi_{p,\varrho}^{EV}(t), \mu_{od,\varrho}^{EV} + \varepsilon_{od,\varrho}^{EV}\right], \forall od \in OD, p \in P_{od}^{EV}, \varrho \in \hat{\mathcal{T}}, t \in \mathcal{T} \quad (10.104)$$

$$\tilde{\phi}_{p,\varrho}^{GV}(t) = \max\left[\phi_{p,\varrho}^{GV}(t), \mu_{od,\varrho}^{GV} + \varepsilon_{od,\varrho}^{GV}\right], \forall od \in OD, p \in P_{od}^{GV}, \varrho \in \hat{\mathcal{T}}, t \in \mathcal{T} \quad (10.105)$$

The modified travel cost $\tilde{\Phi}$ is computed using the given $\mathcal{H}$ as the input of the dynamic network loading, where the travel cost Φ is also associated with the charging price λ^{ch}. Therefore, $\tilde{\Phi}$ can be regarded as a function of $\mathcal{H}$ and λ^{ch}. (10.109) implies that when the projection of the vector $\mathcal{H} - \alpha\tilde{\Phi}(\mathcal{H}, \lambda^{ch})$ onto the space Π is equal to $\mathcal{H}$, the corresponding $\mathcal{H}$ is the boundedly rational dynamic user equilibrium solution.

Equation (10.109) can be further expressed as the following fixed-point problem:

$$\mathcal{H}^* = \arg\min_{\mathcal{H}\in\Pi} \frac{1}{2}\left\|\mathcal{H}^* - \alpha\tilde{\Phi}(\mathcal{H}, \lambda^{ch}) - \mathcal{H}\right\|_2 \quad (10.106)$$

In the above, $\mathcal{H}^*$ denotes the solution under the boundedly rational dynamic user equilibrium.Equation (10.113) indicates that the solution to the fixed-point problem is also a parameter of the original minimization problem. Thus, the boundedly rational dynamic user equilibrium problem itself is a fixed-point problem and can be considered as part of the fixed-point problem of the dynamic network equilibrium. Equation (10.113) can be solved iteratively using the projection method, as follows:

$$\mathcal{H}^{z+1} = \arg\min_{\mathcal{H}\in\Pi} \frac{1}{2}\left\|\mathcal{H}^z - \alpha\tilde{\Phi}(\mathcal{H}^z, \lambda^{ch,k}) - \mathcal{H}\right\|_2 \quad (10.107)$$

Here, the superscript z represents the z-th iteration in solving the BRDUE (inner fixed-point problem), and the superscript k represents the k-th iteration in solving the dynamic power–traffic coupled network equilibrium (outer fixed-point problem). The update rule for the path departure flow in (10.114) is as follows:

$$h_{p,\varrho}^{EV,z+1}(t) = \left[h_{p,\varrho}^{EV,z}(t) - \alpha\tilde{\phi}_{p,\varrho}^{EV,z}(t) + v_{od,\varrho}^{EV,z}\right]_+, \forall od \in OD, p \in P_{od}^{EV}, \varrho \in \hat{\mathcal{T}}, t \in \mathcal{T} \quad (10.108)$$

$$h_{p,\varrho}^{GV,z+1}(t) = \left[h_{p,\varrho}^{GV,z}(t) - \alpha\tilde{\phi}_{p,\varrho}^{GV,z}(t) + v_{od,\varrho}^{GV,z}\right]_+, \forall od \in OD, p \in P_{od}^{GV}, \varrho \in \hat{\mathcal{T}}, t \in \mathcal{T} \quad (10.109)$$

In the formula, $[x]+ = \max(0, x)$; $v^{EV,z}od, \varrho$ and $v_{od,\varrho}^{GV,z}$ are dual variables, satisfying:

$$\sum_{p \in P_{od}^{\mathrm{EV}}} \sum_{t \in T} \left[h_{p,\varrho}^{\mathrm{EV},z}(t) - \alpha \tilde{\phi}_{p,\varrho}^{\mathrm{EV},z}(t) + v_{od,\varrho}^{\mathrm{EV},z} \right]_{+} = Q_{od,\varrho}^{\mathrm{EV}}, \forall od \in OD, \varrho \in \hat{\mathcal{T}} \quad (10.110)$$

$$\sum_{p \in P_{od}^{\mathrm{EV}}} \sum_{t \in T} \left[h_{p,\varrho}^{\mathrm{GV},z}(t) - \alpha \tilde{\phi}_{p,\varrho}^{\mathrm{GV},z}(t) + v_{od,\varrho}^{\mathrm{GV},z} \right]_{+} = Q_{od,\varrho}^{\mathrm{GV}}, \forall od \in OD, \varrho \in \hat{\mathcal{T}} \quad (10.111)$$

When the following convergence criterion is met or $z \geq Z$, the computed solution $\mathcal{H}^{k+1} = \mathcal{H}^{z+1}$ is used as the path departure flow for the $(k+1)$-th iteration of the outer fixed-point problem:

$$\frac{\left\| \mathcal{H}^{z+1} - \mathcal{H}^{z} \right\|_2}{\left\| \mathcal{H}^{z} \right\|_2} \leq \overline{\sigma}_{\mathcal{H}} \quad (10.112)$$

In the formula, $\overline{\sigma}_{\mathcal{H}}$ denotes the convergence threshold for the inner fixed-point problem, and Z is the maximum number of iterations.

10.5 Case Study Analysis

Simulation Setup. The topology of the power-traffic coupled network test system 1 studied in this section is shown in Fig. 10.13. The power network is a 9-node distribution network system, where node 1 is connected to the upper-level main grid, and generators are deployed at nodes 3 and 5. The traffic network is a 17-node system based on the Nguyen transportation network, including four fast-charging station nodes (green nodes), with charging loads supplied by power nodes 9, 7, 3, and 8, respectively. Traffic nodes 1 and 4 serve as origin nodes (blue nodes), and traffic nodes 2 and 3 serve as destination nodes (also blue nodes). The remaining nodes are intermediate nodes (yellow nodes).

There are a total of 26 road segments in the transportation network, among which segments 20, 22, 24, and 26 are fast-charging road segments, and the other 22 are conventional segments. The free-flow travel time for segments 9 and 17 is 0.15 h, for segments 1, 4–8, 14–16, and 19 is 0.1 h, and for all other segments is 0.05 h. Travel demands and travel paths are listed in Tables 10.1 and 10.2, respectively. The main simulation parameters for the transportation network, power network, and solution algorithms are listed in Tables 10.3, 10.4, and 10.5, respectively.

In the multi-period static modeling method, the simulation time step of the transportation network is 1 h; in the dynamic modeling method, the simulation time step is 0.05 h. For the power network, the time step of the multi-period optimal power flow model is uniformly 0.25 h. In simulations, discrepancies in the time resolution

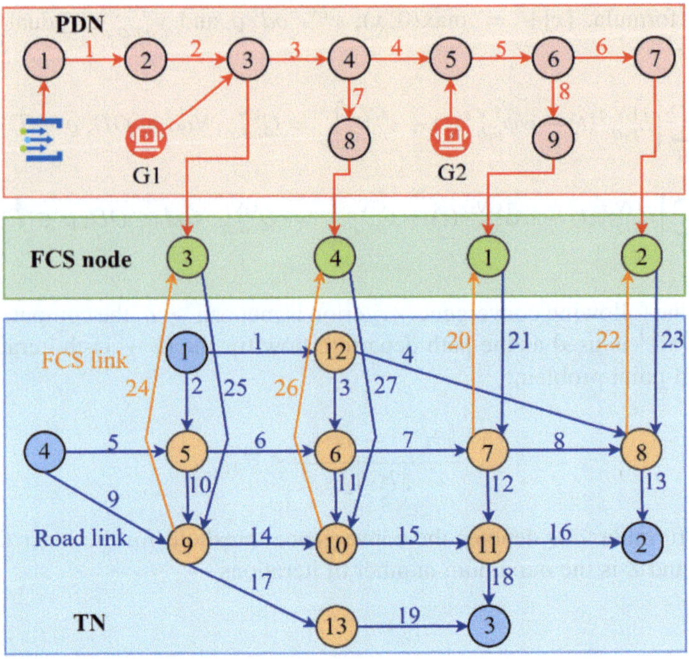

Fig. 10.13 Coupled power-traffic network test system 1

Table 10.1 Travel demand settings

o	d	Q_{od}^{EV}	Q_{od}^{GV}
1	2	750 veh	5250 veh
1	3	750 veh	5250 veh
4	2	750 veh	5250 veh
4	3	750 veh	5250 veh

between traffic flow (charging load) and power flow (nodal marginal electricity price) are aligned through upsampling/downsampling techniques.

The daily distribution curves of travel demand and conventional loads are shown in Fig. 10.14. The conventional loads in the power distribution network, excluding charging loads, are determined by the load distribution curve. The travel demands in Table 10.1 represent daily totals. For the multi-period bounded rational static user equilibrium model, travel demands are allocated into in total of 24 steps (i.e. 1 h/step) according to the daily distribution curve. For the bounded rational dynamic user equilibrium model, daily travel demand is divided into 96 portions at 0.25 h intervals, and each portion's time point serves as the expected arrival time of the corresponding traffic flow.

In the multi-period bounded rational static user equilibrium model, both the number of travel demand periods and the number of time steps are 24. The total

Table 10.2 Travel path settings

o	d	EV			GV	
		#	P_{od}^{EV}	$\upsilon(p)$	#	P_{od}^{GV}
1	2	1	1-4-22-23-13	2	25	1-4-13-0-0
1	2	2	1-3-7-8-22-23-13	2	26	1-3-7-8-13
1	2	3	1-3-7-20-2-1-12-16	1	27	1-3-7-12-16
1	2	4	1-3-11-26-27-15-16	4	28	1-3-11-15-16
1	2	5	2-6-7-8-22-23-13	2	29	2-6-7-8-13
1	2	6	2-6-7-20-21-12-16	1	30	2-6-7-12-16
1	2	7	2-6-11-26-27-15-16	4	31	2-6-11-15-16
1	2	8	2-10-24-25-14-15-16	3	32	2-10-14-15-16
1	3	9	1-3-7-20-21-12-18	1	33	1-3-7-12-18
1	3	10	1-3-11-26-27-15-18	4	34	1-3-11-15-18
1	3	11	2-6-7-20-21-12-18	1	35	2-6-7-12-18
1	3	12	2-6-11-26-27-15-18	4	36	2-6-11-15-18
1	3	13	2-10-24-25-14-15-18	3	37	2-10-14-15-18
1	3	14	2-10-24-25-17-19	3	38	2-10-17-19
4	2	15	5-6-7-8-22-23-13	2	39	5-6-7-12-18
4	2	16	5-6-7-20-21-12-16	1	40	5-6-11-15-18
4	2	17	5-6-11-26-27-15-16	4	41	5-10-14-15-18
4	2	18	5-10-24-25-14-15-16	3	42	9-14-15-18
4	2	19	9-24-25-14-15-16	3	43	9-17-19
4	3	20	5-6-7-20-21-12-18	1	44	5-6-7-12-18
4	3	21	5-6-11-26-27-15-18	4	45	5-6-11-15-18
4	3	22	5-10-14-26-27-15-18	4	46	5-10-14-15-18
4	3	23	9-24-25-14-15-18	3	47	9-14-15-18
4	3	24	9-24-25-17-19	3	48	9-17-19

Table 10.3 Transportation network parameter settings

Parameter	Value	Parameter	Value	Parameter	Value
C_l	720 veh/h	N_l	15 个	τ_l^{q}	0.2 h
τ^{ch}	0.2 h	$\gamma^{\mathrm{EV,t}}/\gamma^{\mathrm{GV,t}}$	5/10 \$/h	P_{ch}	50 kW
$\varepsilon_{od}^{\mathrm{EV}}/\varepsilon_{od}^{\mathrm{GV}}$	1/2 \$	$\gamma^{\mathrm{EV,e}}/\gamma^{\mathrm{GV,e}}$	4/8 \$/h^2	$\gamma^{\mathrm{EV,l}}/\gamma^{\mathrm{GV,l}}$	6/12 \$/h^2
Δt	1/0.05 h	$\Delta\tau$	0.1 h	T	24 h

Table 10.4 Power network parameter settings

Parameter	Value	Parameter	Value	Parameter	Value
a_i	30 $/MW²h	b_i	400 $/MWh	b_0	600 $/MWh
$\overline{P}_{ij}^{\mathrm{L}}/\overline{Q}_{ij}^{\mathrm{L}}$	3 MW	$\overline{P}_i^{\mathrm{G}}/\overline{Q}_i^{\mathrm{G}}$	3 MW	$\overline{P}_0^{\mathrm{G}}$	10 MW
$\underline{U}/\overline{U}$	0.95/1.05 p.u	$\overline{I}_{ij}$	4 p.u	Δt	0.25 h

Table 10.5 Solver algorithm parameter settings

Parameter	Value	Parameter	Value	Parameter	Value
$\overline{\delta}^{\lambda}$	10^{-3}	$\overline{\delta}^{\mathcal{H}}$	10^{-3}	$\overline{\delta}^{P}$	10^{-3}
$\overline{\delta}^{f}$	10^{-3}	$\overline{\sigma}_{\mathcal{H}}$	10^{-3}	α	5
K	30	Z	50		

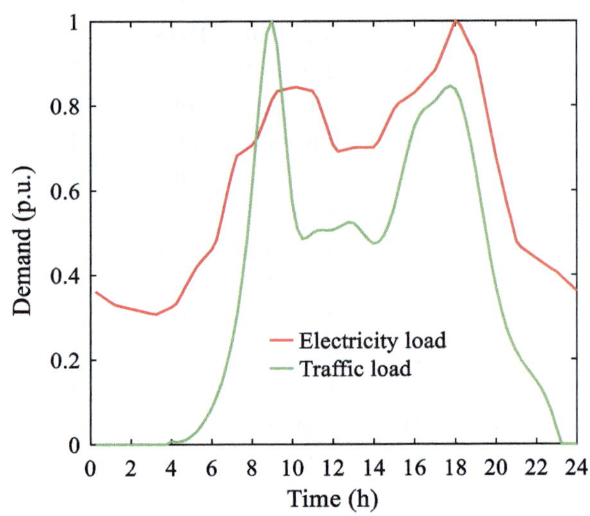

Fig. 10.14 Daily distribution curves of conventional power demand and travel demand

traffic flow within a time slice equals the sum of travel demands for that time step, as traffic flow activities are confined within a single time step. In contrast, for the bounded rational dynamic user equilibrium model, the number of demand periods is 96, which is smaller than the 480 time steps in the model. This is because traffic flow in the dynamic model propagates across multiple time steps, leading to non-fixed traffic volumes in each time slice.

The transportation and power distribution network models in this section are built using MATLAB ver. 2025. The solution to the multi-period bounded rational static user equilibrium model and the multi-period optimal power flow model is based on Gurobi. The built-in functions of Gurobi are used to directly obtain the dual variables of active power balance constraints as nodal marginal electricity prices. The bounded rational dynamic user equilibrium is solved using the iterative algorithm proposed in Sect. 10.4.4.

Analysis of the Multi-Period Static Modeling Approach.

(1) Convergence Analysis.

In the multi-period static model, the fixed-point iteration algorithm (Algorithm 1) is employed for solving the problem. The algorithm satisfies the convergence condition after three iterations, with a total computation time of 43.1 s. The convergence process is shown in Fig. 10.15. Specifically, the nodal marginal price and traffic flow satisfy the convergence criteria after the second and third iterations, respectively, indicating good convergence performance of the iterative coordination mechanism between the power and traffic systems.

(2) Traffic Network Analysis.

The overall spatiotemporal distribution of traffic flow on each road segment is shown in Fig. 10.16, while the charging traffic flow on fast charging station segments is depicted in Fig. 10.17.

In the temporal dimension, the traffic flow distribution aligns closely with the daily travel demand distribution curve in Fig. 10.14. It is primarily concentrated during daytime hours and exhibits a typical "morning-evening peak" bimodal pattern, with 8:00–10:00 as the morning peak and 16:00–19:00 as the evening peak. The charging load of each fast charging station exhibits the same distribution characteristics. Since the BPR road model and M/M/c/K queuing model only describe the increase in travel time with increasing traffic flow without imposing upper bounds on per-period traffic flow, congestion exceeding road capacity occurs during peak hours. For example, during 8:00–9:00, Road 1 carries 1000 vehicles, exceeding its 720-vehicle capacity

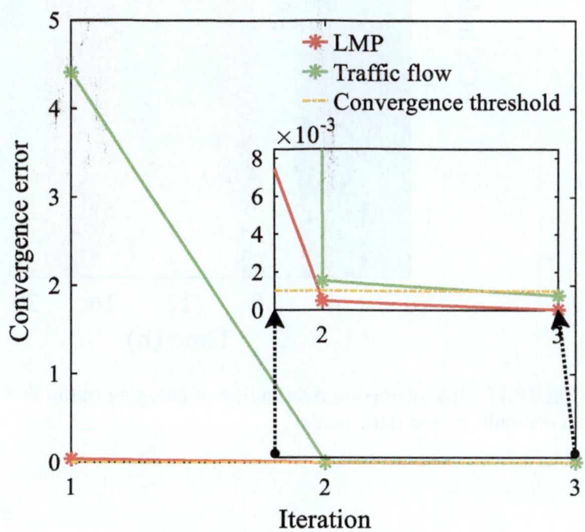

Fig. 10.15 Convergence process of the fixed-point iteration algorithm (Algorithm 1) in the multi-period static model

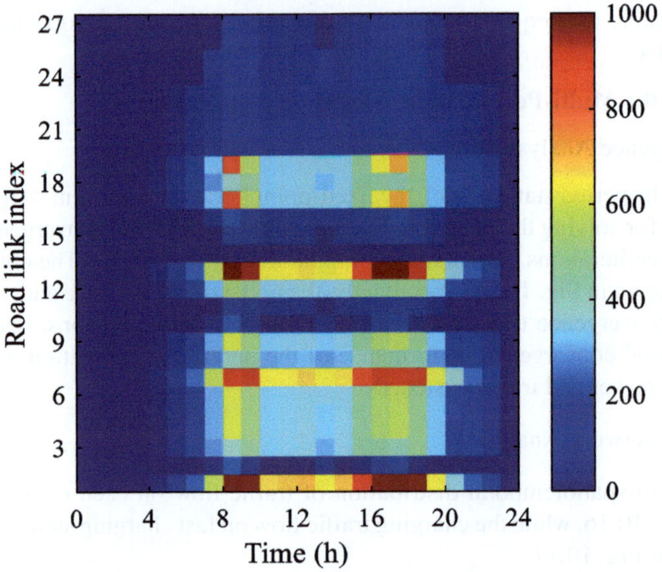

Fig. 10.16 Spatiotemporal distribution of traffic flow on each road segment in the multi-period static model

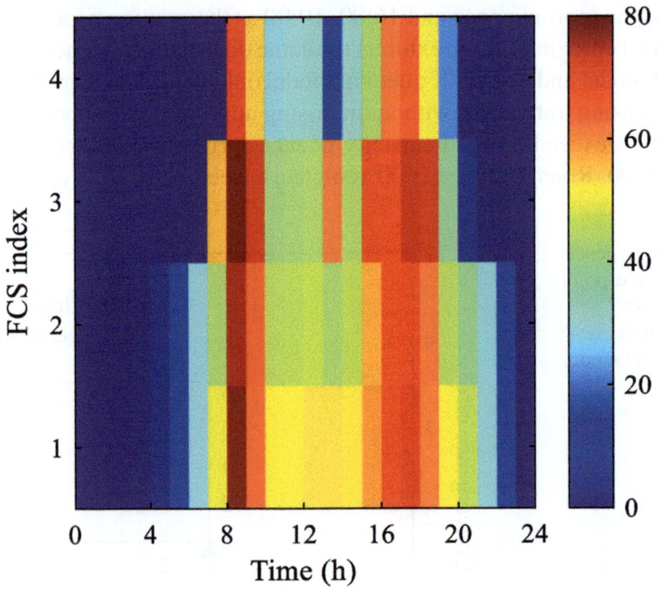

Fig. 10.17 Spatiotemporal distribution of charging traffic flow on fast charging station segments in the multi-period static model

limit. Similarly, the number of electric vehicles (EVs) in Fast Charging Station 3 reaches 80, surpassing its service capacity of 75 vehicles (15 chargers, with each charger serving up to 5 vehicles per hour).

In the spatial dimension, the segments with the highest traffic volumes are Roads 1, 7, 12, 13, 17, 18, and 19. However, due to the limitations of the static model, it is difficult to reflect real-time traffic congestion and queuing dynamics. Additionally, since conventional road segments generally have higher capacities than those of fast charging station segments and the EV penetration rate is relatively low, the traffic flow on conventional segments is significantly greater than on the charging station segments. It is worth noting that due to the time-period flow conservation constraint, the traffic volume entering and exiting each charging station segment (e.g., Roads 20/21, 22/23, 24/25, 26/27) remains equal, making it difficult for the model to capture dwell times and queuing effects within charging stations caused by charging delays.

The travel path selections of electric vehicles and gasoline vehicles (i.e., path-based traffic flows) are shown in Figs. 10.18 and 10.19, respectively.

Taking Paths 20–24 and 44–48 as examples, their only difference lies in whether they pass through a fast charging station; the remaining road segments are identical. Since fast charging stations are distributed across multiple paths, the EV path distribution is more dispersed to avoid excessive queuing at any single station due to concentrated charging demand. It is shown that the EV traffic is distributed across Paths 1, 9, 10, 14, 15, 16, 17, 20, and 24. In contrast, gasoline vehicles do not need to access charging stations, resulting in more concentrated path choices, primarily along Paths 25, 33, 39, and 48.

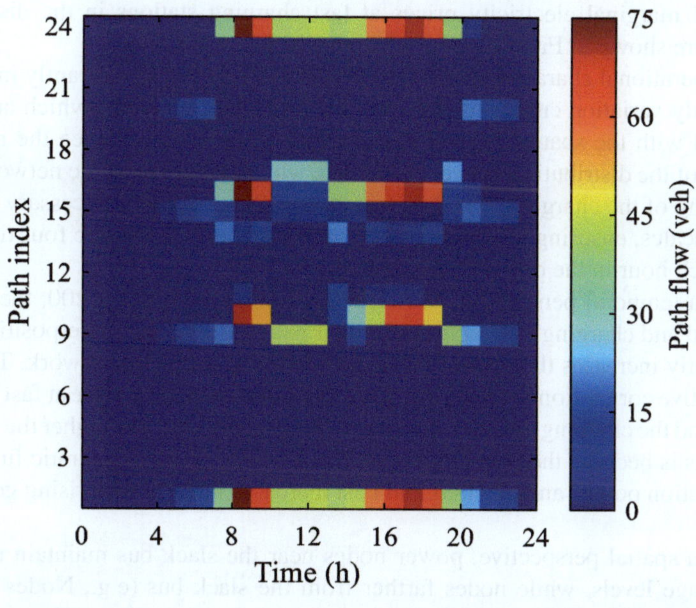

Fig. 10.18 Spatiotemporal distribution of EV path traffic flow in the multi-period static model

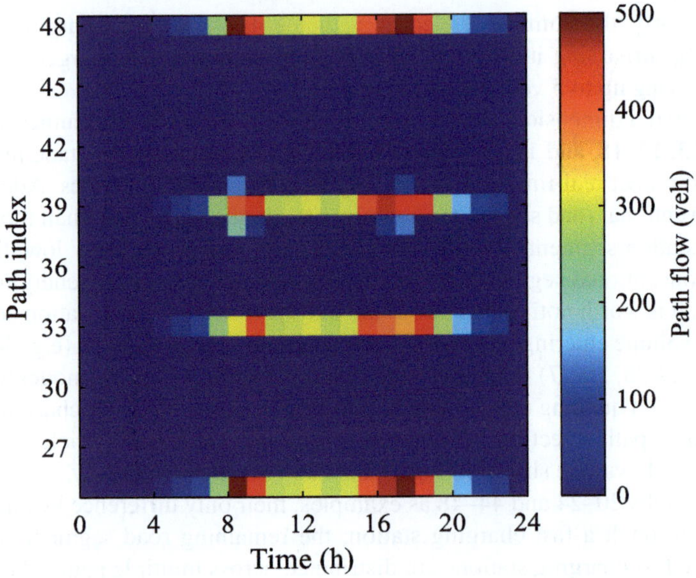

Fig. 10.19 Spatiotemporal distribution of gasoline vehicle path traffic flow in the multi-period static model

(3) Power Network Analysis.

The spatiotemporal distributions of node voltage magnitudes, branch active power, and nodal marginal electricity prices at fast charging stations in the distribution network are shown in Figs. 10.20, 10.21, and 10.22, respectively.

The operational characteristics of the power network are significantly influenced by the daily variation curves of conventional and charging loads, which are highly correlated with the spatiotemporal distribution of traffic flow. Since the modeling time step of the distribution network is 0.25 h, while that of the traffic network is 1 h, upsampling of the charging load data from the traffic network is necessary to match the time scales, ensuring a consistent charging load level across the four time steps within each hour in the distribution network.

From a temporal perspective, during 8:00–10:00 and 16:00–19:00, the conventional load and charging load both reach their peak values. This superposition effect significantly increases the operational stress on the distribution network. There is a clear positive correlation between the nodal marginal electricity price at fast charging stations and the charging load, i.e. the larger the charging load, the higher the marginal price. This is because the generation cost function is a convex quadratic function of the generation power, and the marginal cost increases linearly with rising generation power.

From a spatial perspective, power nodes near the slack bus maintain relatively high voltage levels, while nodes farther from the slack bus (e.g., Nodes 7 and 9, where fast charging stations are located) exhibit noticeable voltage drops. In terms

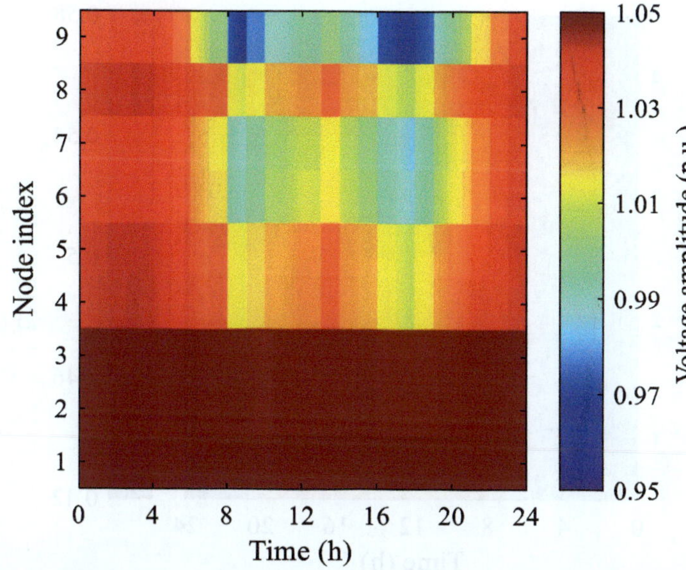

Fig. 10.20 Spatiotemporal distribution of voltage magnitudes at power nodes in the multi-period static model

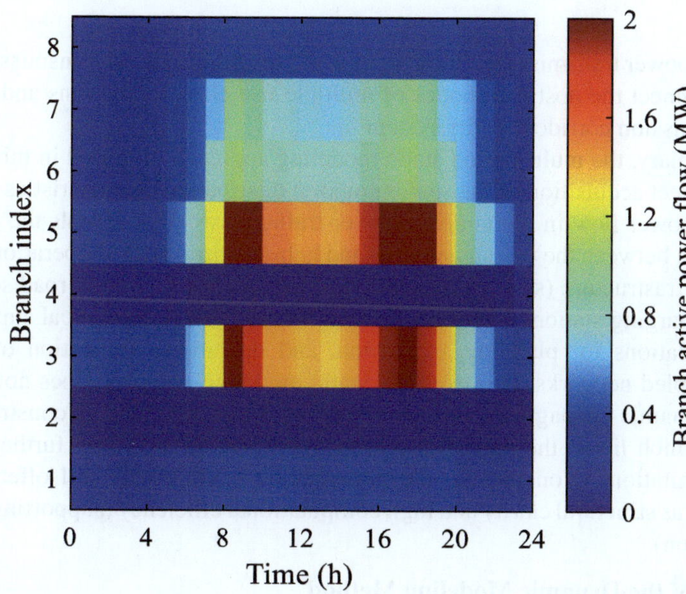

Fig. 10.21 Spatiotemporal distribution of branch active power in the multi-period static model

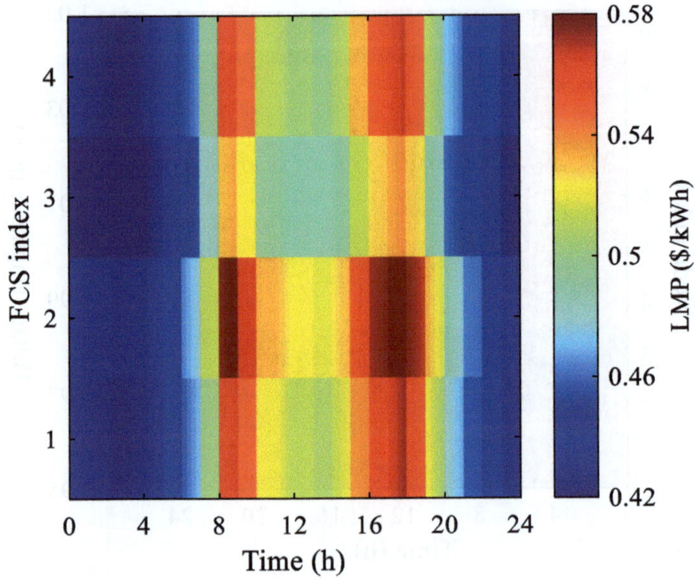

Fig. 10.22 Spatiotemporal distribution of marginal electricity prices at fast charging station nodes in the multi-period static model

of branch power transmission, Branches 3 and 5 bear the heaviest transmission load, as they connect the upstream nodes of multiple fast charging stations and serve as key transmission corridors in the system.

In summary, the multi-period static modeling approach proposed in this chapter enables direct acquisition of the spatiotemporal distribution characteristics of traffic flow and power flow in the coupled power-traffic network. It reveals the coupling mechanism between the two subsystems and helps identify critical operation periods and key infrastructure (such as power nodes, feeders, traffic nodes, road segments, and fast charging stations). This model provides important theoretical support and data foundations for planning, expansion, and operational regulation of power-traffic coupled networks. However, the static modeling approach does not account for the dynamic propagation characteristics of traffic flow and is constrained by (10.19), which limits the ability to reduce the modeling time step further, posing certain limitations. Nonetheless, the multi-period static model still offers advantages such as structural clarity and high computational efficiency (supporting parallel computation).

Analysis of the Dynamic Modeling Method.

(1) Convergence Analysis.

In the dynamic model, the fixed-point iteration algorithm (Algorithm 2) reaches convergence after the 14th iteration, with a total time cost of 306.4 s, which is longer than that of the multi-period static model. The convergence process is illustrated

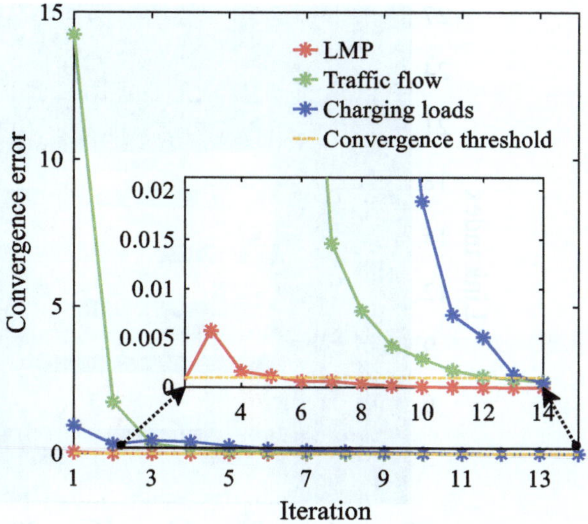

Fig. 10.23 Convergence process of the fixed-point iteration algorithm (Algorithm 2) in the dynamic model

in Fig. 10.23, where the node marginal electricity prices, path departure flows, and charging loads satisfy the convergence criteria after the 6th, 13th, and 14th iterations, respectively.

(2) Traffic Network Analysis.

The spatiotemporal distributions of the inflow traffic Q_l^{in}, queuing traffic S_l, and outflow traffic Q_l^{out} on each road segment are shown in Figs. 10.24, 10.25, and 10.26, respectively. The distributions of the queuing and charging traffic on fast-charging road segments are depicted in Figs. 10.27 and 10.28, respectively.

Temporally, the traffic flows exhibit a similar diurnal pattern to that in Fig. 10.14, mainly concentrated between 07:00 and 20:00, with morning and evening peaks observed from 07:00–09:00 and 16:00–18:00, respectively. Compared to the multi-period static model, the traffic peaks in the dynamic model occur slightly earlier, as users tend to depart earlier to ensure arrival within their desired time windows. Additionally, the dynamic model adopts a finer time resolution (0.05 h vs. 1 h in the static model), enabling a more precise depiction of traffic evolution.

Electric vehicle (EV) users usually choose to charge between 08:00–09:00 and 15:00–17:00 to ensure sufficient energy for the remaining journey. During these periods, fast-charging stations typically operate at full capacity and the maximum number of queuing vehicles reaches 12. The outflow on conventional road segments and charging flows on fast-charging segments are both strictly limited by road capacities and the number of available chargers, verifying the effectiveness of Theorem 2.2.

Chronologically, the inflow, queuing, and outflow traffic flows on each segment follow a sequential propagation pattern, clearly demonstrating the dynamic propagation characteristics of traffic flows and further validating the effectiveness of the improved point queue model proposed in this work.

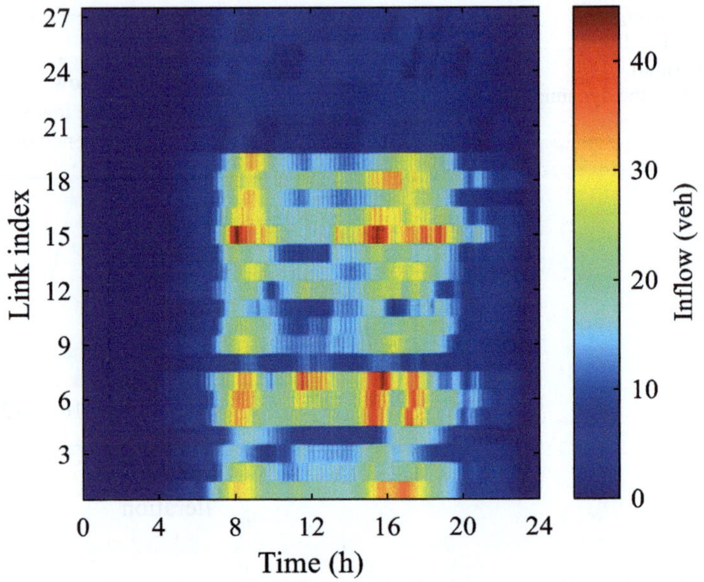

Fig. 10.24 Spatiotemporal distribution of inflow traffic on each road segment in the dynamic model

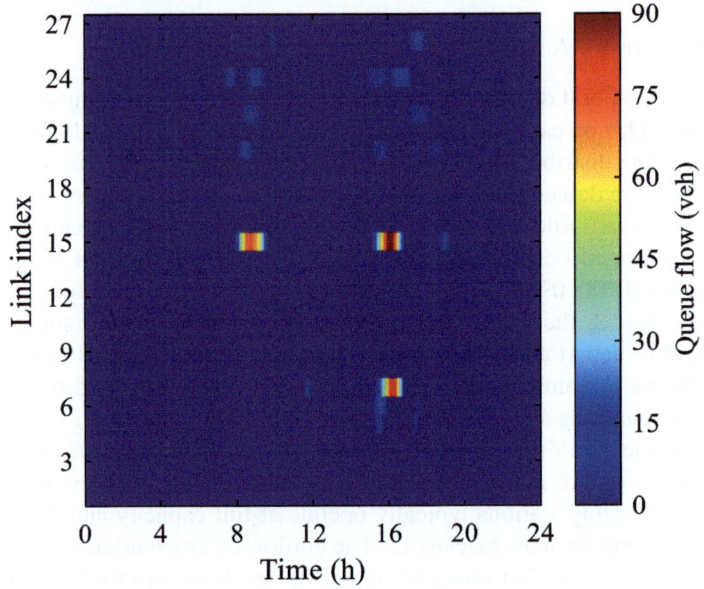

Fig. 10.25 Spatiotemporal distribution of queuing traffic on each road segment in the dynamic model

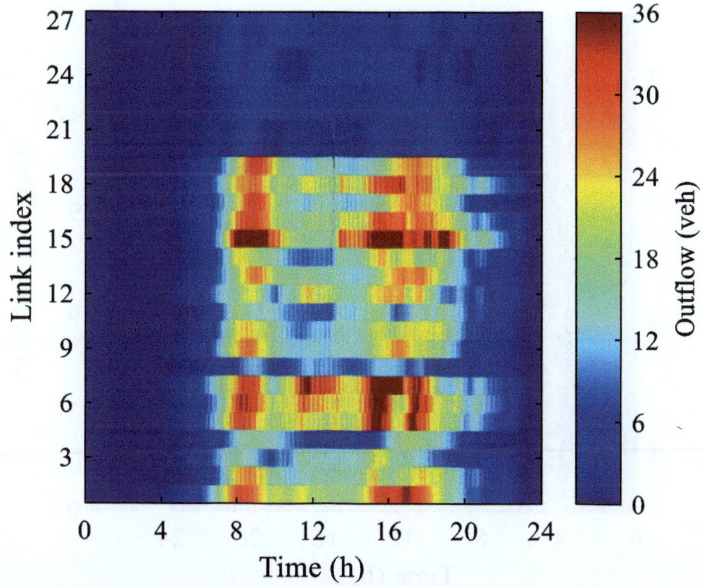

Fig. 10.26 Spatiotemporal distribution of outflow traffic on each road segment in the dynmic model

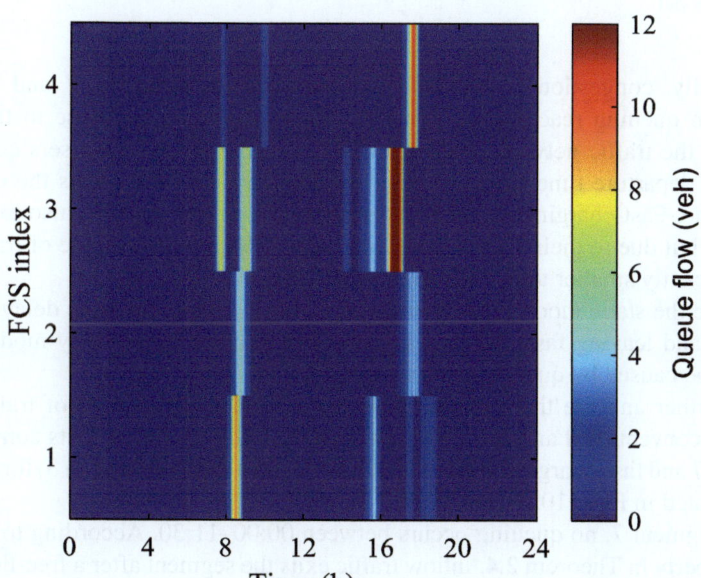

Fig. 10.27 Spatiotemporal distribution of queuing traffic on fast-charging road segments in the dynamic model

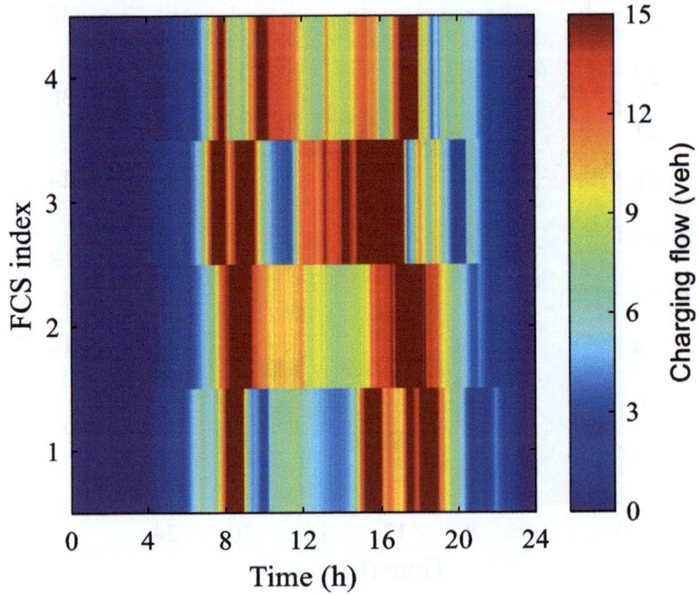

Fig. 10.28 Spatiotemporal distribution of charging traffic on fast-charging road segments in the dynamic model

Spatially, congestion is primarily concentrated on segments 7 and 15, with maximum queuing reaching 90 vehicles. These segments are close to the origin nodes of the traffic network, where traffic tends to aggregate as users commonly choose a departure time based on their expected arrival time minus the estimated travel time. Fast-charging segments (segments 20, 22, 24, and 26) also experience queuing, but due to their access being limited to EVs, the magnitude of traffic flow is significantly smaller than on conventional segments.

Unlike the static model, the dynamic model can capture the time delay between entering and leaving fast-charging segments, thus more realistically modeling the dwell time caused by queuing and charging.

To further analyze the dynamic characteristics and differences of traffic flows between conventional and fast-charging segments, this section selects conventional segment 7 and fast-charging segment 24 (hosting fast-charging station 3) for analysis, as illustrated in Figs. 10.29 and 10.30.

For segment 7, no queuing occurs between 00:00–11:30. According to the free-flow property in Theorem 2.4, inflow traffic exits the segment after a free-flow travel time, resulting in parallel inflow and outflow curves with a time lag equal to 0.1 h. As inflow increases, the outflow also rises until it reaches the segment's capacity (36 vehicles/h), and queuing occurs between 11:30–12:00. During queuing, the outflow curve deviates from the inflow, with a delay greater than the free-flow time, and remains capped at capacity, as validated in Theorem 2.1. Another queuing period is observed from 15:15 to 16:45, with similar dynamics. According to (10.71), the

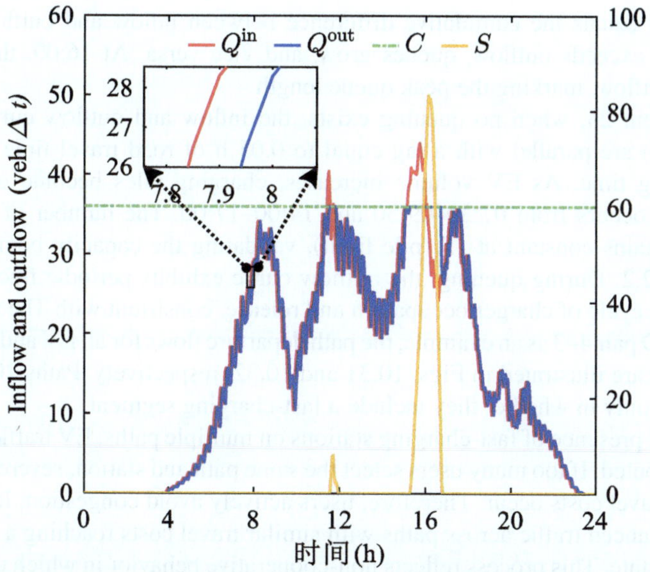

Fig. 10.29 Traffic dynamics on conventional segment 7 in the dynamic model

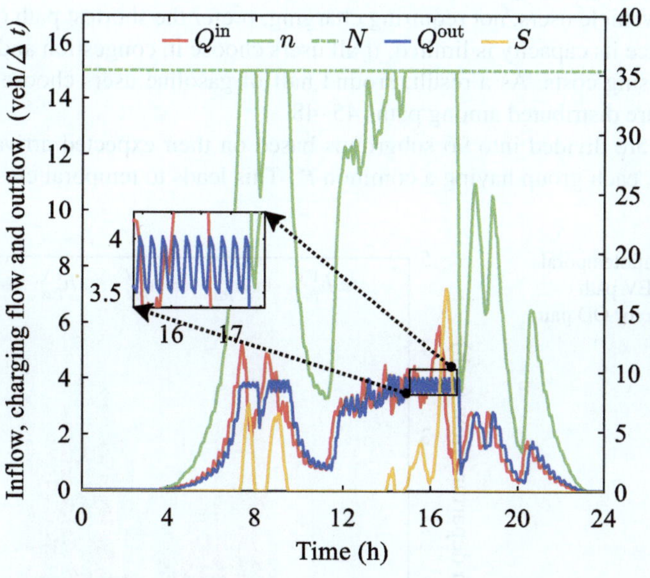

Fig. 10.30 Traffic dynamics on fast-charging segment 24 (station 3) in the dynamic model

queue length equals the cumulative difference between inflow and outflow rates; when inflow exceeds outflow, queues grow, and vice versa. At 16:00, the inflow equals the outflow, marking the peak queue length.

For segment 24, when no queuing exists, the inflow and outflow curves (i.e., Q^{in} and Q^{out}) are parallel with a lag equal to 0.05 h of road travel time plus 0.2 h of charging time. As EV volume increases, charging piles become occupied, and queuing occurs from 07:30–09:30 and 14:00–17:00. The number of vehicles charging remains constant at 15 (pile limit), validating the capacity boundedness in Theorem 2.2. During queuing, the outflow curve exhibits periodic fluctuations, reflecting the cycle of charger occupation and release, consistent with Theorem 2.3.

Taking OD pair 4–3 as an example, the path departure flows for all EV and gasoline vehicle users are illustrated in Figs. 10.31 and 10.32, respectively. Paths 20–24 and 44–48 differ only in whether they include a fast-charging segment.

Due to the presence of fast-charging stations on multiple paths, EV traffic is more evenly distributed. If too many users select the same path and station, severe queuing and higher travel costs occur. Therefore, users actively avoid congestion, leading to relatively balanced traffic across paths with similar travel costs reaching a dynamic equilibrium state. This process reflects non-cooperative behavior in which users rely on navigation apps, charging station apps, and personal experience to determine their travel and charging strategies.

Gasoline vehicle users, not requiring charging, prefer the shortest path (path 44). However, since its capacity is limited, if all users choose it, congestion and queuing occur, increasing costs. As a result, around half of gasoline users choose path 44, and the rest are distributed among paths 45–48.

EV users are divided into 96 subgroups based on their expected arrival time ϱ (i.e., $Q^{\text{EV}}_{3-4,\varrho}$), each group having a common t^*. This leads to temporal clustering in

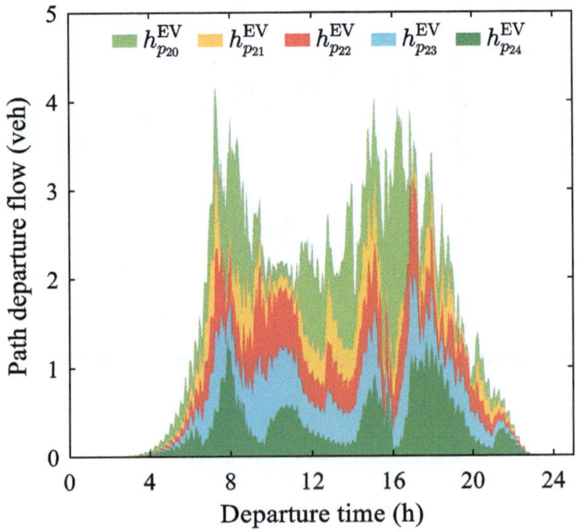

Fig. 10.31 Spatiotemporal distribution of EV path departure flows for OD pair 4–3

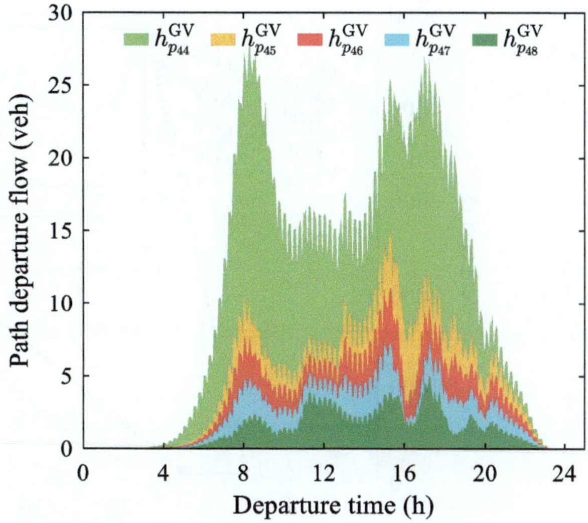

Fig. 10.32 Spatiotemporal distribution of gasoline vehicle path departure flows for OD pair 4–3

path choices, as shown by sharp peaks in Fig. 10.31. A similar pattern is observed for gasoline vehicles in Fig. 10.32.

Further, for EV and gasoline users with $\varrho = 8:00$ (i.e., $Q^{\mathrm{EV}}_{3-4,\varrho}$, with $\varrho = 8:00$), the path travel costs and path departure flows are shown in Figs. 10.33 and 10.34, respectively.

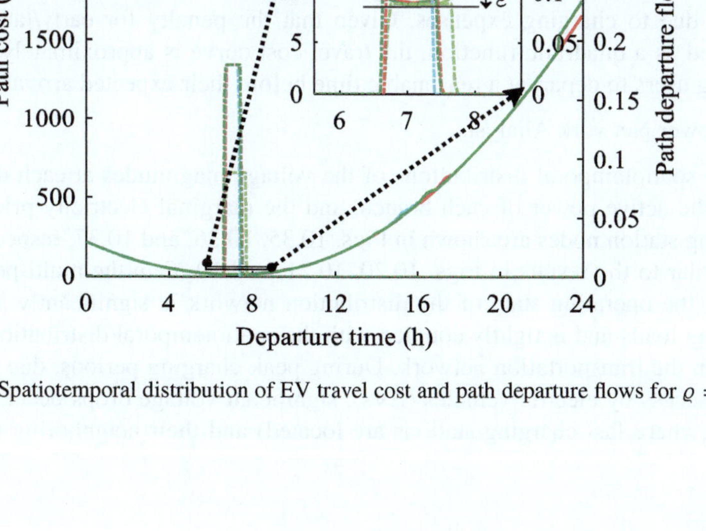

Fig. 10.33 Spatiotemporal distribution of EV travel cost and path departure flows for $\varrho = 8:00$ (OD 4–3)

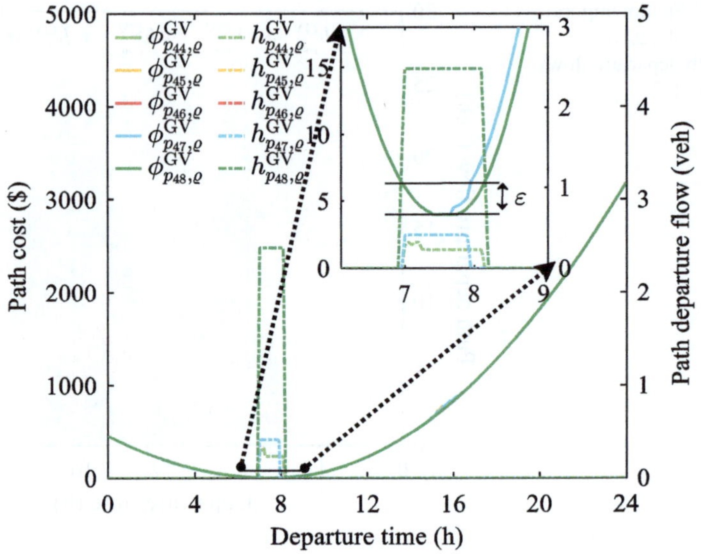

Fig. 10.34 Spatiotemporal distribution of gasoline vehicle travel cost and path departure flows for $\varrho = 8:00$ (OD 4–3)

EV users mainly depart between 06:36–07:48, with more choosing paths 21–24 than path 20. Regions with significant departure flows correspond to nearly optimal travel costs. Due to numerical convergence error, costs range from 8.5–9.6 $/trip, fluctuating within the 1$ tolerance threshold, indicating that users satisfy the bounded rationality condition and reach a dynamic user equilibrium.

Gasoline users mainly depart between 06:54 and 08:12, with most choosing path 48. Similarly, their travel costs range from 4.1 to 6.3 $/trip, aligning with the 2$ tolerance threshold, thus satisfying bounded rationality and dynamic equilibrium.

For users with identical travel demand, EVs generally depart earlier than gasoline vehicles due to mid-journey charging needs. Their total travel cost is also slightly higher due to charging expenses. Given that the penalty for early/late arrival is modeled as a quadratic function, the travel cost curve is approximately parabolic, guiding users to depart at a reasonable time before their expected arrival.

(3) Power Network Analysis.

The spatiotemporal distributions of the voltage magnitudes at each distribution node, the active power of each branch, and the marginal electricity prices at fast-charging station nodes are shown in Figs. 10.35, 10.36, and 10.37, respectively.

Similar to the results in Figs. 10.20, 10.21, and 10.22 in the multi-period static model, the operating state of the distribution network is significantly affected by charging loads and is tightly coupled with the spatiotemporal distribution of traffic flows in the transportation network. During peak charging periods, due to concentrated access by electric vehicles (EVs), significant voltage drops occur at nodes 7 and 9 (where fast-charging stations are located) and their neighboring node 6. To

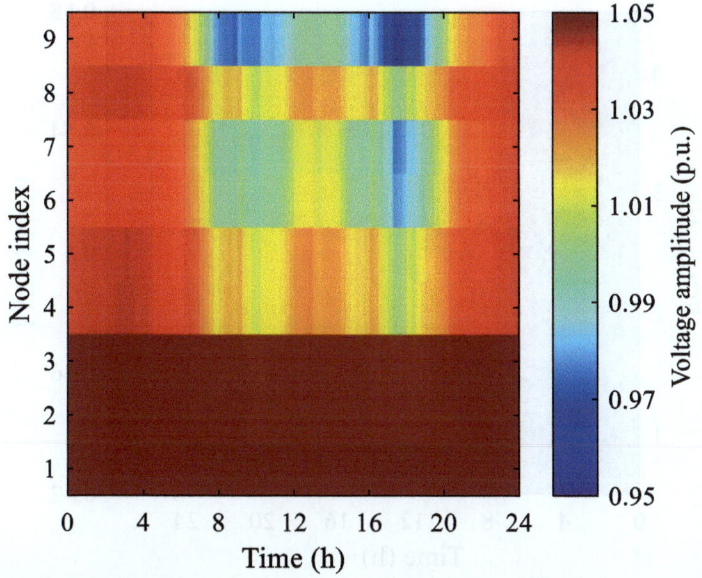

Fig. 10.35 Spatiotemporal distribution of voltage magnitudes at distribution network nodes in the dynamic model

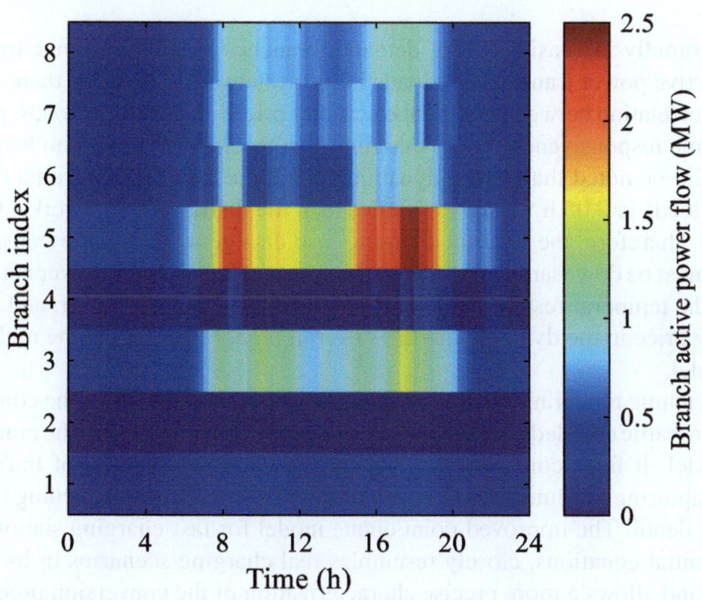

Fig. 10.36 Spatiotemporal distribution of active power on distribution network branches in the dynamic model

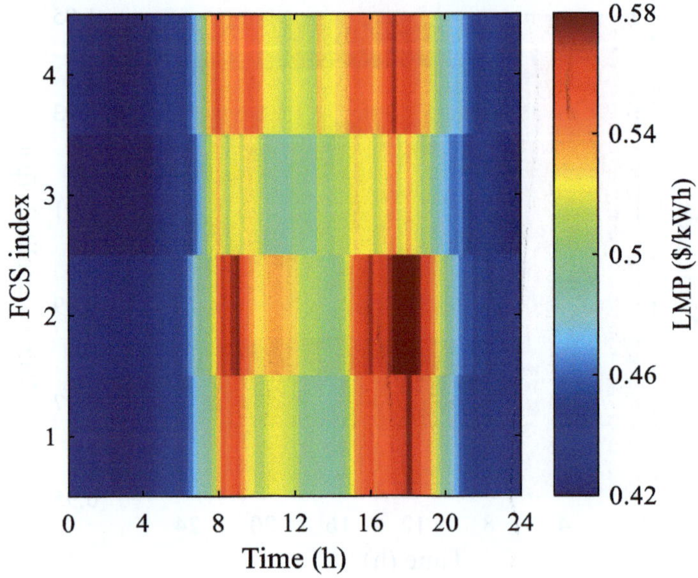

Fig. 10.37 Spatiotemporal distribution of marginal electricity prices at fast-charging station nodes in the dynamic model

meet the rapidly increasing power demand, branches 3 and 5 continue to bear the highest active power transmission loads in the system. Additionally, there is a clear positive correlation between marginal electricity prices and charging loads, reflecting the dynamic responsiveness of the marginal cost mechanism to system load states.

It should be noted that, in the dynamic model, the time step of traffic flows and charging loads is 0.05 h, which is smaller than the distribution network's time step of 0.25 h. Therefore, the original charging load data generated in the transportation network must be downsampled to match the time resolution of the power network. As a result, the temporal resolution of voltage magnitude, branch power, and marginal electricity price in the dynamic model is overall higher than that in the multi-period static model.

The dynamic modeling method proposed in this chapter enables the construction of a power-traffic coupled network model with higher resolution than the multi-period static model. It fully considers the dynamic propagation process of traffic flows, thereby capturing the emergence and evolution of congestion and queuing behaviors in greater detail. The improved point queue model for fast-charging stations, based on differential equations, closely resembles real charging scenarios in its dynamic response and allows a more precise characterization of the conversion process from EV traffic flows to power network charging loads. It reveals the dynamic coupling mechanism between traffic flow and power flow at a micro level. This model provides a solid theoretical foundation and data support for the flexible short-term scheduling and operation of power-traffic coupled systems.

However, compared with the static modeling method, the dynamic modeling approach still suffers from lower computational efficiency, is prone to a higher number of iterations and convergence issues, which pose challenges for large-scale system applications.

10.6 Summary

To evaluate the impact of bounded user rationality on travel and charging behaviors, uncover the dynamic coupling mechanism between traffic flows and power flows at fast-charging stations in power-traffic coupled networks, and provide theoretical support for optimal fast-charging station scheduling, this chapter proposes two modeling approaches from different temporal scales: A multi-period static modeling method based on bounded rationality static user equilibrium; A dynamic modeling method based on bounded rationality dynamic user equilibrium. First, a bounded rationality user decision model is constructed by introducing a user tolerance parameter to quantify the degree of rationality, and comparative analyses are made with a fully rational model to reveal the behavioral impact of bounded rationality on travel and charging decisions. Then, a multi-period static modeling method is proposed, where traffic flows are modeled as a multi-period bounded rationality static user equilibrium without propagation delay. A dynamic modeling method is also presented, where traffic flows are modeled as bounded rationality dynamic user equilibrium with propagation characteristics. In the dynamic model, an improved point queue model based on differential equations is designed to accurately depict the EV's entry, queuing, charging, and departure processes at fast-charging stations. This model enables tracking the transformation of traffic flows into charging loads and thoroughly explores its mathematical properties, physical significance, and consistency with real-world dynamic behavior. In both modeling frameworks, the power system is modeled using a multi-period optimal power flow model. Marginal electricity prices serve as charging prices and are dynamically adjusted based on charging loads. A fixed-point modeling approach is used to characterize the interaction between traffic and power flows, defining a network equilibrium state under coupled conditions. Corresponding iterative algorithms are developed to solve for this equilibrium.

The models and algorithms proposed in this chapter are systematically evaluated and validated through simulation studies, yielding the following conclusions: (1) Both the multi-period static and dynamic modeling methods for power-traffic coupled networks effectively solve the spatiotemporal distributions of traffic and power flows. The static method, with lower time resolution, benefits from a simple model structure and high computational efficiency (including parallel computing capability), making it suitable for fast identification of key periods and critical infrastructure. The dynamic method, considering the propagation of traffic flows, offers higher time accuracy and finer decision support for short-term scheduling, albeit with increased model complexity and computational burden; (2) the improved point queue model based on differential equations accurately describes EV queuing and charging behaviors, enabling detailed tracking of the transformation from traffic flows to distribution network charging loads, thereby revealing the dynamic coupling mechanisms from a microscopic perspective; and finally (3) the fixed-point iterative algorithm effectively solves both static and dynamic network equilibrium problems, and the resulting spatiotemporal distributions of traffic flows satisfy the bounded rationality quasi-optimality condition, demonstrating the practical feasibility and application potential of the proposed modeling approaches for real-world scheduling decisions.

Chapter 11
Dynamic Pricing Strategy for Fast Charging Stations Considering Bounded User Rationality and Market Regulation

Abstract This chapter proposes a dynamic pricing strategy for fast charging stations (FCSs) in coupled power-traffic networks, explicitly considering bounded user rationality and market regulation. Unlike traditional models that assume full rationality and static pricing, this work establishes a tri-level Stackelberg game framework. At the upper level, the distribution network operator determines locational marginal prices (LMPs) via an optimal power flow model. The middle level features an FCS operator who sets dynamic retail prices to maximize profit, subject to a novel regulatory constraint that limits the revenue coefficient to ensure user affordability. At the lower level, electric vehicle (EV) users exhibit bounded rationality, making spatiotemporal charging decisions based on a bounded rationality dynamic user equilibrium (BRDUE) model. To solve this complex hierarchical fixed-point problem, a Gauss–Seidel iterative algorithm with a damping factor is developed, ensuring computational efficiency and convergence. Furthermore, a data-driven adaptive method is introduced to calibrate user tolerance parameters from historical data, enhancing model practicality. Case studies on a coupled 33-bus power and 17-node traffic network demonstrate that the proposed optimization-based dynamic pricing strategy significantly improves FCS operator profits compared to fixed or LMP-based methods. The market regulation mechanism effectively controls user costs, preventing excessive charges. The strategy also successfully shifts charging loads to off-peak periods, reduces distribution network operating costs, and alleviates traffic congestion at charging stations. Sensitivity analysis reveals that moderate bounded rationality can enhance system efficiency by mitigating user overreaction, while the regulatory coefficient is a critical lever for balancing profitability and fairness. This research provides a robust, implementable framework for coordinated FCS pricing in modern coupled infrastructure systems.

Q. Yang et al., *Integration of Massive Electric Vehicles to Smart Grid*, Smart Energy Systems, https://doi.org/10.1007/978-981-92-0557-8_11

11.1 Introduction

In this chapter, it was assumed that fast charging stations (FCSs) and distribution network operators belong to the same asset owner, and therefore, locational marginal prices (LMPs) were directly adopted as the retail charging prices. However, in practical operations, FCS operators are often independent of the distribution network operators, and LMPs merely serve as the wholesale prices at which electricity is purchased from the grid. To ensure profitability, FCS operators must formulate retail charging prices on top of these wholesale costs to provide charging services to electric vehicle (EV) users. A rational pricing strategy not only secures economic returns for the operators but also incentivizes investment in charging infrastructure, thereby promoting EV adoption and the development of green transportation. Hence, optimal pricing strategies for FCSs hold both theoretical significance and practical value.

As a critical tool for regulating the coordinated operation of coupled power–traffic networks (CPTNs), charging prices influence the spatial and temporal distribution of both traffic flow and power flow. An improper pricing strategy may lead to excessive concentration of charging loads in time and space, aggravating the peak–valley gap in the distribution network and jeopardizing grid stability. Moreover, it can also cause clustering of EVs at specific FCSs, intensifying queuing phenomena and even triggering congestion in the surrounding traffic network, thereby reducing overall system efficiency. Therefore, FCSs must consider not only their own profit goals when setting retail prices but also the broader impact on power and traffic systems, using scientific pricing mechanisms to promote efficient coordination in CPTNs.

Existing studies have primarily relied on static modeling approaches to forecast long-term steady-state charging load distributions and design corresponding optimal static pricing strategies. However, static models fail to capture the temporal dynamics of charging demand, limiting their ability to reflect real-time changes in grid prices and user behavior. This undermines operators' ability to implement flexible pricing schemes and reduces revenue potential. Moreover, static pricing lacks the temporal flexibility required to guide load shifting or traffic flow modulation effectively. Thus, there is an urgent need to introduce dynamic modeling frameworks that reflect the short-term evolution of CPTNs and support the development of optimal dynamic pricing strategies for FCSs.

On the other hand, the pricing activities of FCSs fall within the realm of market operations and must be supervised by third-party regulators. Therefore, pricing models should incorporate market regulation policies to ensure legal compliance, operational feasibility, and public acceptability, thereby supporting the orderly development of the charging market and protecting EV user interests. In addition, many current studies assume that EV users respond to price signals with full rationality. In reality, users often exhibit bounded rationality due to limited information and cognitive biases. Therefore, incorporating bounded rationality into pricing models remains a crucial and underexplored research direction.

To address these challenges, this chapter proposes a dynamic pricing strategy for FCSs that jointly considers bounded user rationality and market regulation.

Specifically, based on the dynamic modeling framework for CPTNs established in Chapter 11, a tri-level Stackelberg game model is constructed among the distribution network operator, the FCS operator, and EV users. The distribution network operator determines LMPs via optimal power flow (OPF) models as wholesale prices; the FCS operator sets retail prices to maximize profit; and EV users determine their spatiotemporal charging behavior based on a boundedly rational dynamic user equilibrium (BRDUE) model. To ensure pricing fairness, a regulatory mechanism is introduced into the FCS operator's pricing model, which constrains users' maximum acceptable charging cost via a regulatory coefficient. This ensures pricing reasonableness while incentivizing the operator to shift loads and enhance CPTN coordination via price signals.

To solve the fixed-point problem arising from the tri-level game, this chapter proposes a Gauss–Seidel iterative algorithm that balances computational stability and efficiency. Furthermore, to improve real-world applicability, a data-driven adaptive estimation method is introduced to calibrate users' tolerance levels based on historical behavior. This enhances the responsiveness and practicality of the pricing strategy. Finally, a case study is conducted to verify the effectiveness of the proposed approach and to explore the interaction between the regulatory coefficient and user rationality parameters. Through this research, the chapter aims to provide FCS operators with a scientifically sound, flexible, and implementable dynamic pricing framework, while also offering theoretical and practical guidance for the coordinated optimization of CPTNs and the orderly development of the charging service market.

The structure of this chapter is as follows: Sect. 12.2 constructs the tri-level Stackelberg game model for optimal dynamic pricing of FCSs, characterizing the interaction between the grid operator, FCS operator, and EV users, and introduces regulatory constraints to ensure market fairness. Section 12.3 reformulates the pricing problem as a fixed-point problem and presents a Gauss–Seidel-based iterative solution method, including analysis of the fixed-point solution's existence. Section 12.4 introduces a historical-data-based adaptive estimation and adjustment mechanism for user tolerance parameters. Section 12.5 presents case studies to validate the proposed pricing and solution methods and analyzes the impact of key parameters. Section 12.6 summarizes the key findings of this chapter.

11.2 Optimal Dynamic Pricing Model for Fast Charging Stations Based on a Tri-Level Stackelberg Game

Under the independent operation of fast charging stations (FCSs), their retail pricing strategies simultaneously influence the spatiotemporal distribution of traffic flow and power flow, thereby affecting the operational states and coupling relationships of both the distribution and transportation networks. This section aims to construct a multi-level decision-making game framework among the distribution network operator, the

FCS operator, and EV users, and to propose an optimal dynamic pricing model for FCSs.

Upper Level: Distribution Network Model Based on Optimal Power Flow and Locational Marginal Prices. To model the response behavior of the distribution network operator to charging demand, this section adopts the multi-period optimal power flow (OPF) model introduced in Sect. 10.4.2. The mathematical formulation is omitted here for brevity. Given a set of charging demands, the locational marginal price (LMP) at a node is defined as the incremental operational cost of the distribution network induced by one additional unit of charging load at that node. It is mathematically represented as the partial derivative of the OPF objective function for the charging demand at a node:

$$\lambda_i^{\text{ch}}(t) = \frac{\partial O^{\text{PDN}}}{\partial P_i^{\text{ch}}(t)\Delta t}, \forall v_i \in V^{\text{FCS}}, t \in \mathcal{T} \tag{11.1}$$

When used as the wholesale price for selling electricity to FCS operators, the LMP directly reflects the impact of charging loads on the operational cost of the distribution network. Since local generators in the distribution network typically adopt quadratic cost functions, the LMP increases as the charging load grows. This pricing mechanism helps suppress peak-period load growth and supports the safe and stable operation of the distribution network.

There are three mainstream approaches to computing LMPs: (1) Using commercial solvers to extract dual variables: Some solvers (e.g., Gurobi, Mosek) support simultaneous access to both primal and dual solutions. Thus, by solving the OPF primal problem, one can directly extract the dual variable associated with the nodal active power balance constraint as the LMP; (2) Solving the dual form of the OPF problem: The dual problem, also formulated as a second-order cone programming (SOCP) model, provides LMPs through its dual variables; and (3) Indirect computation via Karush–Kuhn–Tucker (KKT) conditions: One can first solve the primal OPF problem and then substitute the optimal solution into the KKT conditions to solve the resulting linear system for LMPs.

In the tri-level Stackelberg game equilibrium framework proposed in this chapter, the OPF model needs to be solved repeatedly in an iterative process. However, as the scale of the distribution network increases, the computation time for solving the AC-OPF model can grow significantly. To improve computational efficiency, model linearization techniques or the use of DC-OPF approximations can be employed for large-scale networks to reduce the computational burden. Notably, the DC-OPF model ignores line losses and voltage drops during simplification. Despite introducing certain approximation errors, its formulation is a convex quadratic programming (QP) problem with only linear constraints, which enables significantly faster computation compared to the AC model, making it suitable for iterative applications. Moreover, under the DC model, LMPs still increase with charging load and effectively capture the distribution operator's response behavior, making the approach practical and applicable in real-world scenarios.

Middle Level: Optimal Pricing Model for Fast Charging Stations under Market-Regulated Profit Coefficient. Unlike the assumption in Chap. 2, where locational marginal prices (LMPs) are directly used as retail charging prices, this section assumes that all fast charging stations (FCSs) are managed by a single operator to simulate a typical real-world market-oriented operation mode. Under this setting, the LMP serves only as the wholesale electricity price that the operator pays to the distribution network. To ensure profitability and the normal operation and maintenance of charging infrastructure, the FCS operator must determine reasonable retail charging prices for providing services to EV users. The difference between the retail price and the wholesale price constitutes the operator's source of profit. Let the retail charging price be denoted as $\gamma^{\text{ch}} = \{\gamma_i^{\text{ch}}(t)|v_i \in V^{\text{FCS}}, t \in \mathcal{T}\}$. Given a wholesale electricity price and charging demand, the operator's net profit can be expressed as:

$$O^{\text{ch}} = \sum_{v_i \in V^{\text{FCS}}} \sum_{t \in \mathcal{T}} [\gamma_i^{\text{ch}}(t) - \lambda_i^{\text{ch}}(t)] P_i^{\text{ch}}(t) \Delta t \tag{11.2}$$

In existing literature, methods for determining retail charging prices can be categorized into static and dynamic pricing strategies, depending on whether the price varies over time. Static pricing assumes a fixed retail price across all periods and is suitable for relatively stable demand scenarios. Dynamic pricing allows the retail price to vary over time, enabling the operator to adjust prices in response to temporal variations in LMPs and charging demand. Dynamic pricing methods can be further divided into two subcategories: LMP-based pricing: The retail price is determined by multiplying the LMP by a pricing coefficient, which must be greater than or equal to 1. Optimization-based pricing: An optimization model is constructed to maximize the profit function defined in (11.2), directly solving for the optimal retail price. The mathematical formulations of these three pricing strategies are introduced as follows:

(1) Fixed Pricing Method (Static Pricing Method).

$$\gamma^{\text{fix}} = \{\gamma_i^{\text{ch}}(t)|\gamma_i^{\text{ch}}(t) = \gamma \geq \lambda_i^{\text{ch}}(t), v_i \in V^{\text{FCS}}, t \in \mathcal{T}\} \tag{11.3}$$

where $\boldsymbol{\gamma}$ is a constant that exceeds the nodal marginal price.

(2) Nodal Marginal Price-Based Method (Dynamic Pricing Method).

$$\gamma^{\text{LMP}} = \{\gamma_i^{\text{ch}}(t)|\gamma_i^{\text{ch}}(t) = \alpha^{\text{p}} \lambda_i^{\text{ch}}(t), v_i \in V^{\text{FCS}}, t \in \mathcal{T}\} \tag{11.4}$$

α^{p} is the pricing coefficient, which is greater than or equal to 1.

(3) Optimization-Based Method (Dynamic Pricing Method)

$$\gamma^{\text{opt}} = \arg\max_{\gamma^{\text{ch}}} O^{\text{ch}} = \sum_{v_i \in V^{\text{FCS}}} \sum_{t \in \mathcal{T}} [\gamma_i^{\text{ch}}(t) - \lambda_i^{\text{ch}}(t)] P_i^{\text{ch}}(t) \Delta t \tag{11.5}$$

$$\lambda_i^{\text{ch}}(t) \leq \underline{\gamma} \leq \gamma_i^{\text{ch}}(t) \leq \overline{\gamma}, \forall v_i \in V^{\text{FCS}}, t \in \mathcal{T} \tag{11.6}$$

$$\frac{\sum_{v_i \in V^{\mathrm{FCS}}} \sum_{t \in \mathcal{T}} \gamma_i^{\mathrm{ch}}(t)}{|V^{\mathrm{FCS}}||\mathcal{T}|} <= \frac{\underline{\gamma} + \overline{\gamma}}{2} \tag{11.7}$$

(11.6) defines the upper and lower bounds of the retail charging price, where $\underline{\gamma}$ and $\overline{\gamma}$ represent the upper and lower limits of the retail price, respectively. (11.7) imposes a constraint on the average value of the retail charging price, with γ^{av} denoting the specified mean price. Given the nodal marginal prices and charging demand, this problem constitutes a linear programming model.

To ensure comparability across the three pricing strategies, their average retail prices are set equal to γ^{av}. For the fixed pricing method, this implies $\gamma = \gamma^{\mathrm{av}}$. For the nodal marginal price-based method, the pricing coefficient

$$\mathrm{Average}(\gamma^{\mathrm{LMP}}) = \frac{\sum_{v_i \in V^{\mathrm{FCS}}} \sum_{t \in \mathcal{T}} \alpha^{\mathrm{p}} \lambda_i(t)}{|V^{\mathrm{FCS}}||\mathcal{T}|} = \gamma^{\mathrm{av}} \tag{11.8}$$

α^{p} can be derived from:

$$\alpha^{\mathrm{p}} = \frac{\gamma^{\mathrm{av}} |V^{\mathrm{FCS}}||\mathcal{T}|}{\sum_{v_i \in V^{\mathrm{FCS}}} \sum_{t \in \mathcal{T}} \lambda_i(t)} \geq 1 \tag{11.9}$$

In this study, it is assumed that all fast charging stations within the distribution network are operated by a single operator. Meanwhile, electric vehicle (EV) users exhibit inelastic demand for charging services, resulting in an oligopolistic market dominated by the charging operator. EV users, lacking bargaining power, act merely as price takers. To regulate orderly market operations, a third-party regulatory agency is introduced in the model to supervise and restrict the pricing behavior of the charging station operator. This aims to prevent excessively high retail prices and protect EV users' legitimate interests in an oligopolistic environment.

Although constraints (11.6) and (11.7) impose certain limitations on the operator's pricing decisions, they primarily regulate the pricing process itself. These constraints cannot directly control the actual charging costs borne by users. Consequently, the dominant operator may still manipulate the retail pricing strategy in a way that indirectly increases users' overall expenditure on charging services.

In this study, it is assumed that all fast charging stations within the distribution network are independently operated by a single charging station operator. Meanwhile, electric vehicle (EV) users exhibit inelastic charging demand. As a result, the market structure resembles an oligopoly dominated by the operator, in which EV users act merely as price takers with limited bargaining power. To ensure the orderly operation of the charging market, a third-party regulatory agency is introduced to supervise and constrain the operator's pricing behavior, thereby preventing the imposition of excessively high retail prices and protecting users' legitimate rights in such an oligopolistic environment.

Although constraints (11.6) and (11.7) can partially regulate the operator's pricing strategy, they focus primarily on the pricing process itself, rather than directly controlling the actual charging costs borne by EV users. Thus, the dominant operator may still manipulate pricing strategies to indirectly increase users' expenditures.

To directly regulate EV users' charging expenditures, the following market regulation constraint is proposed:

$$1 < \chi = \frac{\sum\limits_{v_i \in V^{\mathrm{FCS}}} \sum\limits_{t \in \mathcal{T}} \gamma_i^{\mathrm{ch}}(t) P_i^{\mathrm{ch}}(t) \Delta t}{\sum\limits_{v_i \in V^{\mathrm{FCS}}} \sum\limits_{t \in \mathcal{T}} \gamma^{\mathrm{av}} P_i^{\mathrm{ch}}(t) \Delta t} \le \overline{\chi} \tag{11.10}$$

Here, χ denotes the revenue coefficient of the fast charging station operator. The numerator represents the operator's revenue under the current pricing strategy, while the denominator is the revenue baseline when the retail charging price is fixed at a reference value γ^{av}, $\overline{\chi}$ is the upper bound of the revenue coefficient set by the regulatory agency. Since the operator's total revenue is equivalent to the aggregate charging expenditure of all EV users, constraining the revenue coefficient effectively regulates the overall charging cost in the coupled power-transportation network, ensuring it remains within a reasonable range.

In the denominator of (11.10), the total EV charging load is summed and then multiplied by the mean retail price. Given that the number of EV users is fixed in the transportation network, the total EV charging demand is constant and related to the travel demand $Q_{od,\varrho}^{\mathrm{EV}}$.. Therefore, constraint (11.10) can be equivalently reformulated as:

$$1 \le \chi = \frac{\sum\limits_{v_i \in V^{\mathrm{FCS}}} \sum\limits_{t \in \mathcal{T}} \gamma_i^{\mathrm{ch}}(t) P_i^{\mathrm{ch}}(t) \Delta t}{\gamma^{\mathrm{av}} P_{\mathrm{ch}} \tau^{\mathrm{ch}} \sum\limits_{od \in OD} \sum\limits_{\varrho \in \hat{\mathcal{T}}} Q_{od,\varrho}^{\mathrm{EV}}} \le \overline{\chi} \tag{11.11}$$

By integrating the market regulation constraint in (11.10) with the conventional pricing constraints in (11.6) and (11.7), the pricing behavior of the fast charging station operator can be further regulated. Accordingly, two types of dynamic pricing methods under market regulation are proposed and formulated based on the aforementioned strategies.

(4) Node Marginal Price Method with Market Regulation (Dynamic Pricing Method)

$$\gamma^{\overline{\mathrm{LMP}}} = \left\{ \gamma_i^{\mathrm{ch}}(t) \middle| \gamma_i^{\mathrm{ch}}(t) = \alpha^{\mathrm{p}} \lambda_i(t), v_i \in V^{\mathrm{FCS}}, t \in \mathcal{T} \right\} \tag{11.12}$$

$$\alpha^{\mathrm{p}} = \min \left[\frac{\gamma^{\mathrm{av}} \left| V^{\mathrm{FCS}} \right| \left| \mathcal{T} \right|}{\sum\limits_{v_i \in V^{\mathrm{FCS}}} \sum\limits_{t \in \mathcal{T}} \lambda_i(t)}, \frac{\overline{\chi} \gamma^{\mathrm{av}} P_{\mathrm{ch}} \tau^{\mathrm{ch}} \sum\limits_{od \in OD} \sum\limits_{\varrho \in \hat{\mathcal{T}}} Q_{od,\varrho}^{\mathrm{EV}}}{\sum\limits_{v_i \in V^{\mathrm{FCS}}} \sum\limits_{t \in \mathcal{T}} \lambda_i(t) P_i^{\mathrm{ch}}(t) \Delta t} \right] \ge 1 \tag{11.13}$$

When the pricing coefficient α^{p} is given by (11.9), the resulting revenue coefficient of the fast charging station operator may exceed the upper bound specified by the market regulation constraint. Therefore, the coefficient α^{p} must be adjusted downward according to (11.13) to ensure compliance with regulatory requirements.

$$\gamma^{\overline{\mathrm{opt}}} = \arg\max_{\gamma^{\mathrm{ch}}} O^{\mathrm{ch}} = \sum_{v_i \in V^{\mathrm{FCS}}} \sum_{t \in \mathcal{T}} \left[\gamma_i^{\mathrm{ch}}(t) - \lambda_i^{\mathrm{ch}}(t)\right] P_i^{\mathrm{ch}}(t) \Delta t \tag{11.14}$$

$$\lambda_i^{\mathrm{ch}}(t) \le \underline{\gamma} \le \gamma_i^{\mathrm{ch}}(t) \le \overline{\gamma}, \forall v_i \in V^{\mathrm{FCS}}, t \in \mathcal{T} \tag{11.15}$$

$$\frac{\sum\limits_{v_i \in V^{\mathrm{FCS}}} \sum\limits_{t \in \mathcal{T}} \gamma_i^{\mathrm{ch}}(t)}{\left|V^{\mathrm{FCS}}\right| \left|\mathcal{T}\right|} \le \gamma^{\mathrm{av}} = \frac{\underline{\gamma} + \overline{\gamma}}{2} \tag{11.16}$$

$$1 \le \chi = \frac{\sum\limits_{v_i \in V^{\mathrm{FCS}}} \sum\limits_{t \in \mathcal{T}} \gamma_i^{\mathrm{ch}}(t) P_i^{\mathrm{ch}}(t) \Delta t}{\gamma^{\mathrm{av}} P_{\mathrm{ch}} \tau^{\mathrm{ch}} \sum\limits_{od \in OD} \sum\limits_{\varrho \in \hat{\mathcal{T}}} Q_{od,\varrho}^{\mathrm{EV}}} \le \overline{\chi} \tag{11.17}$$

The market regulation constraint (11.17) is linear and does not introduce additional complexity to the model. Therefore, the above optimization problem remains a linear programming (LP) problem and can be efficiently solved.

Lower Level: Traffic Network Model Based on Bounded Rationality Dynamic User Equilibrium. Given the retail charging prices, the electric vehicle (EV) traffic flow responds accordingly, altering the spatiotemporal distribution of traffic. This adjustment also involves interactions with internal combustion engine (ICE) vehicles, ultimately leading to changes in charging demand at fast charging stations.

Based on the retail prices set by charging station operators, the spatiotemporal distribution of charging demand can be determined using the bounded rationality dynamic user equilibrium (BR-DUE) model proposed in Sect. 10.4.2. Different from the case considered in Sect. 10.4.1, where charging prices are equal to node marginal prices, the prices in this section are the retail prices determined by the charging station operator. Therefore, (10.58), (10.105), (10.109), (10.113), and (10.114) should be modified as follows:

$$\theta_p^{\mathrm{EV}}(t) = \sum_{t' \in T_p^{\mathrm{ch}}(t)} \gamma_{\upsilon(p)}^{\mathrm{ch}}(t') P^{\mathrm{ch}} \Delta t, \forall od \in OD, p \in P_{od}^{\mathrm{EV}}, t \in \mathcal{T} \tag{11.18}$$

$$\mathcal{H} = f^{\mathrm{BRDUE}}(\gamma^{\mathrm{ch}}) \tag{11.19}$$

$$\mathcal{H} = P_{\Pi}\left[\mathcal{H} - \alpha \tilde{\Phi}(\mathcal{H}, \gamma^{\mathrm{ch}})\right] \tag{11.20}$$

$$\mathcal{H}^* = \arg\min_{\mathcal{H} \in \Pi} \frac{1}{2} \left\| \mathcal{H}^* - \alpha \tilde{\Phi}(\mathcal{H}, \gamma^{\mathrm{ch}}) - \mathcal{H} \right\|_2 \tag{11.21}$$

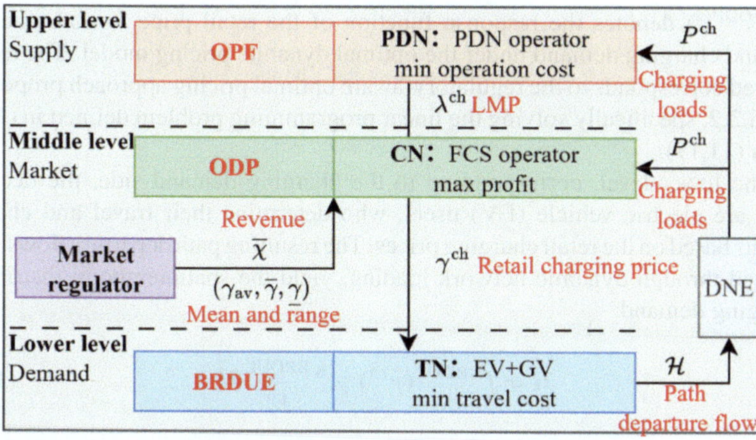

Fig. 11.1 Three-level Stackelberg game model

$$\mathcal{H}^{z+1} = \arg\min_{\mathcal{H}\in\Pi} \frac{1}{2}\left\|\mathcal{H}^z - \alpha\tilde{\Phi}(\mathcal{H}^z, \gamma^{\mathrm{ch},k}) - \mathcal{H}\right\|_2 \tag{11.22}$$

Three-Level Stackelberg Game Model. Based on the above analysis, the distribution network operator, fast charging station operator, third-party regulatory agency, and traffic users together form a three-level Stackelberg game in the coupled power-traffic network, as illustrated in Fig. 11.1. In each pairwise game between adjacent layers, the upper layer acts as the leader, while the lower layer follows.

The upper level corresponds to the electricity supply side. The decision-maker is the distribution network operator, who determines the node marginal prices through the optimal power flow model based on charging demand:

$$\lambda^{\mathrm{ch}} = f^{\mathrm{OPF}}(P^{\mathrm{ch}}):P^{\mathrm{ch}} \xrightarrow[\text{QP + dual}]{\text{OPF}} \lambda^{\mathrm{ch}} \tag{11.23}$$

where $f^{\mathrm{OPF}}(\cdot)$ is the price response function to charging demand. When the direct current optimal power flow (DCOPF) model is used as an approximation, this problem becomes a quadratic programming problem. Through dual analysis, the marginal prices can be obtained, consistent with Sect. 11.4.3.

At the middle level, corresponding to the charging market side, the decision-maker is the fast charging station operator under market regulation, who determines the optimal retail charging price strategy under the dynamic pricing model based on node marginal prices, charging demand, and market regulatory constraints:

$$\gamma^{\mathrm{ch}} = f^{\mathrm{ODP}}(\lambda^{\mathrm{ch}}, P^{\mathrm{ch}}):(\lambda^{\mathrm{ch}}, P^{\mathrm{ch}}) \xrightarrow[\text{LP}]{\text{opt}} \gamma^{\mathrm{ch}} \tag{11.24}$$

Here, $f^{\mathrm{ODP}}(\cdot)$ denotes the response function of the retail price to node marginal prices and charging demand under the optimal dynamic pricing model. The method employed corresponds to the regulatory-aware optimal pricing approach proposed in Sect. 12.2.2, specifically solving the linear programming problem defined in (11.14) through (11.17).

At the lower level, corresponding to the charging demand side, the decision-makers are electric vehicle (EV) users, who determine their travel and charging decisions based on the retail charging prices. The resulting path departure flows, when processed through dynamic network loading, yield the spatiotemporal distribution of charging demand:

$$\mathcal{H} = f^{\mathrm{BRDUE}}(\gamma^{\mathrm{ch}}): \gamma^{\mathrm{ch}} \xrightarrow[\mathrm{FP}]{\mathrm{BRDUE}} \mathcal{H} \tag{11.25}$$

$$P^{\mathrm{ch}} = f^{\mathrm{DNL}}(\mathcal{H}) : \mathcal{H} \xrightarrow{\mathrm{DNL}} P^{\mathrm{ch}} \tag{11.26}$$

In these equations, $f^{\mathrm{BRDUE}}(\cdot)$ represents their response function to retail charging prices, rather than the node marginal prices used in Sect. 10.4.3. This response is formulated as a fixed-point problem under the bounded rationality dynamic user equilibrium (BR-DUE) framework. During this process, interactions between internal combustion engine (ICE) vehicles and EV users are considered. Therefore, in Fig. 11.1, ICE vehicles are also treated as decision-makers at the lower level. The function $f^{\mathrm{DNL}}(\cdot)$ reflects how the charging demand at each fast charging station evolves spatiotemporally in response to the given departure flows, consistent with the model in Sect. 11.4.3.

Compared to the dynamic network equilibrium model of the coupled power–traffic network established in Sect. 11.4.3, the model proposed in this section introduces fast charging stations as an independent intermediate layer between the power and transportation networks, enabling a more granular and market-oriented pricing and decision-making mechanism.

11.3 Solution Strategy for the Optimal Dynamic Pricing Fixed-Point Problem Based on the Gauss–Seidel Iteration Method

Fixed-Point Model of the Optimal Dynamic Pricing Problem. In the three-level Stackelberg game model established in Sect. 11.2.4, given the node marginal prices at the upper level and the charging demand at the lower level, the optimal dynamic pricing problem at the intermediate level can be reduced to a linear programming (LP) problem. However, when the fast charging station operator adjusts the retail charging price, EV users will correspondingly revise their original charging behavior, shifting demand toward time periods with lower prices. This shift in charging demand will, in turn, trigger changes in the node marginal prices within the distribution network, thereby altering the parameters of the objective function in the optimal pricing model. To maintain profit maximization, the operator must then re-optimize the retail pricing strategy accordingly.

Due to the hierarchical structure of the three-level Stackelberg game, the optimal dynamic pricing problem can be approached using two major categories of solution strategies: (1) exact mathematical optimization methods, and (2) numerical iterative approximation methods. The optimization-based approach typically involves reformulating the multi-level game into a single-level solvable model. Specifically, the pricing problem at the intermediate level is a linear program and can be replaced with its KKT (Karush–Kuhn–Tucker) conditions to reduce the three-level game to a bi-level structure.

However, at the lower level, the traffic network model employs a bounded rationality dynamic user equilibrium (BR-DUE) framework, which is non-convex, nonlinear, and non-closed. This complexity prevents the derivation of the corresponding KKT conditions and thus precludes further reduction to a single-level model within polynomial time. To overcome this modeling and computational challenge, this section proposes a numerical solution strategy based on the Gauss–Seidel iterative method, which sequentially passes response characteristics among the distribution network operator, charging station operator, and EV users to gradually converge to a Stackelberg equilibrium. The final result is an optimal dynamic pricing strategy for fast charging stations.

By combining and substituting the relationships among the model components, the optimal dynamic pricing problem in Sect. 11.2.4 can be reformulated as the following fixed-point model:

$$
\begin{aligned}
\gamma^{\text{ch}} &= f^{\text{ODP}}\left(f^{\text{OPF}}(P^{\text{ch}}), f^{\text{DNL}}(\mathcal{H})\right) \\
&= f^{\text{ODP}}\left(f^{\text{OPF}}\left(f^{\text{DNL}}(\mathcal{H})\right), f^{\text{DNL}}\left(f^{\text{BRDUE}}(\gamma^{\text{ch}})\right)\right) \\
&= f^{\text{ODP}}\left(f^{\text{OPF}}\left(f^{\text{DNL}}\left(f^{\text{BRDUE}}(\gamma^{\text{ch}})\right)\right), f^{\text{DNL}}\left(f^{\text{BRDUE}}(\gamma^{\text{ch}})\right)\right) \\
&= f^{\text{ODP - FP}}(\gamma^{\text{ch}})
\end{aligned}
\tag{11.27}
$$

where $f^{\text{OP - FP}}(\cdot)$ denotes the self-response function of the retail charging price in the optimal dynamic pricing fixed-point model. As analyzed in Sect. 10.4.4, the BR-DUE model embedded within $f^{\text{BRDUE}}(\cdot)$ is also a fixed-point problem. Thus, (11.27) constitutes a bi-level fixed-point problem.

Solution Method. To efficiently solve the fixed-point problem defined in (11.27), this section builds upon the iterative methods discussed in Sects. 11.3.4 and 11.4.4 and develops a Gauss–Seidel-based solution strategy for optimal dynamic pricing.

Algorithm 3: Gauss–Seidel-Based Solution Strategy for Optimal Dynamic Pricing

Step 1	Set the maximum number of iterations K, convergence thresholds $\overline{\delta}^{\lambda}, \overline{\delta}^{\lambda}, \overline{\delta}^{\mathcal{H}}, \overline{\delta}^{P}$, iteration counter $k = 0$; initialize the node marginal price $\lambda^{\mathrm{ch},0}$, retail charging price $\gamma^{\mathrm{ch},0}$, path departure flow $\mathcal{H}^{0}$, charging load $P^{\mathrm{ch},0}$, and damping factor $\beta \in [0, 1]$
Step 2	Given the charging load at iteration k, solve the multi-period optimal power flow model to update the node marginal price for iteration $k+1$, i.e., $\lambda^{\mathrm{ch},k+1} = (1-\beta^{k})\lambda^{\mathrm{ch},k} + \beta^{k} f^{\mathrm{OPF}}(P^{\mathrm{ch},k})$
Step 3	Given the node marginal price at iteration $k+1$ and the charging load at iteration k, solve the optimal dynamic pricing model to update the retail charging price for iteration $k+1$, i.e., $\gamma^{\mathrm{ch},k+1} = (1-\beta^{k})\gamma^{\mathrm{ch},k} + \beta^{k} f^{\mathrm{ODP}}(\lambda^{\mathrm{ch},k+1}, P^{\mathrm{ch},k})$
Step 4	Given the retail charging price at iteration $k+1$, , solve the bounded rationality dynamic user equilibrium (BR-DUE) model to update the path departure flow for iteration $k+1$, i.e., $\mathcal{H}^{k+1} = (1-\beta^{k})\mathcal{H}^{k} + \beta^{k} f^{\mathrm{BRDUE}}(\gamma^{\mathrm{ch},k+1})$
Step 5	Given the updated path departure flow at iteration $k+1$, perform dynamic network loading to update the charging load for iteration $k+1$, i.e., $P^{\mathrm{ch},k+1} = (1-\beta^{k})P^{\mathrm{ch},k} + \beta^{k} f^{\mathrm{DNL}}(\mathcal{H}^{k+1})$
Step 6	Calculate the convergence errors $\delta^{\lambda}_{k+1} = \frac{\lVert \lambda^{\mathrm{ch},k+1} - \lambda^{\mathrm{ch},k} \rVert_2}{\lVert \lambda^{\mathrm{ch},k} \rVert_2}$, $\delta^{\gamma}_{k+1} = \frac{\lVert \gamma^{\mathrm{ch},k+1} - \gamma^{\mathrm{ch},k} \rVert_2}{\lVert \gamma^{\mathrm{ch},k} \rVert_2}$, $\delta^{\mathcal{H}}_{k+1} = \frac{\lVert \mathcal{H}^{k+1} - \mathcal{H}^{k} \rVert_2}{\lVert \mathcal{H}^{k} \rVert_2}$, $\delta^{P}_{k+1} = \frac{\lVert P^{\mathrm{ch},k+1} - P^{\mathrm{ch},k} \rVert_2}{\lVert P^{\mathrm{ch},k} \rVert_2}$ 。 If $\delta^{\lambda}_{k+1} \le \overline{\delta}^{\lambda}, \delta^{\gamma}_{k+1} \le \overline{\delta}^{\gamma}, \delta^{\mathcal{H}}_{k+1} \le \overline{\delta}^{\mathcal{H}}, \delta^{P}_{k+1} \le \overline{\delta}^{P}$, then output the optimal dynamic pricing strategy $\gamma^{\mathrm{ch},k+1}$; Otherwise, set $k = k+1$, and return to **Step 2**

Compared to the conventional iterative algorithms adopted in Sects. 11.3.4 and 11.4.4, the Gauss–Seidel iterative algorithm proposed in this section utilizes the most recently updated results from the current iteration in each step, rather than relying on historical values from the previous round. This strategy improves the update efficiency of each iteration. In addition, an exponential damping factor is introduced to control the step size during the iteration process: relatively large step sizes are adopted in the early stages to accelerate convergence, while gradually decreasing step sizes are applied in later stages to enhance the stability and precision of convergence. To ensure that the coupled power-transportation network ultimately reaches a dynamically balanced state, the designed convergence criterion comprehensively considers the convergence of key variables such as node marginal prices, path departure flows, and charging loads, thereby evaluating system stability from multiple dimensions.

11.4 Estimation and Adaptive Correction of User Tolerance Based on Historical Data

User tolerance directly determines the degree of user rationality and is one of the most critical parameters in bounded rationality decision-making models. It reflects users' sensitivity to charging prices. If the set tolerance level is too high, the resulting pricing strategy may become overly aggressive, which can weaken users' willingness to charge and lead to the loss of EV customers, thereby undermining the competitiveness of fast charging station operators. On the other hand, if the tolerance level is set too low, the pricing strategy will become overly conservative, limiting profit growth and potentially triggering vicious price competition among multiple operators. Therefore, accurately estimating the user tolerance parameter is a prerequisite for fast charging station operators to develop reasonable pricing strategies.

However, existing studies generally assume the user tolerance level to be a known constant, which is often difficult to satisfy in practical applications. To address this issue, this section proposes a user tolerance estimation and adaptive correction method based on historical data.

It is assumed that fast charging station operators can obtain historical travel data within a given region through transportation statistics and surveys. This data may include: the user's origin location, departure time, travel path, selected charging station, destination location, and actual arrival time. Based on this information, the following approach can be used to estimate the user tolerance parameter.

Algorithm 4: Estimation and Adaptive Correction of User Tolerance Based on Historical Data	
Step 1	Let the historical operating profit of the fast charging station be denoted as $O^{\text{ch},0}$, the correction iteration index as $k = 1$, and the maximum number of corrections as K. Set the initial value of the EV user tolerance coefficient as $\varepsilon_{od}^{\text{EV},0} = 0$, and the correction coefficient as $a_{\varepsilon} > 0$
Step 2	Collect the departure flow data $h_{p,\varrho}^{\text{EV}}(t)$ and $h_{p,\varrho}^{\text{GV}}(t)$, as well as their associated travel costs $\phi_{p,\varrho}^{\text{EV}}(t)$ and $\phi_{p,\varrho}^{\text{GV}}(t)$ for each OD pair. Assume that traffic users arrive at their destinations within their desired arrival time window. Therefore, for EV users, extract **time cost** $\varphi_p^{\text{EV}}(t)$ and **charging cost** $\theta_p^{\text{EV}}(t)$; For gasoline vehicle users, extract **time cost** $\varphi_p^{\text{GV}}(t)$ only
Step 3	Estimate the historical user tolerance based on the following expression: $\varepsilon_{od}^{\text{EV},1} = \max_{p\in P_{od}^{\text{EV}}} \max_{\varrho\in\hat{T}} \max_{t\in T} \phi_{p,\varrho}^{\text{EV}}(t) - \min_{p\in P_{od}^{\text{EV}}} \min_{\varrho\in\hat{T}} \min_{t\in T} \phi_{p,\varrho}^{\text{EV}}(t), \forall od \in OD$ $\varepsilon_{od}^{\text{GV}} = \max_{p\in P_{od}^{\text{GV}}} \max_{\varrho\in\hat{T}} \max_{t\in T} \phi_{p,\varrho}^{\text{GV}}(t) - \min_{p\in P_{od}^{\text{GV}}} \min_{\varrho\in\hat{T}} \min_{t\in T} \phi_{p,\varrho}^{\text{GV}}(t), \forall od \in OD$ Output the gasoline vehicle users' tolerance coefficient $\varepsilon^{\text{GV}} = \{\varepsilon_{od}^{\text{GV}} \mid od \in OD\}$。
Step 4	Using the current EV user tolerance coefficient $\varepsilon^{\text{EV},k}$ and ε^{GV}, apply the optimal dynamic pricing method proposed in Sect. 12.3 to solve for the pricing strategy. Let the resulting fast charging station profit be denoted as O^k

(continued)

(continued)

Algorithm 4: Estimation and Adaptive Correction of User Tolerance Based on Historical Data	
Step 5	Adaptively update the EV user tolerance coefficient using: $\varepsilon_{od}^{\mathrm{EV},k+1} = \varepsilon_{od}^{\mathrm{EV},k} + a_\varepsilon(O^{\mathrm{ch},k} - O^{\mathrm{ch},k-1}), \forall od \in T$
Step 6	If $\left\|\varepsilon^{\mathrm{EV},k+1} - \varepsilon^{\mathrm{EV},k}\right\|_2 \leq \sigma_\varepsilon$ or $k \geq K$, terminate the adaptive correction process and output the final EV user tolerance coefficient $\varepsilon^{\mathrm{EV},k+1} = \{\varepsilon_{od}^{\mathrm{EV},k+1} \mid od \in OD\}$ Otherwise, increment $k = k + 1$ and return to Step 4

The estimation method in **Step 3** can be derived from the *bounded rationality suboptimality condition* proposed in Sect. 12.2.2. Since only non-zero path departure flows can be observed in **Step 2**, the corresponding travel costs should satisfy the bounded rationality condition, i.e., lie within the user tolerance interval $[\mu_{od,\varrho}^{\mathrm{EV}}, \mu_{od,\varrho}^{\mathrm{EV}} + \varepsilon_{od}^{\mathrm{EV}}]$. Therefore, the width of this interval, estimated by the difference between the maximum and minimum observed travel costs, can be used to approximate the user tolerance level $\varepsilon_{od}^{\mathrm{EV},1}$. Considering that charging prices directly affect only the behavior of electric vehicle (EV) users, and that the influence of EV flows on gasoline vehicle flows is relatively minor, the travel behavior of gasoline vehicle users can be approximately assumed to be independent of charging prices. As such, the tolerance coefficient estimated for gasoline vehicle users in **Step 3** can be regarded as a reasonable proxy for the true user tolerance level and used as the final estimation result.

In **Step 5**, since the operating profit of the fast charging station is positively correlated with the tolerance level of EV users under a given pricing strategy, the operator can observe the profit variation after implementing the new strategy. Based on the direction and magnitude of profit changes, the user tolerance parameter can be adaptively corrected, thereby dynamically optimizing the pricing strategy.

11.5 Case Study

Simulation Setup. The topology of the power–transportation coupled network Test System 2 used in this section is illustrated in Fig. 11.2. The power network is based on the IEEE 33-bus distribution system [175], where Bus 1 is connected to the upstream main grid, and distributed generators are installed at Buses 2, 3, 6, 9, 13, 25, and 30. The transportation network adopts a 17-node system derived from the Nguyen transportation network, with the structure and parameters consistent with those presented in Sect. 11.5. Four fast charging stations are deployed in the system and powered by Buses 29, 15, 4, and 7 of the power network. The travel demand matrix is presented in Table 11.1, while the other key simulation parameters are listed in Table 11.2. Parameters identical to those in Sect. 11.5 will not be repeated here. In the power network, the time step for calculating nodal marginal electricity prices is set to 0.25 h, whereas the time step for fast charging station retail pricing and traffic flow modeling is 0.05 h. The software platforms used for model construction and solution in this section are the same as those adopted in Sect. 11.5 and will not be detailed again.

Performance Analysis of the Gauss–Seidel Iterative Method. The proposed Gauss–Seidel iterative algorithm (Algorithm 3) reaches convergence at the 17th iteration, with a total computation time of 289.5 s. The convergence trajectory is illustrated in Fig. 11.3. Specifically, the marginal electricity price converges at the 17th iteration, the retail charging price converges rapidly by the 2nd iteration, while the path departure flow and charging load reach convergence by the 10th and 16th iterations, respectively.

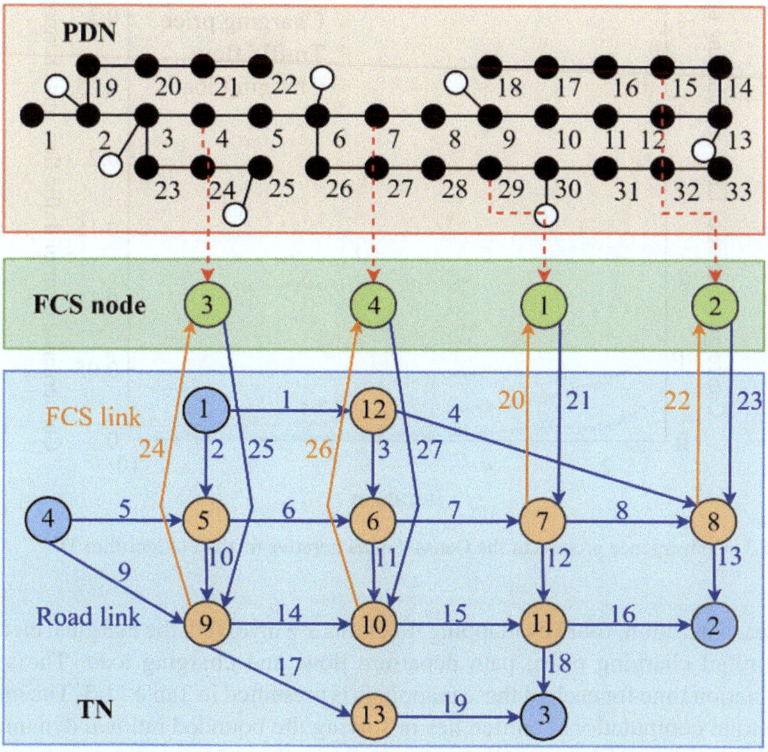

Fig. 11.2 Coupled power–transportation network test system 2

Table 11.1 Travel demand settings in coupled power–transportation network test system 2

Origin o	Destination d	EV demand Q_{od}^{EV}	GV demand Q_{od}^{GV}
1	2	1500 veh	2500 veh
1	3	1500 veh	2500 veh
4	2	1500 veh	2500 veh
4	3	1500 veh	2500 veh

Table 11.2 Parameter settings in the heterogeneous traffic network test system 2

Parameter	Value	Parameter	Value	Parameter	Value
Δt	0.05/0.25 h	N_l	30	$\overline{\chi}$	1.05
$\underline{\gamma}$	0.67 \$/kWh	$\overline{\gamma}$	0.87 \$/kWh	γ^{av}	0.77 \$/kWh
β	0.95	$\overline{\delta}^{P}$	10^{-2}	$\overline{\delta}^{\mathcal{H}}$	10^{-2}
C_l	432 veh/h				

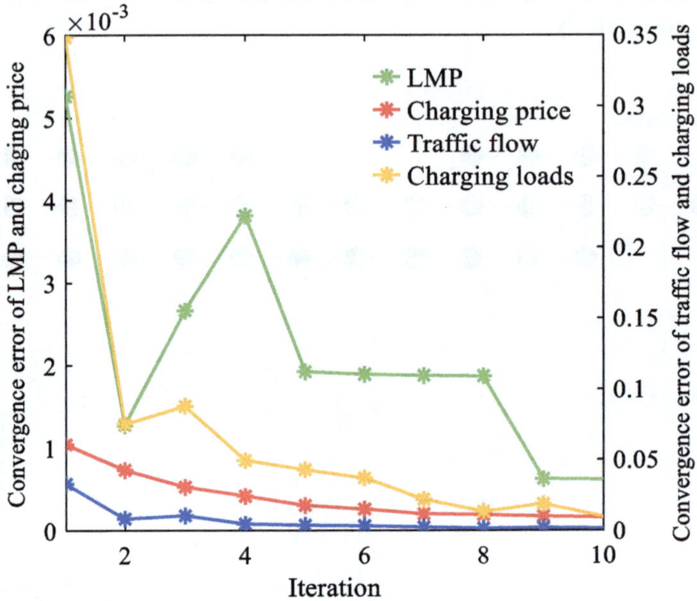

Fig. 11.3 Convergence process of the Gauss–Seidel iterative method (Algorithm 3)

In each iteration, four key mapping functions are involved: the marginal electricity price, retail charging price, path departure flow, and charging load. The average computation time for each of these mappings is presented in Table 11.3. The majority of the total computational burden lies in solving the bounded rational dynamic user equilibrium (BR-DUE) model. As discussed in Sect. 11.4.4, this model itself is formulated as a nested fixed-point problem, capturing multi-path, time-dependent traffic behavior, thereby resulting in substantial computational complexity.

Table 11.3 Average computation time of each mapping function per iteration

$f^{\mathrm{OPF}}(\cdot)$	$f^{\mathrm{ODP}}(\cdot)$	$f^{\mathrm{BRDUE}}(\cdot)$	$f^{\mathrm{DNL}}(\cdot)$	$f^{\mathrm{OP\text{-}FP}}(\cdot)$
0.6 s	0.2 s	15.7 s	0.5 s	17.0 s

In contrast, the optimal power flow (OPF) model of the distribution network and the dynamic pricing model of the fast charging stations correspond to convex optimization problems, i.e. quadratic programming (QP) and linear programming (LP), respectively, and can be efficiently solved using commercial solvers. The dynamic network loading process is relatively less computationally intensive. Based on the path departure flow as the boundary input, the link-based cumulative queue dynamics are updated over discrete time steps through forward recursion. This process primarily involves basic algebraic operations, making it computationally efficient.

In summary, the proposed Gauss–Seidel iterative method demonstrates both convergence reliability and computational efficiency. It is thus well-suited for practical applications in large-scale coupled power-traffic systems.

Economic Performance Comparison of Different Charging Pricing Strategies. The five charging pricing strategies discussed in Sect. 12.2.2 are respectively integrated into the proposed model. Once the power–traffic coupled network reaches a dynamic equilibrium, the operating profit of fast charging station operators, the operating cost of the distribution network operator, and the travel cost of electric vehicle (EV) users are illustrated in Fig. 11.4. Figure 11.5 further details the composition of travel costs for EV and internal combustion engine vehicle (ICEV) users under different pricing strategies. The EV travel cost includes time cost, schedule delay penalty cost, and charging cost, while the ICEV travel cost only consists of time and schedule delay penalties.

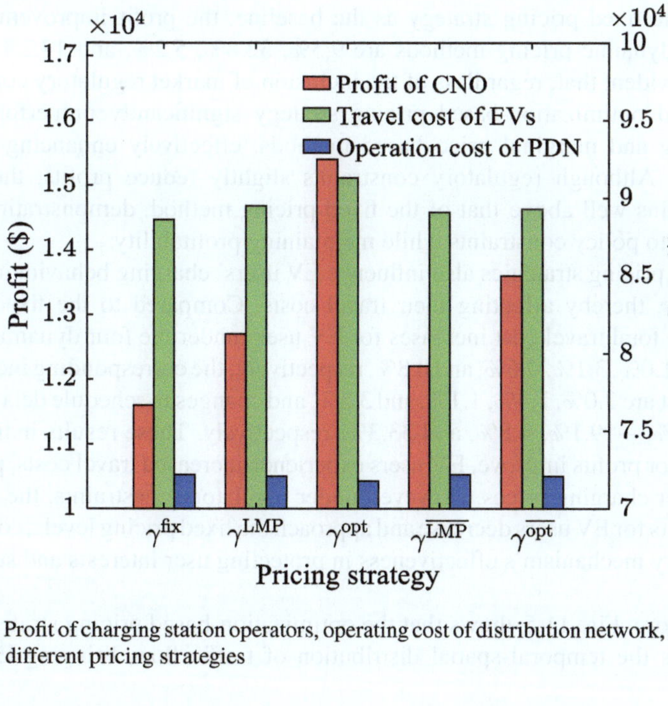

Fig. 11.4 Profit of charging station operators, operating cost of distribution network, and user travel cost under different pricing strategies

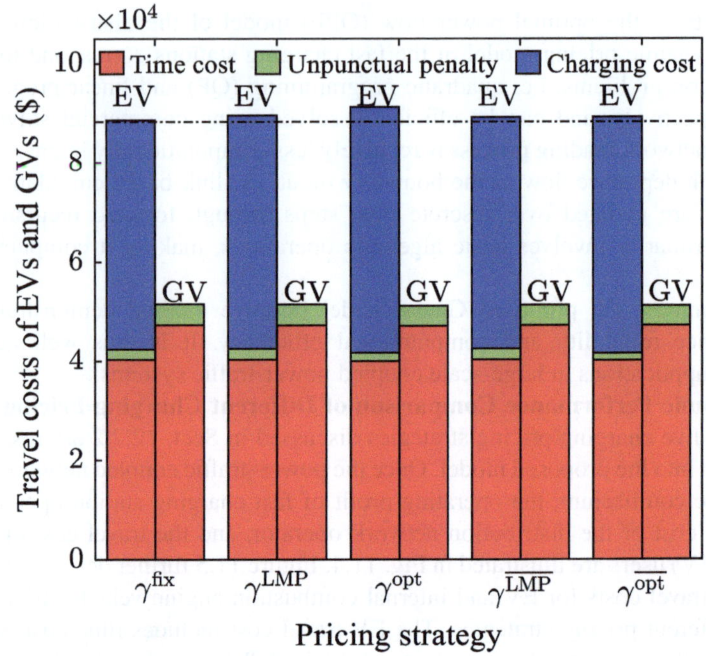

Fig. 11.5 Composition of travel costs for EV and ICEV users under different pricing strategies

Taking the fixed pricing strategy as the baseline, the profit improvement rates of the four dynamic pricing methods are 9.5%, 32.8%, 5.2%, and 17.2%, respectively. It is evident that, regardless of the inclusion of market regulatory constraints, the proposed optimization-based pricing strategy significantly outperforms both fixed pricing and marginal price–based methods, effectively enhancing operator profitability. Although regulatory constraints slightly reduce profits, the performance remains well above that of the fixed pricing method, demonstrating strong adaptability to policy constraints while maintaining profitability.

Different pricing strategies also influence EV users' charging behavior and travel path choices, thereby affecting their travel costs. Compared to the fixed pricing baseline, the total travel cost increases for EV users under the four dynamic pricing methods by 1.0%, 3.1%, 0.6%, and 0.8%, respectively; the corresponding increases in charging cost are 2.0%, 7.4%, 1.1%, and 3.3%, and changes in schedule delay penalty cost are 4.8%, −19.1%, 4.8%, and 33.3%, respectively. These results indicate that while operator profits improve, EV users experience increased travel costs, primarily due to higher charging prices. However, under regulatory constraints, the total and charging costs for EV users decrease and approach the fixed pricing levels, confirming the regulatory mechanism's effectiveness in protecting user interests and stabilizing prices.

Furthermore, Fig. 11.5 shows that the optimization-based pricing strategy effectively guides the temporal-spatial distribution of traffic flow, bringing EV users' actual arrival times closer to their expected times, thus reducing schedule delay penalties and improving overall travel efficiency. Since ICEV travel costs are not directly affected by charging prices, different pricing strategies have little impact on them. As analyzed in Sect. 11.4.3, EV and ICEV traffic flows interact; although charging prices do not directly affect ICEV behavior, they can indirectly influence ICEV distribution by altering EV traffic patterns. Nonetheless, due to the relatively small proportion of EVs in the current network, this indirect effect is limited.

Changes in EV charging behavior also impact the spatial–temporal distribution of charging loads, thereby influencing distribution network operating costs. Relative to the fixed pricing strategy, the cost variations under the four dynamic pricing methods are -0.2%, -0.4%, -0.2%, and -0.2%, respectively. By adjusting prices to guide load redistribution, the dynamic strategies generally reduce network operating costs. According to the definition in (11.1), the nodal marginal price inherently promotes cost minimization in the distribution network, encouraging load shifting from peak to off-peak periods and flattening demand curves. Moreover, to maximize profits, the optimization-based pricing strategy not only boosts electricity sales but also strategically adjusts charging demand toward lower-cost time slots, thereby reducing wholesale electricity procurement costs and enhancing overall economic efficiency.

In summary, the proposed optimization-based pricing strategy with the market regulation can effectively enhance the profitability of fast charging station operators, maintain reasonable cost burdens for EV users, and improve the spatial–temporal structure of traffic flows. It also enables more rational load distribution, achieving significant reductions in distribution network operating costs. This strategy fosters a virtuous interaction among distribution network operators, charging station operators, and users, thus significantly promoting synergistic efficiency and orderly operation within the power–traffic coupled network.

Impact of Different Charging Pricing Strategies on Traffic Flow and Load Distribution. To analyze the impact of different charging pricing strategies on the spatiotemporal distribution of traffic flows and corresponding charging loads at fast charging stations, this section compares the dynamic characteristics under two strategies: the fixed pricing strategy and the optimization-based pricing strategy incorporating market regulatory constraints. Figure 11.6 illustrates the dynamic changes in charging flow, queuing flow, nodal marginal electricity price, and retail charging price at each charging station under the fixed pricing strategy. Figure 11.7 presents the same variables under the optimization-based pricing strategy with market regulation. Figure 11.8 further compares the total charging and queuing flows across all fast charging stations under both strategies.

Under the fixed pricing strategy, a uniform retail electricity price is applied across all periods. The lack of temporal price differentiation leads to a limited ability to guide the spatial and temporal distribution of traffic flows. As a result, charging demand largely follows the natural diurnal pattern of travel demand. Charging peaks emerge during the periods of 7:00–9:00 and 15:00–18:00, during which the number of EVs approaches or exceeds the service capacity of the charging stations. All four fast charging stations experience queuing between 12:00 and 17:45. Specifically, Station 1 peaks at 14:12 with a queue of 12.2 vehicles, Station 2 peaks at 14:42 with

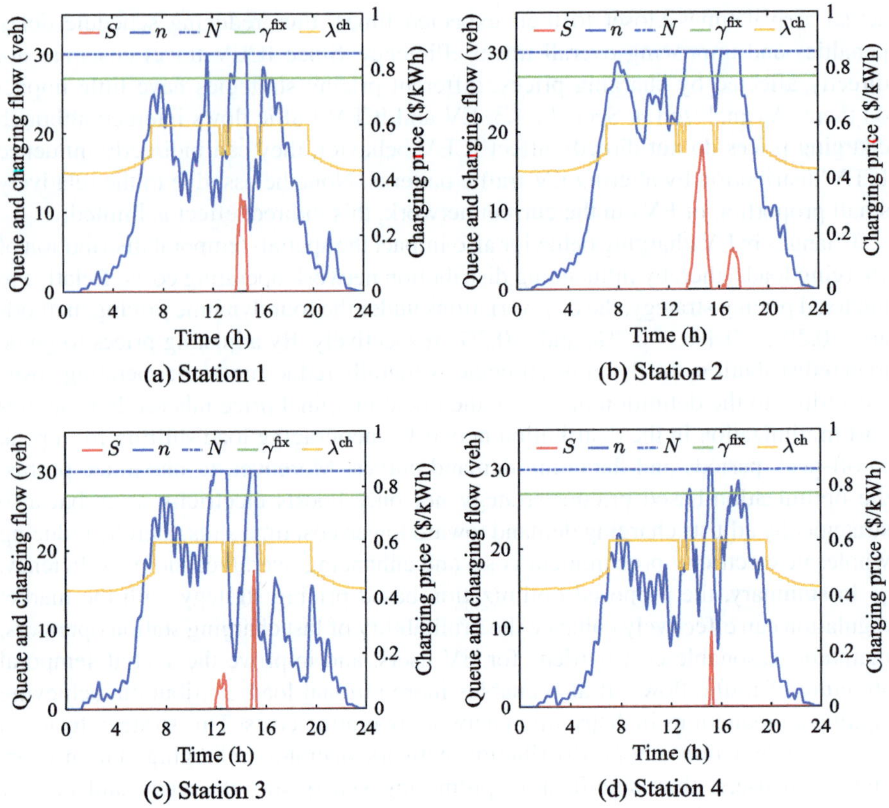

Fig. 11.6 Charging and queuing flow, nodal marginal price, and retail price dynamics under fixed pricing strategy

18.1 vehicles, Station 3 at 15:03 with 17.4 vehicles, and Station 4 at 15:12 with 9.4 vehicles. The queues last about 1 h at Stations 1 and 3, and over 2 h at Station 2.

As daytime charging load increases significantly, the marginal price at the nodes where the charging stations are connected rises from 0.45 $/kWh at night to approximately 0.60 $/kWh during the day, indicating that the distribution network must rely on upstream grid supply. In this context, since the marginal cost of distributed generation is higher than the grid purchase price, the additional load is primarily supplied by the main grid.

Under the optimization-based pricing strategy with market regulation, retail electricity prices vary across time periods, with higher prices generally occurring during the daytime. The resulting price signals effectively influence user behavior, leading to a more balanced temporal and spatial distribution of charging traffic. As shown in Fig. 11.7, compared to the fixed pricing case, charging and queuing flows under the optimized pricing strategy are more evenly distributed, with shorter and more dispersed peaks replacing the previously concentrated high-demand periods.

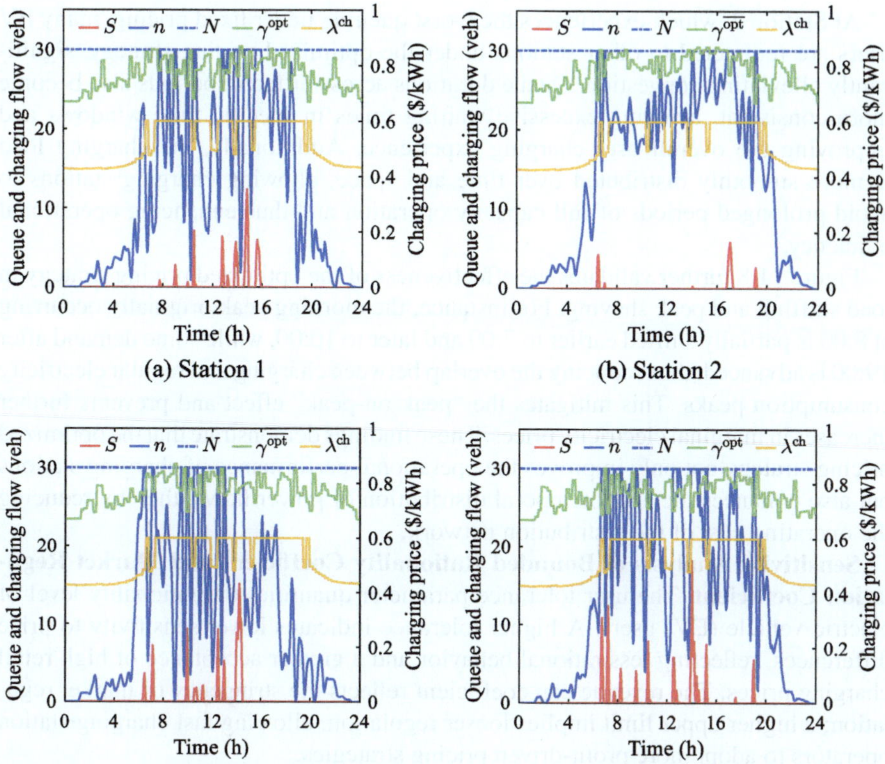

(a) Station 1 (b) Station 2

(c) Station 3 (d) Station 4

Fig. 11.7 Charging and queuing flow, nodal marginal price, and retail price dynamics under optimization-based pricing strategy with regulation

Fig. 11.8 Dynamic charging and queuing flow across all charging stations under both pricing strategies

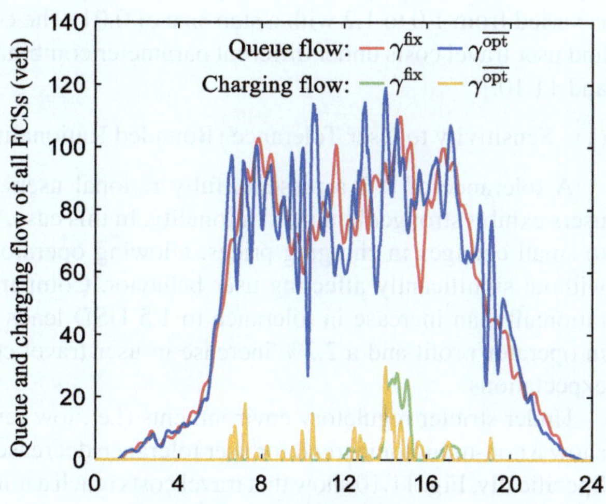

At Station 2, which experiences the worst queuing under fixed pricing, many EV users are redirected to other stations under the optimized pricing strategy, significantly alleviating congestion. Queue durations across different periods also become more consistent, avoiding excessive waiting times in specific time windows and improving the overall user charging experience. Additionally, the charging load is more smoothly distributed over time and space, allowing charging stations to avoid prolonged periods of full capacity operation and thus enhancing operational efficiency.

Figure 11.8 further validates the effectiveness of the optimized pricing strategy in load shifting and peak shaving. For instance, the morning peak originally occurring at 8:00 is partially shifted earlier to 7:00 and later to 10:00, while some demand after 19:00 is advanced, thus reducing the overlap between charging and regular electricity consumption peaks. This mitigates the "peak-on-peak" effect and prevents further increases in marginal electricity prices. These findings demonstrate that the optimized pricing strategy not only improves the operational performance of charging stations but also enhances the spatiotemporal distribution of power flows, thereby reducing the operating cost of the distribution network.

Sensitivity Analysis of Bounded Rationality Coefficient and Market Regulation Coefficient. The user tolerance parameter quantifies the rationality level of electric vehicle (EV) users. A higher tolerance indicates lower sensitivity to price differences, reflecting less rational behavior and a greater acceptance of high retail charging prices. The revenue cap coefficient reflects the stringency of market regulation: a higher upper limit implies looser regulation, allowing fast charging station operators to adopt more profit-driven pricing strategies.

To evaluate the impact of the bounded rationality coefficient and revenue cap coefficient on the profitability of charging station operators and the travel cost of EV users, a joint sensitivity analysis is conducted. The EV user tolerance parameter is varied from 0 to 1.5 USD with a step size of 0.1 USD, and the revenue cap coefficient is varied from 1.0 to 1.2 with a step size of 0.01. The corresponding operator profits and user travel costs under different parameter combinations are shown in Figs. 11.9 and 11.10.

(1) Sensitivity to User Tolerance (Bounded Rationality Coefficient).

A tolerance of 0 $ represents fully rational users. As the tolerance increases, users exhibit stronger bounded rationality. In this case, users become less responsive to small changes in charging prices, allowing operators to moderately raise prices without significantly affecting user behavior. Compared to the baseline of perfect rationality, an increase in tolerance to 1.5 USD leads to an average 1.9% increase in operator profit and a 2.2% increase in user travel cost, aligning with theoretical expectations.

Under stricter regulatory environments (i.e., low revenue caps), user travel costs show a non-monotonic trend for user tolerance decreases initially and then increases. Specifically, Fig. 11.10 show that travel costs reach a minimum when user tolerance is in the range of 0.3 and 0.5 USD. In this range, moderate bounded rationality helps alleviate excessive competition among EV users in the traffic network. Since user route

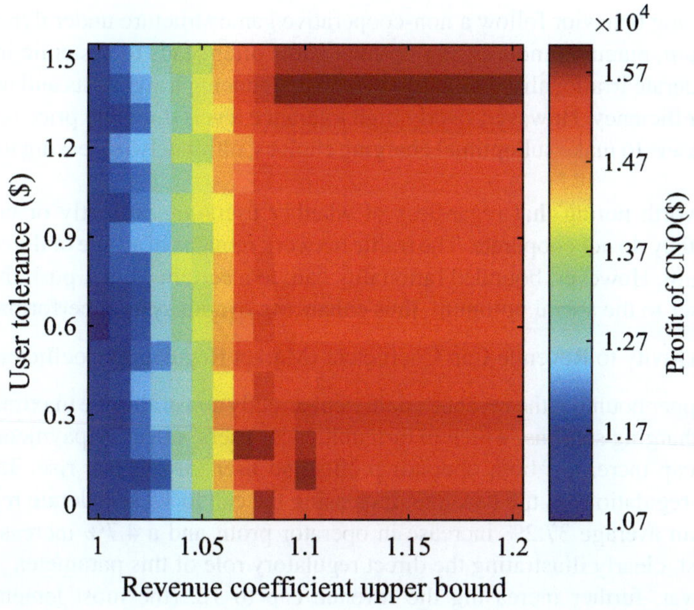

Fig. 11.9 Profit of fast charging station operators under different user tolerance levels and revenue cap coefficients

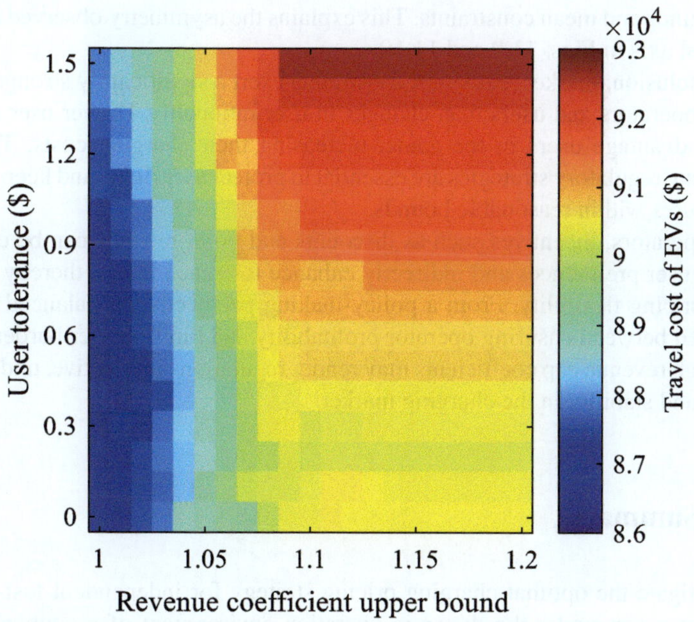

Fig. 11.10 Total travel cost of electric vehicle users under different user tolerance levels and revenue cap coefficients

and charging behavior follow a non-cooperative game structure under dynamic user equilibrium, pursuing individually optimal paths often leads to systemic inefficiencies. Moderate irrationality mitigates this effect, reducing travel costs and improving network efficiency. However, overly high tolerance levels diminish price sensitivity, causing users to make suboptimal charging choices, ultimately increasing total travel costs.

It is worth noting that regardless of whether users are perfectly or boundedly rational, they do not cooperate. The traffic network remains in a state of dynamic user equilibrium. However, bounded rationality can, to a certain extent, push the system state closer to the social optimum, thus enhancing overall system performance.

(2) Sensitivity to Revenue Cap Coefficient (Market Regulation Coefficient).

The upper bound of the revenue coefficient directly constrains the maximum profit level of charging stations, which in turn influences users' effective payments. As the revenue cap increases, both operator profits and user travel costs rise. Taking 1.0 (strictest regulation) as the baseline, increasing the cap to 1.1 (moderate regulation) leads to an average 37.2% increase in operator profit and a 4.7% increase in user travel cost, clearly illustrating the direct regulatory role of this parameter.

However, further increasing the revenue cap to 1.2 (the most lenient regulation) results in a stabilization of both operator profit and user travel cost. At this stage, the regulatory constraint defined in (11–17) no longer binds the optimization problem. The pricing model degenerates into a simpler form with only price upper/lower bounds and mean constraints. This explains the asymmetry observed along the horizontal axis in Figs. 11.9 and 11.10.

In conclusion, market regulation constraints exert a significantly stronger impact on both operators and users than changes in user rationality. Higher user tolerance may disadvantage users in the game, increasing their charging costs. Therefore, appropriate regulatory strategies are essential to protect user rights and keep charging expenditures within reasonable bounds.

For operators, incentives such as discounts and point rewards can be employed to shape user preferences and indirectly enhance tolerance levels, thereby enabling greater pricing flexibility. From a policy-making perspective, a balanced trade-off is required between ensuring operator profitability and limiting user burden. Excessively high revenue cap coefficients may render regulations ineffective, undermining fairness and stability in the charging market.

11.6 Summary

To investigate the optimal charging pricing strategy for independent fast-charging station operators under the dynamic operation environment of a coupled power–transportation network, and to analyze the mechanism through which pricing signals regulate both traffic flows and power flows, this chapter proposes an optimized dynamic pricing method for fast-charging stations that comprehensively considers user bounded rationality and market regulation. First, building upon the dynamic modeling framework of the coupled power–transportation network presented in Chap. 2, a three-level Stackelberg game model is established. In this framework, the distribution network operator determines nodal marginal electricity prices (wholesale prices) using an optimal power flow model; the fast-charging station operator aims to maximize profit by setting retail electricity prices; and electric vehicle (EV) users determine their charging demand's spatiotemporal distribution based on a bounded-rationality dynamic user equilibrium model. Second, to protect user interests and facilitate system-wide coordination, a regulatory mechanism for charging prices is designed.

In this chapter, by introducing a revenue regulation coefficient, the mechanism constrains operators' pricing behavior, ensuring reasonable EV user costs while guiding charging loads toward off-peak periods and enhancing peak-shaving and valley-filling performance. Next, a Gauss–Seidel–based fixed-point iterative method is developed to solve the formulated three-level game problem, and an iterative algorithm is proposed to ensure efficient convergence. Furthermore, to improve the model's adaptability and practical applicability, a historical-data-driven self-adaptive estimation method for the user tolerance parameter is introduced. By dynamically adjusting the rationality coefficient according to actual operational data, this method enhances the model's responsiveness to real-world conditions.

Through numerical simulations, the effectiveness of the proposed model and algorithm is validated. The main conclusions are as follows: (1) Compared with fixed pricing and nodal marginal price–based strategies, the proposed optimized dynamic pricing method significantly improves the profits of fast-charging station operators. Although EV users may face higher charging costs, the introduction of regulatory constraints effectively reduces these costs, demonstrating the regulator's role in protecting user interests; (2) the optimized dynamic pricing strategy with market regulation exerts a stronger guiding influence than static pricing strategies. It shortens the duration of peak traffic congestion caused by charging and queuing, thereby improving the operational efficiency of charging stations; (3) the strategy successfully guides charging demand toward periods with lower electricity prices, reducing operational costs of the distribution network. Additionally, it enhances the spatiotemporal distribution of power flows and strengthens the coordinated operation of power and transportation systems; (4) the user rationality coefficient significantly affects trip costs. Moderate levels of bounded rationality help alleviate excessive competition among users, thereby improving the overall efficiency of the transportation system; and (5) the market regulation coefficient directly influences both operator profit and user cost. If the revenue coefficient is set too high, it may weaken the effectiveness of regulatory measures. Hence, regulatory agencies should carefully define an appropriate range for this coefficient to balance market order and system performance.

Chapter 12
Energy Optimization and Control Strategy for Fast Charging Station Aggregators Under Electricity Market Environment

Abstract This chapter proposes a stochastic optimization-based energy regulation method for fast charging station aggregators participating in two-stage electricity markets, addressing the challenges of multidimensional uncertainties and bounded rationality in EV user behavior. A path merging and flow splitting strategy is first introduced to simplify the bounded rationality dynamic user equilibrium model, enabling efficient prediction of spatiotemporal charging load distributions in large-scale networks while significantly reducing computational complexity. A group-based charging scheduling strategy is then developed, clustering EVs with similar charging requirements to substantially decrease decision variables. An aggregator model incorporating local renewable energy and storage is formulated, with K-means clustering used to generate representative scenarios for travel demand, real-time prices, and renewable output uncertainties. The stochastic optimization framework determines optimal day-ahead bidding and real-time charging strategies. Case studies based on empirical data from a provincial electricity market in China demonstrate that the simplified model reduces iterations by 68.8% and computation time by 94.9% while maintaining high accuracy. The proposed strategies increase aggregator profits by up to 14.7% through peak load shifting, reduced deviation penalties, and enhanced renewable integration. The method exhibits excellent scalability for large-scale EV applications, providing theoretical and technical support for aggregator participation in electricity markets.

12.1 Introduction

In Chap. 11, the charging power was assumed to be constant, focusing primarily on the response of charging demand to pricing strategies. However, in actual operation, fast charging stations are not only subject to pricing strategies but can also manage charging load through direct control of charging power. With the large-scale integration of electric vehicles (EVs) into the grid, uncoordinated charging behavior may pose significant risks to the security and stability of distribution systems, potentially

Q. Yang et al., *Integration of Massive Electric Vehicles to Smart Grid*, Smart Energy Systems, https://doi.org/10.1007/978-981-92-0557-8_12

leading to issues such as voltage violations, line congestion, and equipment overloading. Since the ability of price signals to regulate behavior is inherently limited, relying solely on indirect user guidance is insufficient to address the challenges posed by charging loads. Therefore, on top of existing pricing mechanisms, the development of direct control methods for charging power is of great significance to improve flexibility and responsiveness in load management.

Against this backdrop, EV aggregators have emerged as new market participants capable of coordinating the charging power of multiple EVs, enabling orderly integration of charging loads, reducing peak load impacts, and lowering system peak-to-valley differences. When the number of aggregated EVs reaches a certain scale, aggregators can also participate in electricity market transactions. With appropriate bidding strategies, they can effectively reduce electricity costs. As such, EV aggregators are regarded as a key approach to managing the large-scale group behavior of EVs.

Current research on aggregators predominantly focuses on slow-charging scenarios during nighttime. However, with the increasing popularity of public fast charging facilities, it has become urgent to study the orderly integration of fast charging loads. Fast charging loads are characterized by high power, short duration, and strong randomness. In "charge-and-go" mode, their flexibility is limited, requiring energy delivery within a short time frame. As a result, the schedulability of fast charging is much lower than that of slow charging, placing higher demands on the time resolution and computational efficiency of scheduling algorithms. To distinguish from traditional aggregators in community or residential slow-charging contexts, this chapter defines aggregators composed of fast charging stations, renewable energy facilities, and energy storage resources as fast charging station aggregators and focuses on their energy optimization and control strategies under electricity market environments.

Taking a two-stage electricity market (day-ahead and real-time) as an example, fast charging station aggregators need to forecast charging loads and formulate bidding plans during the day-ahead stage, and dynamically adjust EV charging power in real-time to ensure consistency between actual electricity consumption and day-ahead declarations. Efficient and accurate forecasting of the spatiotemporal distribution of cluster charging loads is critical, especially in large-scale traffic networks, where efficient modeling and simulation of traffic flow are key challenges. Additionally, in formulating day-ahead strategies, aggregators must address multidimensional uncertainties arising from user travel behavior, variations in charging demand, user rationality, and electricity price volatility. If the station is equipped with renewable generation or storage, the inherent variability of renewable energy output must also be considered, further increasing the complexity of energy control.

To tackle these challenges, this chapter proposes an energy optimization and control method for fast charging station aggregators in electricity market environments. First, based on the bounded rationality dynamic user equilibrium model established in Chap. 10, a path merging and flow splitting strategy is introduced to simplify traffic network topology and focus on traffic flows with charging demand, thus enabling efficient prediction of cluster load. Second, a group-based charging scheduling strategy is designed to cluster EVs with similar charging needs into scheduling units, thereby significantly reducing the decision-making problem's dimensionality. Furthermore, an aggregator model including local renewable energy is formulated, and scenario-based stochastic optimization methods are applied to model uncertainties in load, price, and renewable generation. Based on this, an energy control strategy is proposed for the aggregator under the two-stage electricity market, providing optimal solutions for day-ahead bidding and real-time scheduling. Finally, numerical simulations are conducted to validate the effectiveness of the proposed method. The goal of this chapter is to provide a high-efficiency, flexible energy control framework for fast charging station aggregators participating in electricity markets, offering theoretical support and practical guidance for the orderly integration of EV charging loads and secure operation of power systems.

The remainder of this chapter is organized as follows: Sect. 12.1 introduces the path merging and flow splitting strategy to simplify the bounded rationality dynamic user equilibrium model, enabling rapid load prediction in large-scale networks. Section 12.2 designs a group-based charging scheduling strategy that reduces decision complexity by integrating similar charging demands. Section 12.3 establishes an aggregator energy control model incorporating local renewables, applies scenario-based modeling for uncertainty analysis, and proposes a stochastic optimization strategy for bidding and scheduling in a two-stage market. Section 12.4 validates the proposed models and methods through case studies. Section 12.5 summarizes the key findings of the chapter.

12.2 Simplified Bounded Rationality Dynamic User Equilibrium Model Based on Path Merging and Flow Splitting

In large-scale urban traffic networks, nodes and road segments are densely distributed, while the number of fast charging stations is relatively small. EV traffic flows must traverse many conventional road segments to reach a charging station. As network scale increases, the number of feasible paths grows exponentially. As a result, the bounded rationality dynamic user equilibrium model presented in Sect. 10.4.1 becomes computationally inefficient when applied to large-scale networks. A significant portion of the computational burden lies in modeling the dynamics of conventional road segments, i.e. the paths from origins to fast charging stations and from charging stations to destinations.

The purpose of modeling the travel from origin to charging station is to estimate travel time and determine which station the EVs will choose and their expected arrival time. Similarly, modeling the travel from the charging station to the destination aims to calculate users' actual arrival time and evaluate whether it satisfies the desired time window, thus determining the tardiness penalty cost in the total travel cost.

where $\tau_{p,\varrho}^{\mathrm{EV}}(t)$ denotes the travel time at time t along path p from station f to destination d for an electric vehicle user with expected arrival time ϱ; $\Pi_{fd,t}^{\tau,\mathrm{EV}}$ is the set of such $\tau_{p,\varrho}^{\mathrm{EV}}(t)$; P_{fd}^{EV} denotes the set of all EV paths from station f to destination d; for any $p \in P_{fd}^{\mathrm{EV}}$, $\upsilon(p) = f$; and $FD = \{fd\}$ is the set of all station–destination pairs.

The average travel time of equivalent paths from station f to destination d is defined as:

$$\hat{\tau}_{fd}^{\mathrm{EV}}(t) = \frac{1}{\left|\Pi_{fd,t}^{\tau,\mathrm{EV}}\right|} \sum_{\tau_{p,\varrho}^{\mathrm{EV}}(t) \in \Pi_{fd,t}^{\tau,\mathrm{EV}}} \tau_{p,\varrho}^{\mathrm{EV}}(t), \forall fd \in FD, t \in \mathcal{T} \tag{12.10}$$

Combining (10.61) and (10.63), we have:

$$\begin{aligned} &\hat{\xi}_{p'}^{\mathrm{EV}}(t) \triangleq \hat{\xi}_{p'}^{\mathrm{EV}*}(t) + \hat{\tau}_{fd}^{\mathrm{EV}}\left[\hat{\xi}_{p'}^{\mathrm{EV}*}(t)\right] \approx \xi_{p}^{\mathrm{EV}}(t), \\ &\forall o \in O, fd \in FD, p' \in P_{od}^{\mathrm{equi}}, p \in P_{od}^{\mathrm{EV}}, \upsilon(p') = \upsilon(p) = f, t \in \mathcal{T} \end{aligned} \tag{12.11}$$

where Here, $\hat{\xi}_{p'}^{\mathrm{EV}}(t)$ denotes the estimated travel time of the equivalent path p' at time t.

For electric vehicle users, their total travel cost is directly determined by two key components: (1) the charging phase duration (determined by the end time of charging $\xi_{p}^{\mathrm{EV}*}(t)$ in (10.63)), and (2) the time to reach the destination (determined by the travel time $\xi_{p}^{\mathrm{EV}}(t)$ in (10.61)). Based on the path merging strategy proposed in this section, the complex nested functions in (10.61) and (10.63) can be simplified into explicit expressions. The equivalent route replaces multiple consecutive route segments with a single integrated route. Further, through the transformation from a segment-based model to a path-based model, the original traffic network composed of segments is simplified into an equivalent path-based structure. In this simplified model, intermediate nodes and their corresponding flow conservation constraints are no longer maintained. Only the origin node, fast charging station node, and destination node are retained, as illustrated in Fig. 12.1.

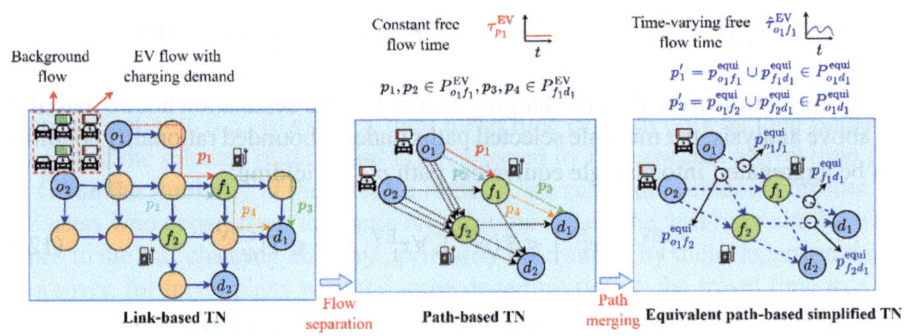

Fig. 12.1 Illustration of path merging and flow separation strategy

The path merging strategy proposed in this section significantly reduces the computational complexity and burden associated with modeling traffic flows on conventional road segments in large-scale transportation networks, while ensuring the modeling accuracy of fast charging stations. As a result, the limited rationality-based dynamic user equilibrium model becomes feasible and practical for estimating the spatiotemporal distribution of charging loads across clusters of fast charging stations.

It should be noted that the above path merging and equivalent approximation analyses are based on the assumption of known distributions of electric and fuel-powered vehicle traffic flows. When these traffic flow distributions undergo significant changes, the corresponding path travel times will also adjust accordingly. Therefore, further investigation is required to explore how traffic flow perturbations affect the validity and effectiveness of the proposed path merging strategy.

Mixed Traffic Flow Separation Strategy. The distribution of path travel times is closely related to the distribution of electric vehicle (EV) and internal combustion engine vehicle (ICEV) traffic flows. On conventional road segments, due to congestion effects, EVs and ICEVs compete for limited road resources. However, in most urban traffic networks, EVs with mid-trip fast charging needs account for a much smaller proportion compared to ICEVs. Consequently, the marginal delay caused by EV traffic on conventional roads is relatively minor. In other words, the additional delays caused by EVs on conventional road segments are generally negligible. As a result, path travel times are primarily determined by ICEV traffic flows, and the influence of EV traffic can be approximated as constant, with both flows exhibiting similar spatiotemporal distributions.

Based on this assumption, this section proposes a mixed traffic flow separation strategy. The core idea is as follows: first, independently solve for the spatiotemporal distribution of ICEV traffic flows; then, based on this distribution, calculate the travel times for each path in the previously defined sets P_{of}^{EV} and P_{fd}^{EV}; the obtained path travel times are then used to estimate the expected travel time functions $\hat{\tau}_{of}^{\mathrm{EV}}(t)$ and $\hat{\tau}_{fd}^{\mathrm{EV}}(t)$ for path merging; the ICEV traffic distribution is treated as background traffic, assumed to be fixed. Under this assumption of travel time equivalence, only EV traffic affected by pricing strategies and marginal delays at fast charging stations is modeled to estimate its spatiotemporal distribution. A schematic of this approach is shown in Fig. 12.1.

Within this framework, EV and ICEV traffic flows are modeled separately to achieve computational decoupling. ICEV traffic is treated as background input, and only the marginal delays it causes at each location are incorporated into EV path travel time estimation. Thus, the equivalent travel time of each path is determined by the departure time of EV users but does not vary with changes in the ICEV traffic distribution. In practical applications, if navigation platforms or transportation management systems can provide traffic distribution data, the expected travel times from origins to fast charging stations f (or from stations to destinations d) and corresponding path recommendations can be used directly to estimate EV travel paths.

This enables the development of simplified EV traffic flow models and improves overall computational efficiency and scalability.

Assuming that the congestion conditions of each road segment in the transportation network can be quantified by a time-varying congestion coefficient $\vartheta(t) \geq 1$, the travel times of the equivalent paths $\hat{\tau}_{of}^{\mathrm{EV}}(t)$ and $\hat{\tau}_{fd}^{\mathrm{EV}}(t)$ can be estimated respectively as:

$$\hat{\tau}_{of}^{\mathrm{EV}}(t) = \tau_{of,0}^{\mathrm{EV}}\vartheta(t), \forall of \in OF, t \in \mathcal{T} \tag{12.12}$$

$$\hat{\tau}_{fd}^{\mathrm{EV}}(t) = \tau_{fd,0}^{\mathrm{EV}}\vartheta(t), \forall fd \in FD, t \in \mathcal{T} \tag{12.13}$$

In these equations, the congestion coefficient $\vartheta(t) = 1$ indicates that the transportation network is uncongested, while $\vartheta(t) > 1$ indicates the presence of congestion. A higher value of $\vartheta(t)$ implies more severe congestion. $\tau_{of,0}^{\mathrm{EV}}$ represents the baseline uncongested travel time from the origin o to the fast charging station f, while $\tau_{fd,0}^{\mathrm{EV}}$ denotes the baseline uncongested travel time from the fast charging station f to the destination d. Under uncongested traffic conditions, these baseline values can be obtained based on the statistical results from (12.5) and (12.10).

User Equilibrium Model with Bounded Rationality in Simplified Transportation Networks. By leveraging the aforementioned path aggregation and traffic flow separation strategies, the originally complex transportation network structure can be transformed into a simplified network composed of equivalent paths. On this basis, this section introduces a user equilibrium model with bounded rationality for dynamic traffic flow in such a simplified transportation network.

(1) *Equivalent Path Traffic Flow Model.* After applying the path aggregation and flow separation strategies, it is only necessary to model the temporal and spatial distribution of electric vehicle (EV) traffic flow at three types of nodes in the transportation network, namely: origin nodes, fast charging station nodes, and destination nodes. The modeling only considers the following three states for EV traffic flow in the network: (1) traveling along the equivalent path p_{of}^{equi} from origin o to charging station f; (2) waiting in the queue to charge at station f (excluding the actual charging state); (3) traveling along the equivalent path p_{fd}^{equi} from station f to destination d.

Since the travel time on the equivalent paths p_{of}^{equi} and p_{fd}^{equi} is known and time-varying, there is no need to model the dynamics along conventional road segments using node-link queuing models. Instead, it is sufficient to improve node-based queuing models to describe the dynamics of EV flow within fast charging stations, thereby greatly reducing the computational complexity of the model.

At this point, the dynamics of EV traffic flow along the equivalent path p_{of}^{equi} can be simplified as follows:

$$Q_{of}^{\text{in}}(t) = \sum_{d \in D} \sum_{p' \in P_{od}^{\text{equi}}} \Lambda_{p'}^{of} Q_{of,p'}^{\text{in}}(t), \forall of \in OF, t \in \mathcal{T} \tag{12.14}$$

In this equation, $Q_{of}^{\text{in}}(t)$ denotes the EV inflow into the equivalent path p_{of}^{equi} at time t; $Q_{of,p'}^{\text{in}}(t)$ denotes the inflow into the equivalent sub-path p' that is part of p_{of}^{equi} at time t; $\Lambda_{p'}^{of}$ is a 0–1 indicator variable, which identifies whether sub-path p' belongs to the set of equivalent paths p_{of}^{equi}. The calculation method is as follows:

$$\Lambda_{p'}^{of} = \begin{cases} 0, \text{ Otherwise} \\ 1, \; p' \in P_{od}^{\text{equi}}, \upsilon(p') = f \end{cases}, \forall of \in OF, d \in D \tag{12.15}$$

The equivalent path decomposition weight $Q_{of,p'}^{\text{in}}(t)$ can be further expressed directly using the path departure flow $h_{p,\varrho}^{\text{EV}}(t)$. However, due to path aggregation, $Q_{of,p'}^{\text{in}}(t)$ may include departure flows from multiple original paths p, which all depart from the same origin o, pass through fast charging station f for charging, and eventually reach destination d. These different paths are ultimately aggregated into the equivalent path p' for modeling. Thus, $Q_{of,p'}^{\text{in}}(t)$ can be represented as:

$$Q_{of,p'}^{\text{in}}(t) = \sum_{\varrho \in \hat{\mathcal{T}}} \sum_{p \in P_{od}^{\text{EV}}} \Lambda_{of}^{p} h_{p,\varrho}^{\text{EV}}(t), \forall of \in OF, d \in D, p' \in P_{od}^{\text{equi}}, t \in \mathcal{T} \tag{12.16}$$

In the above expression, Λ_{of}^{p} is a 0–1 indicator variable denoting whether the segment from origin to fast charging station in path p belongs to the equivalent path p_{of}^{equi}. The calculation method is as follows:

$$\Lambda_{of}^{p} = \begin{cases} 0, \text{ Otherwise} \\ 1, \; p \in P_{od}^{\text{EV}}, \upsilon(p) = f \end{cases}, \forall of \in OF, d \in D \tag{12.17}$$

To simplify, the path departure flow of the grouped equivalent path can be defined as $h_{p',\varrho}^{EV}(t)$, namely:

$$h_{p',\varrho}^{\text{EV}}(t) = \sum_{p \in P_{od}^{\text{EV}}} \Lambda_{of}^{p} h_{p,\varrho}^{\text{EV}}(t), \forall of \in OF, d \in D, p' \in P_{od}^{\text{equi}}, \varrho \in \hat{\mathcal{T}}, t \in \mathcal{T} \tag{12.18}$$

Thus, (12.16) can be rewritten as:

$$Q_{of,p'}^{\text{in}}(t) = \sum_{\varrho \in \hat{\mathcal{T}}} h_{p',\varrho}^{\text{EV}}(t), \forall of \in OF, d \in D, p' \in P_{od}^{\text{equi}}, \Lambda_{p'}^{of} = 1, t \in \mathcal{T} \tag{12.19}$$

The path departure traffic flow on the equivalent path p_{of}^{equi} can be expressed as:

$$Q_{of}^{\text{out}}(t) = \sum_{d \in D} \sum_{p' \in P_{od}^{\text{equi}}} \Lambda_{p'}^{of} Q_{of,p'}^{\text{out}}(t), \forall of \in OF, t \in \mathcal{T} \tag{12.20}$$

In this expression, $Q_{of}^{\text{out}}(t)$ denotes the path departure traffic flow on equivalent path p_{of}^{equi} at time t (i.e., traffic entering the fast charging station f); $Q_{of,p'}^{\text{out}}(t)$ denotes the decomposition weight of $Q_{of,p'}^{\text{out}}(t)$ along equivalent path p' and the path inflow decomposition weight $Q_{of,p'}^{in}(t)$ can be simplified as:

At this point, there is no need to explicitly consider capacity constraints on conventional road segments, as the congestion effect caused by capacity constraints has already been modeled through time-varying travel impedance parameters. Thus:

$$Q_{of}^{\text{out}}\left[t + \hat{\tau}_{of}^{\text{EV}}(t)\right] = Q_{of}^{\text{in}}(t), \forall of \in OF, t \in \mathcal{T} \tag{12.21}$$

According to the first-in-first-out condition (2) in (10.69), the temporal relationship between the path departure traffic flow decomposition weight $Q_{of,p'}^{\text{out}}(t)$ and the path inflow decomposition weight $Q_{of,p'}^{\text{in}}(t)$ can be simplified as:

$$Q_{of,p'}^{\text{out}}\left[t + \hat{\tau}_{of}^{\text{EV}}(t)\right] = Q_{of,p'}^{\text{in}}(t), \forall of \in OF, d \in D, p' \in P_{od}^{\text{equi}}, \Lambda_{p'}^{of} = 1, t \in \mathcal{T} \tag{12.22}$$

The boundary condition is:

$$Q_{of,p'}^{\text{out}}(t) = 0, \forall of \in OF, d \in D, p' \in P_{od}^{\text{equi}}, \Lambda_{p'}^{of} = 1, t \in \left[0, \hat{\tau}_{of}^{\text{EV}}(0)\right) \tag{12.23}$$

Meanwhile, the first-in-first-out condition (1) in (10.66) ensures monotonicity of the mapping function $t + \hat{\tau}_{of}^{\text{EV}}(t)$, namely:

$$t_1 + \hat{\tau}_{of}^{\text{EV}}(t_1) \le t_2 + \hat{\tau}_{of}^{\text{EV}}(t_2), \forall of \in OF, 0 \le t_1 \le t_2 \le T \tag{12.24}$$

Similarly, the path departure traffic flow on equivalent path p_{fd}^{equi} can be expressed as:

$$Q_{fd}^{\text{in}}(t) = \sum_{o \in O} \sum_{p' \in P_{od}^{\text{equi}}} \Lambda_{p'}^{fd} Q_{fd,p'}^{\text{in}}(t), \forall fd \in FD, t \in \mathcal{T} \tag{12.25}$$

$$\Lambda_{p'}^{fd} = \begin{cases} 0, \text{ Otherwise} \\ 1, \ p' \in P_{od}^{\text{equi}}, \ \upsilon(p') = f \end{cases}, \forall o \in O, fd \in FD \tag{12.26}$$

$$Q_{fd}^{\text{out}}(t) = \sum_{o \in O} \sum_{p' \in P_{od}^{\text{equi}}} \Lambda_{p'}^{fd} Q_{fd,p'}^{\text{out}}(t), \forall fd \in FD, t \in \mathcal{T} \tag{12.27}$$

$$Q_{fd,p'}^{\text{out}}\left[t+\hat{\tau}_{fd}^{\text{EV}}(t)\right]=Q_{fd,p'}^{\text{in}}(t),\forall o\in O,fd\in FD,p'\in P_{od}^{\text{equi}},\Lambda_{p'}^{fd}=1,t\in\mathcal{T} \tag{12.28}$$

$$Q_{fd,p'}^{\text{out}}(t)=0,\forall o\in O,fd\in FD,p'\in P_{od}^{\text{equi}},\Lambda_{p'}^{fd}=1,t\in\left[0,\tau_{fd}^{\text{EV}}(0)\right) \tag{12.29}$$

$$t_1+\hat{\tau}_{fd}^{\text{EV}}(t_1)\le t_2+\hat{\tau}_{fd}^{\text{EV}}(t_2),\forall fd\in FD,0\le t_1\le t_2\le T \tag{12.30}$$

In (12.25), $Q_{fd}^{\text{in}}(t)$ denotes the inflow traffic on equivalent path p_{fd}^{equi} at time t; $Q_{fd,p'}^{\text{in}}(t)$ denotes the decomposition weight of equivalent path p' along the grouped equivalent path. Equation (12.26) defines the 0–1 coefficient $\Lambda_{p'}^{fd}$ indicating whether path p_{fd}^{equi} includes the grouped equivalent path p'. In (12.27), $Q_{fd}^{\text{out}}(t)$ denotes the departure traffic flow (i.e., arriving at terminal node d) on equivalent path p_{fd}^{equi} at time t; $Q_{fd,p'}^{\text{out}}(t)$ denotes the decomposition weight of $Q_{fd}^{\text{out}}(t)$ along the grouped equivalent path p'. Equation (12.28) represents the temporal relationship between the inflow decomposition weight $Q_{fd,p'}^{\text{in}}(t)$ and the outflow decomposition weight $Q_{fd,p'}^{\text{out}}(t)$, corresponding to the first-in-first-out condition (2). Equation (12.29) provides the boundary condition of the traffic flow decomposition weight $Q_{fd,p'}^{\text{in}}(t)$. Equation (12.30) describes the first-in-first-out condition (1) for traffic flow on equivalent path p_{fd}^{equi}.

$$\sum_{p'\in P_{od}^{\text{equi}}}\sum_{\varrho\in\hat{\mathcal{T}}}\sum_{t\in\mathcal{T}}h_{p',\varrho}^{\text{EV}}(t)=\sum_{\varrho\in\hat{\mathcal{T}}}Q_{od,\varrho}^{\text{EV}},\forall od\in OD \tag{12.31}$$

(2) *Fast Charging Station Node Model Based on Modified Queue.* We now discuss the fast charging station model between equivalent paths p_{of}^{equi} and p_{fd}^{equi}. In this case, the fast charging station is no longer modeled as a segment, but instead as a node. The corresponding traffic flow dynamics of inflow and outflow at the fast charging station are expressed as follows:

$$Q_f^{\text{in}}(t)=\sum_{od\in OD}\sum_{p'\in P_{od}^{\text{equi}}}\Lambda_{p'}^{f}Q_{f,p'}^{\text{in}}(t)=\sum_{o\in O}Q_{of}^{\text{out}}(t),\forall f\in F,t\in\mathcal{T} \tag{12.32}$$

$$Q_f^{\text{out}}(t)=\sum_{od\in OD}\sum_{p'\in P_{od}^{\text{equi}}}\Lambda_{p'}^{f}Q_{f,p'}^{\text{out}}(t)=\sum_{d\in D}Q_{fd}^{\text{in}}(t),\forall f\in F,t\in\mathcal{T} \tag{12.33}$$

$$\Lambda_{p'}^{f}=\begin{cases}0, & \upsilon(p')\neq f\\ 1, & \upsilon(p')=f\end{cases},\forall f\in F,od\in OD,p'\in P_{od}^{\text{equi}} \tag{12.34}$$

In (12.32), $Q_f^{\text{in}}(t)$ represents the inflow traffic at fast charging station f at time t, which equals the sum of traffic arriving at f from all origins. $Q_{f,p'}^{\text{in}}(t)$ denotes the

$$Q_{f,p'}^{\text{out}}\left[t+\hat{\varsigma}_f(t)\right]=\begin{cases}\frac{Q_{f,p'}^{\text{in}}(t)}{Q_f^{\text{in}}(t)}Q_f^{\text{out}}\left[t+\hat{\varsigma}_f(t)\right], & Q_f^{\text{in}}(t)\neq 0\\ 0, & Q_f^{\text{in}}(t)=0\end{cases}, \forall f\in F, t\in\mathcal{T} \quad (12.42)$$

Based on the above analysis and considering the path-node structure of the transportation network, the fast-charging station node is the only intermediate node for these paths, i.e., the only traffic node. Therefore, we only need to consider the disaggregated flow conservation relationship at the fast-charging station node.

From this, it can be derived that the disaggregated flow on the equivalent path p_{od}^{equi} is equal to the disaggregated inflow on the same path before entering the fast-charging station:

$$Q_{of,p'}^{\text{out}}(t)=Q_{f,p'}^{\text{in}}(t), \forall of\in OF, d\in D, p'\in P_{od}^{\text{equi}}, \Lambda_{p'}^{of}=1, t\in\mathcal{T} \quad (12.43)$$

Likewise, the outflow on path p_{fd}^{equi} after leaving the fast-charging station is equal to the disaggregated outflow on that path:

$$Q_{f,p'}^{\text{out}}(t)=Q_{fd,p'}^{\text{in}}(t), \forall o\in O, fd\in FD, p'\in P_{od}^{\text{equi}}, \Lambda_{p'}^{fd}=1, t\in\mathcal{T} \quad (12.44)$$

Equations (12.43) and (12.44) represent flow conservation constraints at the fast-charging station node for disaggregated flows, ensuring that all path flows entering and exiting the station remain consistent.

(3) *Equivalent Path Travel Cost and Simplified Bounded Rational User Equilibrium Model.* In the simplified transportation network described above, the travel cost calculation method for the equivalent path corresponding to the electric vehicle path flow emission is as follows:

$$\phi_{p',\varrho}^{\text{EV}}(t)=\varphi_{p'}^{\text{EV}}(t)+\rho_{p',\varrho}^{\text{EV}}(t)+\theta_{p'}^{\text{EV}}(t), \forall od\in OD, p'\in P_{od}^{\text{equi}}, \varrho\in\hat{\mathcal{T}}, t\in\mathcal{T} \quad (12.45)$$

$$\varphi_{p'}^{\text{EV}}(t)=\gamma^{\text{EV,t}}\left[\hat{\xi}_{p'}^{\text{EV}}(t)-t\right], \forall od\in OD, p'\in P_{od}^{\text{equi}}, t\in\mathcal{T} \quad (12.46)$$

$$\rho_{p',\varrho}^{\text{EV}}(t)=\gamma^{\text{e}}\min\left[0,\hat{\xi}_{p'}^{\text{EV}}(t)-\varrho+\Delta\tau\right]+\gamma^{\text{l}}\max\left[0,\hat{\xi}_{p'}^{\text{EV}}(t)-\varrho-\Delta\tau\right],$$
$$\forall od\in OD, p'\in P_{od}^{\text{equi}}, \varrho\in\hat{\mathcal{T}}, t\in\mathcal{T} \quad (12.47)$$

$$\theta_{p'}^{\text{EV}}(t)=\sum_{t'\in T_{p'}^{\text{ch}}(t)}\gamma_{\upsilon(p')}^{\text{ch}}(t')P^{\text{ch}}\Delta t, \forall od\in OD, p'\in P_{od}^{\text{equi}}, t\in\mathcal{T} \quad (12.48)$$

$$T_{p'}^{\text{ch}}(t) = \left(\hat{\xi}_{p'}^{\text{EV}*}(t) - \tau^{\text{ch}}, \hat{\xi}_{p'}^{\text{EV}*}(t)\right], \forall od \in OD, p' \in P_{od}^{\text{equi}}, t \in \mathcal{T} \quad (12.49)$$

In these formulas, $\phi_{p',\varrho}^{\text{EV}}(t)$ denotes the travel cost of electric vehicle path flow $h_{p',\varrho}^{\text{EV}}(t)$, and the cost incurred when electric vehicle user ϱ departing from OD pair od selects the equivalent path p' at time t; $\varphi_{p'}^{\text{EV}}(t)$ is the time cost, $\rho_{p',\varrho}^{\text{EV}}(t)$ is the penalty for early or delayed arrival, $\theta_{p'}^{\text{EV}}(t)$ is the electricity cost, and $T_{p'}^{\text{ch}}(t)$ is the charging duration.

The user's electricity cost depends on the expected charging duration and actual completion time. Charging cost is determined by the average electricity price over the expected charging period, rather than by the real-time price during charging, as shown in (12.48). When the electricity demand is known in advance, the adjustment of charging power output rate does not affect the user's electricity expense, nor does it alter the temporal distribution of EV traffic. Instead, it affects the allocation of electricity consumption across the fast-charging station and its impact on the local grid's electricity cost.

In short, the grid-side cost savings brought by power control will be fully reflected in the operating benefit of the fast-charging station cluster. For the EV users, this process is imperceivable. Therefore, under the assumption of zero electricity price uncertainty, the electricity demand of EV users can be regarded as a known parameter, which can directly be used in energy dispatch models.

According to the research on electricity pricing for fast-charging in Chap. 11, the electricity price for EVs at fast-charging stations is affected by the electricity market prices. This can be modeled using the Stackelberg game model. In practical applications, fast-charging stations often participate in the competitive electricity market through aggregators. Therefore, the price at which they procure electricity from the grid is no longer fixed, but dynamically determined based on real-time and forecasted prices, as discussed in Chap. 11.

By simplifying the multi-layer Stackelberg game model proposed in Chap. 11 to consider only the interaction between fast-charging station aggregators and EV users, the optimal retail electricity price can still be obtained using the Gauss–Seidel iterative algorithm from Chap. 11.

Since this study focuses on the flexible power adjustment capability of fast-charging station aggregators and does not rely on the detailed retail pricing mechanism, the retail electricity price is treated as an exogenous parameter, and the discussion on its determination process is omitted.

To simplify the problem, suppose the retail price is linearly correlated with the real-time market price. That is:

$$\gamma_f^{\text{ch}}(t) = \alpha^{\text{p}} \lambda^{\text{RT}}(t), \forall f \in F, t \in \mathcal{T} \quad (12.50)$$

On this basis, the user equilibrium condition for the simplified transportation network can be expressed as:

$$\forall od \in \mathrm{O}, p' \in P_{od}^{\text{equi}}\varrho \in \hat{\mathcal{T}}, t \in \mathcal{T} \Rightarrow \begin{cases} h_{p',\varrho}^{\text{EV}}(t) > 0 \Rightarrow \phi_{p',\varrho}^{\text{EV}}(t) \in [\mu_{od,\varrho}^{\text{EV}}, \mu_{od,\varrho}^{\text{EV}} + \varepsilon_{od}^{\text{EV}}] \\ \phi_{p',\varrho}^{\text{EV}}(t) > \mu_{od,\varrho}^{\text{EV}} + \varepsilon_{od}^{\text{EV}} \Rightarrow h_{oc,\varrho}^{\text{EV}}(t) = 0 \end{cases} \tag{12.51}$$

$$\mu_{od,\varrho}^{\text{EV}} = \min_{p' \in P_{od}^{\text{equi}}} \min_{t \in \mathcal{T}} \phi_{od,\varrho}^{\text{EV}}(t), \forall od \in \mathrm{O}, \varrho \in \hat{\mathcal{T}} \tag{12.52}$$

Due to the traffic queuing and waiting time involved in the computation, the resistance delay and time cost in the charging phase correspond closely to the approximation function in (12.11) and (12.40). Therefore, the bounded rational user equilibrium model proposed in this section is consistent with the lightweight characteristics of this chapter.

To conclude, with the route merging and flow decomposition strategy proposed in this chapter, the simplified bounded rational user equilibrium model can be applied to estimate large-scale charging demand at fast-charging station clusters in transportation networks.

Aggregated Charging Demand Regulation Strategy for Fast-Charging Stations. In the management of large-scale fast-charging station clusters, if each charging pile independently adjusts its power output, it introduces a large number of decision variables. When the number of piles is large, it significantly increases the computational complexity and solution burden of the scheduling model. To improve the scheduling efficiency of large-scale clusters, this section proposes a strategy based on aggregated charging demand regulation to achieve simplified load management and optimization for fast-charging stations.

Considering the urgency of fast-charging, most EV users' dwell time does not exceed 1 h, and their state of charge (SoC) is typically below 30% when entering the station, resulting in a tendency to perform rapid charging. At present, the battery capacities of mainstream EVs do not differ significantly, so the aggregated charging demand can be treated in a unified manner. Based on this observation, during the modeling of charging behavior, the user's charging demand can be approximated as completing a fixed charging amount E^{ch} within a certain charging duration t^{ch}. This abstraction method effectively avoids the computational complexity brought by modeling pile-level detail and provides a unified load input for subsequent grid-side energy regulation models.

At time t_s, the number of EV users predicted to charge at station f is denoted as $Q_f^{\text{pre}}(t_s)$. Since traffic flow is a macroscopic concept, $Q_f^{\text{pre}}(t_s)$ is allowed to take any value in the continuous range $[0, N_f]$. Let Q be the number of charging piles used by traffic flow $Q_f^{\text{pre}}(t_s)$, then for convenience of analysis, define Q as the ceiling of $Q_f^{\text{pre}}(t_s)$, that is:

$$Q = \text{ceil}\left[Q_f^{\text{pre}}(t_s)\right], \forall f \in F, t_s \in \mathcal{T} \tag{12.53}$$

Here, ceil(·) denotes the upward rounding operator. For any EV user q in $Q_f^{\text{pre}}(t_s)$, its charging demand is subject to the following constraint:

$$0 \le P_{f,q,t_s}^{\text{ch}}(t) \le \overline{P}^{\text{ch}}, \forall f \in F, q \in [1, \ldots, Q], t \in (t, t+\tau^{\text{ch}}], t_s \in \mathcal{T} \tag{12.54}$$

$$P_{f,q,t_s}^{\text{ch}}(t) = 0, \forall f \in F, q \in [1, \ldots, Q], t \in [0, t] \cup (t+\tau^{\text{ch}}, t], t_s \in \mathcal{T} \tag{12.55}$$

$$\sum_{t \in (t_s, t_s+\tau^{\text{ch}}]} P_{f,q,t_s}^{\text{ch}}(t)\Delta t = E^{\text{ch}}, \forall f \in F, q \in [1, \ldots, Q], t_s \in \mathcal{T} \tag{12.56}$$

In (12.54), $P_{f,q,t}^{\text{ch}}(t)$ denotes the charging power of EV user q who starts charging at fast-charging station f at time t_s and is actively charging at time t; P^{ch} is the upper bound of the charging pile power.

Equations (12.54) and (12.55) specify that charging power is only allowed to be positive when EV user q is connected to the charging pile.

The non-charging period fixes the power output to zero. Equation (12.56) specifies that the total amount of energy charged during the charging period must satisfy the user's energy demand. Due to the homogeneity of EV traffic charging demand, and the consistency of the constraints among users starting to charge at the same time, users in $Q_f^{\text{pre}}(t_s)$ can be grouped together and treated with a unified charging power regulation strategy, that is:

$$P_{f,t_s}^{\text{ch}}(t) = P_{f,1,t_s}^{\text{ch}}(t) = \cdots = P_{f,Q,t_s}^{\text{ch}}(t), \forall f \in F, t \in T, t_s \in \mathcal{T} \tag{12.57}$$

In this strategy, $P_{f,q,t}^{\text{ch}}(t)$ is shared across users in the same group starting at t_s, simplifying the charging constraints as follows:

$$0 \le P_{f,t_s}^{\text{ch}}(t) \le \overline{P}^{\text{ch}}, \forall f \in F, t \in (t, t+\tau^{\text{ch}}], t_s \in \mathcal{T} \tag{12.58}$$

$$P_{f,t_s}^{\text{ch}}(t) = 0, \forall f \in F, t \in [0, t] \cup (t+\tau^{\text{ch}}, t], t_s \in \mathcal{T} \tag{12.59}$$

$$\sum_{t \in (t_s, t_s+\tau^{\text{ch}}]} P_{f,t_s}^{\text{ch}}(t)\Delta t = E^{\text{ch}}, \forall f \in F, t_s \in \mathcal{T} \tag{12.60}$$

At time t, after weighting the power across all groups, the total power demand of fast-charging station f is:

$$P_f^{\text{ch}}(t) = \sum_{t_s \in \mathcal{T}} Q_f^{\text{pre}}(t_s) P_{f,t_s}^{\text{ch}}(t), \forall t \in \mathcal{T} \tag{12.61}$$

According to the properties of $P_{f,q}^{\text{ch}}(t, t_s)$ defined in (12.54) and (12.55), (12.61) can be further simplified as:

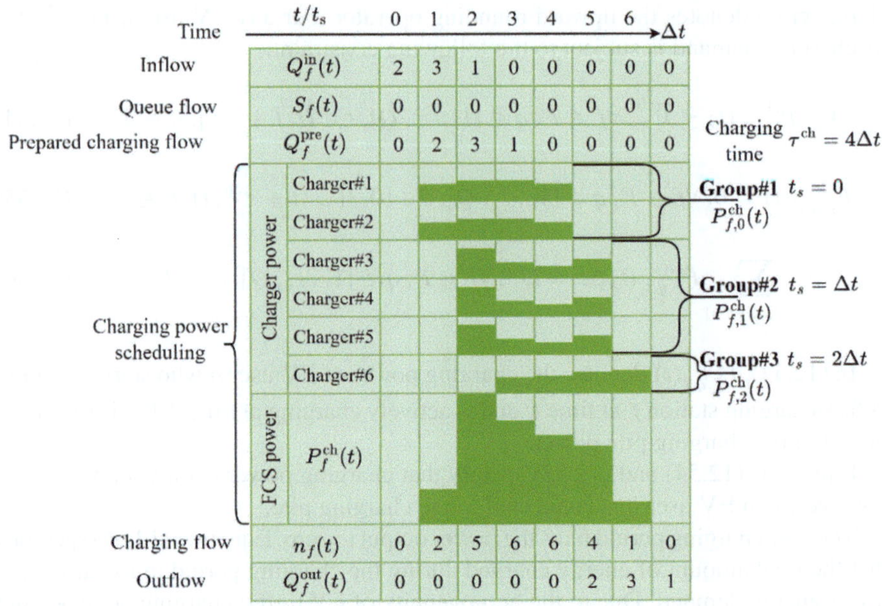

Fig. 12.2 Illustration of the group-based aggregated charging scheduling algorithm

$$P_f^{ch}(t) = \sum_{t_s \in [\max(t-\tau^{ch},0),t)} Q_f^{pre}(t_s) P_{f,t_s}^{ch}(t), \forall t \in \mathcal{T} \tag{12.62}$$

An illustrative example of the group-based charging regulation strategy is shown in Fig. 12.2. The fast-charging station has six charging piles, and the charging duration is $4\Delta t$. At $t = 0/1/2\Delta t$, 2, 3, and 1 EVs arrive at the fast-charging station, respectively, forming three groups. Each group follows a uniform charging power control strategy, whereby the total real-time charging power of the fast-charging station equals the sum of the real-time power of all charging piles.

Provided the user's charging demand is met, the above group-based charging control strategy can reduce the number of decision variables and constraints for power control to only $1/Q$ of the original amount, thereby significantly improving the aggregation and management efficiency for large-scale fast-charging station clusters.

Moreover, due to the abstract nature of traffic flow modeling, the number of centralized control decision variables in this strategy is fixed and only related to the number of charging stations $|F|$, time discretization steps $|\mathcal{T}|$, and the charging period length τ^{ch}, satisfying:

$$\left|\Pi^{ch}\right| = |F||\mathcal{T}|\frac{\tau^{ch}}{\Delta t} \tag{12.63}$$

where Π^{ch} denote the set of decision variables related to charging scheduling, defined as:

$$\Pi^{\text{ch}} = \left\{ P_{f,t_s}^{\text{ch}}(t) | f \in F, t \in (t, t + \tau^{\text{ch}}], t_s \in \mathcal{T} \right\} \tag{12.64}$$

Further analysis indicates that when no electric vehicle user is connected to a charging station f at time t_s (i.e., $Q_f^{\text{begin}}(t_s) = 0$), the corresponding group charging power $P_{f,t_s}^{\text{ch}}(t)$ remains fixed at zero. The associated decision variable is considered inactive and may be removed from the variable set. The set of active charging decision variables is denoted by $\tilde{\Pi}^{\text{ch}}$, defined as:

$$\tilde{\Pi}^{\text{ch}} = \left\{ P_{f,t_s}^{\text{ch}}(t) | Q_f^{\text{pre}}(t_s) > 0, f \in F, t \in (t, t + \tau^{\text{ch}}], t_s \in \mathcal{T} \right\} \tag{12.65}$$

It is evident that:

$$\left| \tilde{\Pi}^{\text{ch}} \right| \leq \left| \Pi^{\text{ch}} \right| = |F||\mathcal{T}| \frac{\tau^{\text{ch}}}{\Delta t} \tag{12.66}$$

In the above scheduling model, the number of activated decision variables has a well-defined upper bound. Since a single $Q_f^{\text{begin}}(t_s)$ can accommodate a maximum flow of N_f electric vehicles, the total modeled electric vehicle quantity $Q_{od,\theta}^{\text{EV}}$ increases without significantly increasing the number of non-zero $Q_f^{\text{begin}}(t_s)$. As such, the scale of the problem does not expand proportionally with the number of EVs, implying that the proposed modeling approach remains tractable and suitable for large-scale electric vehicle access scenarios.

Coordinated Power Control Model for Fast Charging Stations Incorporating On-Site Renewable Energy and Energy Storage Systems. In the process of constructing a green, low-carbon, and sustainable integrated transportation and energy system, it is essential not only to promote the electrification of transportation systems, but also to comprehensively integrate distributed renewable energy sources to achieve a clean power supply. To fully utilize distributed renewable energy in urban areas, fast charging stations can be equipped with rooftop photovoltaic panels, small wind turbines, and other energy-generating systems during site deployment. This enables local clean and low-carbon charging services for electric vehicles. Such an approach enhances the consumption capacity of distributed renewable energy and mitigates the grid stress caused by concentrated charging loads, thereby reducing the operational pressure on the power grid, lowering "peak shaving" risks, and significantly cutting the comprehensive electricity cost of fast charging stations. It also improves the autonomy and economic efficiency of on-site renewable energy systems.

The coordinated power control model for fast charging stations incorporating on-site renewable energy and energy storage systems is expressed as follows:

$$P_f^{\mathrm{ch}}(t) = P_f^{\mathrm{grid}}(t) + P_f^{\mathrm{PV}}(t) + P_f^{\mathrm{wind}}(t) + P_f^{\mathrm{ESS}}(t), \forall f \in F, t \in \mathcal{T} \tag{12.67}$$

$$0 \le P_f^{\mathrm{ch}}(t) \le N_f P_{\mathrm{ch}}, \forall f \in F, t \in \mathcal{T} \tag{12.68}$$

$$0 \le P_f^{\mathrm{grid}}(t) \le \overline{P}^{\mathrm{grid}}, \forall f \in F, t \in \mathcal{T} \tag{12.69}$$

$$0 \le P_f^{\mathrm{PV}}(t) \le \overline{P}_f^{\mathrm{PV}}(t), \forall f \in F, t \in \mathcal{T} \tag{12.70}$$

$$0 \le P_f^{\mathrm{wind}}(t) \le \overline{P}_f^{\mathrm{wind}}(t), \forall f \in F, t \in \mathcal{T} \tag{12.71}$$

$$P_f^{\mathrm{ESS}}(t) = P_f^{\mathrm{ESS-}}(t) - P_f^{\mathrm{ESS+}}(t), \forall f \in F, t \in \mathcal{T} \tag{12.72}$$

$$0 \le P_f^{\mathrm{ESS-}}(t) \le u_f(t)\overline{P}_f^{\mathrm{ESS-}}, \forall f \in F, t \in \mathcal{T} \tag{12.73}$$

$$0 \le P_f^{\mathrm{ESS+}}(t) \le [1 - u_f(t)]\overline{P}_f^{\mathrm{ESS+}}, \forall f \in F, t \in \mathcal{T} \tag{12.74}$$

$$\begin{aligned} \mathrm{SoC}_f^{\mathrm{ESS}}(t + \Delta t) = {} & \mathrm{SoC}_f^{\mathrm{ESS}}(t) \\ & + \left(P_f^{\mathrm{ESS+}}(t)\Delta t \eta^{+} - P_f^{\mathrm{ESS-}}(t)\Delta t/\eta^{-}\right)/E_f^{\mathrm{ESS}}, \forall f \in F, t \in \mathcal{T} \end{aligned} \tag{12.75}$$

$$\underline{\mathrm{SoC}}_f^{\mathrm{ESS}} \le \mathrm{SoC}_f^{\mathrm{ESS}}(t) \le \overline{\mathrm{SoC}}_f^{\mathrm{ESS}}, \forall f \in F, t \in \mathcal{T} \tag{12.76}$$

$$\mathrm{SoC}_f^{\mathrm{ESS}}(0) = \mathrm{SoC}_f^{\mathrm{ESS}}(T) = \mathrm{SoC}_f^{\mathrm{ESS,init}}, \forall f \in \mathcal{T} \tag{12.77}$$

where (12.67) represents the power balance constraint at a fast charging station f, where the total charging power $P_f^{\mathrm{ch}}(t)$ at time t is provided by the sum of the power delivered by the power grid $P_f^{\mathrm{grid}}(t)$, photovoltaic generation $P_f^{\mathrm{PV}}(t)$, wind generation $P_f^{\mathrm{wind}}(t)$, and the energy output from energy storage $P_f^{\mathrm{ESS}}(t)$. Equation (12.68) through (12.71) specify the constraints for total charging power, grid power, PV power, and upper and lower bounds for wind power, respectively. Here, $N_f P_{\mathrm{ch}}$, $\overline{P}^{\mathrm{grid}}$, $\overline{P}_f^{\mathrm{PV}}(t)$, and $\overline{P}_f^{\mathrm{wind}}(t)$ represent the corresponding upper bounds of power output. Since the maximum power output of PV and wind generation is subject to solar irradiance and wind speed, these quantities are time-dependent. Equation (12.72) defines the power output $P_f^{ESS}(t)$ from energy storage, which accounts for losses during charging and discharging. To accurately compute the output, the state of charge (SoC) of the storage system is introduced. The discharging power $P_f^{\mathrm{ESS-}}(t)$ and charging power $P_f^{\mathrm{ESS+}}(t)$ are defined, and the net output is obtained as their difference. Equations (12.73) and (12.74) constrain the maximum charging and discharging rates, where

$u_f(t)$ is a binary variable (0–1) indicating whether discharging is permitted at time t, and $\overline{P}_f^{\mathrm{ESS}+}$ and $\overline{P}_f^{\mathrm{ESS}-}$ represent the upper bounds for charging and discharging power, respectively. Equation (12.75) describes the SoC computation for storage. $\mathrm{SoC}_f^{\mathrm{ESS}}(t)$ denotes the SoC of the storage device at time t; η^+ and η^- are the charging and discharging efficiencies, and E_f^{ESS} is the energy capacity of the storage system. Equation (12.76) defines the permissible operating range of SoC, where $\underline{\mathrm{SoC}}_f^{\mathrm{ESS}}$ and $\overline{\mathrm{SoC}}_f^{\mathrm{ESS}}$ are the upper and lower bounds, respectively. Finally, Eq. (12.77) imposes periodic constraints on storage operation, requiring that the initial and terminal SoC within the scheduling horizon T be equal to a predetermined initial value $\mathrm{SoC}_f^{\mathrm{ESS,init}}$.

12.3 Stochastic Optimization Method for Energy Regulation of Fast Charging Station Aggregators Participating in the Two-Stage Electricity Market

Based on the previously proposed energy regulation model for fast charging station clusters, the charging stations can participate in electricity market transactions in an aggregated manner through aggregators, thereby enabling orderly grid integration of large-scale charging loads in the coupled electricity-transportation network. This section introduces the modeling methods of uncertainty scenarios faced by fast charging station aggregators in the two-stage electricity market, along with the corresponding stochastic optimization strategies for energy regulation.

Multidimensional Uncertainty Scenario Modeling Based on Clustering. The aforementioned models are formulated under deterministic assumptions. However, in actual operation, fast charging station aggregators, when organizing station clusters to participate in electricity market trading, encounter various operational uncertainties from multiple dimensions, including user charging demands (determined by travel behavior), electricity market prices, and renewable generation within stations. Following the approach, this study assumes that the day-ahead electricity price is known at the day-ahead bidding stage. Thus, the primary uncertainty to be addressed in market bidding and regulation decisions lies in the real-time electricity price for the following day.

To effectively handle the multidimensional uncertainties in energy regulation optimization, it is first necessary to characterize the statistical distribution of each uncertain parameter. Accordingly, this section employs the K-means clustering method to perform scenario clustering on historical data from different uncertainty sources. A representative set of typical scenarios is extracted and used as inputs to the stochastic programming model, thereby reducing problem complexity while preserving essential uncertainty characteristics.

Based on these representative scenarios, a stochastic optimization framework is developed in the subsequent section to jointly determine the optimal day-ahead electricity bidding strategy and real-time charging scheduling strategy for fast charging station aggregators in the two-stage electricity market, to enhance both economic efficiency and system robustness.

Taking a certain stochastic parameter (e.g., traffic demand, real-time electricity price, upper bounds of photovoltaic and wind generation) as an example, let R denote the historical dataset over I days. Each day is regarded as an individual scenario. Denote the sampling interval set for each day as $\mathcal{T}$, and the sampled value at time t on day i as $R_i(t)$. Then, the time series of sampled values for day i is represented as R_i, namely:

$$R_i = \{R_i(t)|t \in \mathcal{T}\}, \forall i \in I \tag{12.78}$$

The scenario reduction method based on K-means clustering is introduced as follows:

Algorithm 12.1 Scenario Reduction Based on K-means Clustering

Step 1	Set the number of clusters K, the current iteration index $z = 0$, and the maximum number of iterations Z
Step 2	Randomly select K daily sample time series R_i as the initial cluster centers, forming the set $\hat{R}^z = \hat{R}_k^z \vert k \in K$
Step 3	Calculate the Euclidean distance between each daily sample time series R_i and each cluster center $\hat{R}_k^z$: $d(R_i, \hat{R}_k^z) = \Vert R_i - \hat{R}_k^z \Vert_2 = \sqrt{\sum_{t \in T} [R_i(t) - \hat{R}_k^z(t)]^2}, \forall i \in I, k \in K$ Assign each R_i to the cluster k whose center $\hat{R}_k^z$ is closest in terms of Euclidean distance. Define the cluster assignment function $g(R_i)$ as: $g(R_i) = \arg\min_{k \in K} d(R_i, \hat{R}_k^z), \forall i \in I$
Step 4	Update the cluster G_k^{z+1} by grouping all time series R_i assigned to cluster k, i.e., $G_k^{z+1} = \{R_i \vert g(R_i) = k, i \in I\}, \forall k \in K$ Compute the centroid of each cluster G_k^{z+1} to form the new cluster centers: $\hat{R}^{z+1} = \{\hat{R}_k^{z+1} \vert k \in K\}$ $\hat{R}_k^{z+1} = \frac{1}{\vert G_k^{z+1} \vert} \sum_{R_i \in G_k^{z+1}} R_i$
Step 5	If $\hat{R}^{z+1}$ is equal to $\hat{R}^z$, or if $z \geq Z$, then output the final clustering result $\hat{R}^{z+1}$; otherwise, increment $z = z + 1$ and return to Step 3

Using the above method, K representative scenario sets can be obtained separately for traffic demand, real-time electricity price, photovoltaic generation power upper bound, and wind power generation power upper bound. These are denoted as $\{Q_k^{\text{EV}}\}$, $\{\lambda_k^{\text{RT}}\}$, $\{\overline{P}_k^{\text{PV}}\}$, and $\{\overline{P}_k^{\text{wind}}\}$, respectively. By combining representative scenarios of these uncertain factors, the set of uncertain operating scenarios for the aggregated fast charging station operator can be formulated as $\mathcal{W} = w$, containing a total of K^4

operating scenarios. Specifically, the wth scenario corresponds to the combination of traffic demand, real-time electricity price, photovoltaic power upper bound, and wind power upper bound represented by $Q_{k_1}^{\mathrm{EV}}$, $\lambda_{k_2}^{\mathrm{RT}}$, $\overline{P}_{k_3}^{\mathrm{PV}}$, and $\overline{P}_{k_4}^{\mathrm{wind}}$, respectively. The mapping relationship satisfies:

$$w = 1 + k_1 + K(k_2 - 1) + K^2(k_3 - 1) + K^3(k_4 - 1), \forall w \in W \tag{12.79}$$

Accordingly, the index w can be used to represent the operating scenario number (k_1, k_2, k_3, k_4).

Power Trading Model of Aggregated Fast Charging Stations. The architecture of the aggregated fast charging station operator is illustrated in Fig. 12.3. The aggregator centrally coordinates charging piles, renewable energy resources, and energy storage systems across all fast charging stations. Under the premise of satisfying individual EV users' charging demands, the aggregator enables the station cluster to participate in the two-stage competitive electricity spot market. This facilitates not only the orderly grid integration of charging loads and local consumption of distributed renewable energy but also reduces the overall electricity cost for the station cluster.

Taking the two-stage competitive electricity spot market in Guangdong Province as an example, the aggregator must submit the forecasted electricity demand for the next day in the day-ahead stage. Specifically, this involves reporting the power

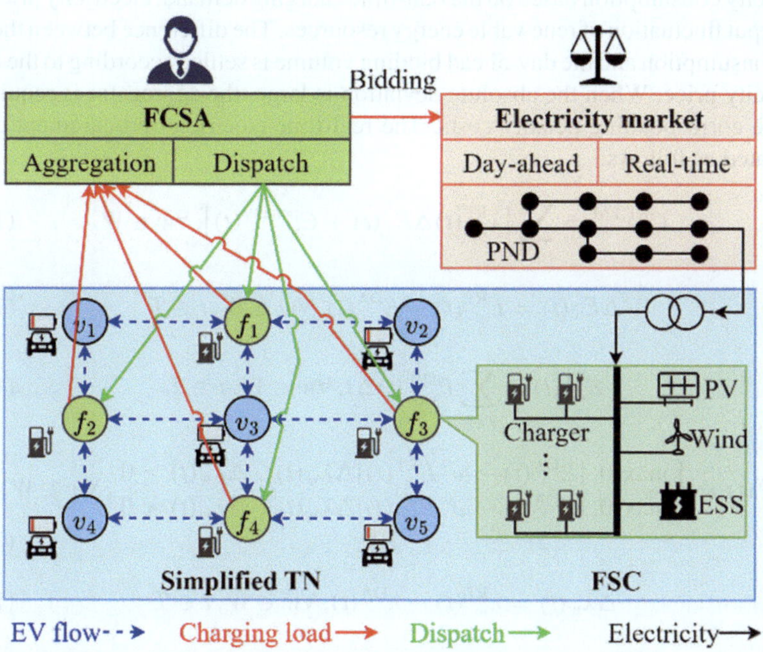

Fig. 12.3 Aggregator model for fast charging stations

consumption curve of grid electricity across all time intervals for the next day. The aggregator pays the corresponding day-ahead bidding cost based on the day-ahead prices. The calculation method is as follows:

$$C^{\mathrm{DA,EM}} = \sum_{t \in \mathcal{T}} \lambda^{\mathrm{DA}}(t) E^{\mathrm{DA}}(t) \tag{12.80}$$

$$E^{\mathrm{DA}}(t) \geq 0, \forall t \in \mathcal{T} \tag{12.81}$$

$$\underline{E}^{\mathrm{DA}} \leq \sum_{t \in \mathcal{T}} E^{\mathrm{DA}}(t) \leq \overline{E}^{\mathrm{DA}} \tag{12.82}$$

In (12.80), $C^{\mathrm{DA,EM}}$ denotes the day-ahead electricity bidding cost for the fast charging station aggregator; $\lambda^{\mathrm{DA}}(t)$ represents the day-ahead electricity price at time t; $E^{\mathrm{DA}}(t)$ is the forecasted day-ahead electricity demand submitted by the aggregator for each time period. Equation (12.81) specifies that the fast charging station aggregator is only allowed to purchase electricity from the grid within a bounded range. It constrains the total submitted day-ahead electricity volume E^{DA} within a lower and upper bound $\underline{E}^{\mathrm{DA}}$ and $\overline{E}^{\mathrm{DA}}$, respectively. These bounds represent the qualified bidding range for participants in the two-stage competitive electricity spot market.

In the real-time stage, the fast charging station aggregator can adjust the actual grid electricity consumption based on the real-time charging demand, electricity price, and the output fluctuation of renewable energy resources. The difference between the real-time consumption and the day-ahead bidding volume is settled according to the actual electricity price. When the absolute deviation is large, the aggregator is required to pay the corresponding penalty costs. The real-time electricity procurement cost is calculated as follows:

$$C_w^{\mathrm{RT,EM}} = \sum_{t \in \mathcal{T}} \left[\lambda_w^{\mathrm{RT}}(t) \Delta E_w(t) + C_w^{\mathrm{penalty}}(t)\right], \forall w \in W \tag{12.83}$$

$$\Delta E_w(t) = E_w^{\mathrm{RT}}(t) - E^{\mathrm{DA}}(t), \forall w \in W, t \in \mathcal{T} \tag{12.84}$$

$$E_w^{\mathrm{RT}}(t) = \sum_{f \in F} P_{w,f}^{\mathrm{grid}}(t) \Delta t, \forall w \in W, t \in \mathcal{T} \tag{12.85}$$

$$C_w^{\mathrm{penalty}}(t) = \begin{cases} \max\{0, [E^{\mathrm{DA}}(t) - \omega^+ E_w^{\mathrm{RT}}(t)] \Delta\lambda_w(t)\}, \ \Delta\lambda_w(t) \geq 0 \\ \max\{0, [E^{\mathrm{DA}}(t) - \omega^- E_w^{\mathrm{RT}}(t)] \Delta\lambda_w(t)\}, \ \Delta\lambda_w(t) < 0 \end{cases}, \forall w \in W, t \in \mathcal{T} \tag{12.86}$$

$$\Delta\lambda_w(t) = \lambda_w^{\mathrm{RT}}(t) - \lambda^{\mathrm{DA}}(t), \forall w \in W, t \in \mathcal{T} \tag{12.87}$$

In (12.83), $C_w^{\mathrm{RT,EM}}$ denotes the real-time electricity procurement cost of the fast charging station aggregator under scenario w, including both the electricity quantity deviation cost and the penalty cost; $\lambda^{\mathrm{RT}}(t)$ represents the real-time electricity price at time t under scenario w; $\Delta E_w(t)$ denotes the deviation between real-time electricity usage and day-ahead bid electricity usage under scenario w for the aggregator. In (12.84), the real-time electricity consumption under scenario w is defined as $E_w^{\mathrm{RT}}(t)$, and the difference from the day-ahead electricity bid $E^{\mathrm{DA}}(t)$ at time t determines whether there is an excess ($\Delta E_w(t) > 0$) or shortfall ($\Delta E_w(t) < 0$) relative to the day-ahead bid. In (12.85), $E_{w,f}^{\mathrm{RT}}(t)$ denotes the grid electricity consumption of fast charging station f at time t under scenario w, and $C_w^{\mathrm{penalty}}(t)$ represents the penalty cost incurred due to the deviation of real-time electricity usage, as calculated in (12.86); ω^+ and ω^- are the penalty coefficients, typically set to 1.05 and 0.95, respectively; $\Delta\lambda_w(t)$ represents the difference between real-time and day-ahead electricity prices at time t under scenario w, as calculated in (12.87). To control the penalty cost, the fast charging station aggregator is incentivized to align the real-time electricity consumption with the day-ahead bid as closely as possible. In cases where real-time prices are significantly higher or lower than day-ahead prices, the aggregator is encouraged to reduce or increase real-time electricity consumption accordingly, thus avoiding large-scale arbitrage behaviors that exploit price deviations and ensuring the orderly operation of the electricity market.

Energy Bidding and Real-time Charging Scheduling Strategy Based on Stochastic Programming. Considering the uncertainty in traffic demand $Q_{w,od}^{\mathrm{EV}}$ and zero-carbon electricity price γ_c, the stochastic nature of charging demand can be characterized through multiple solutions to representative scenarios using simplified dynamic equilibrium methods. At this point, the charging cost coefficient matrix of electric vehicle users derived from (12.31) and (12.48) is revised as follows:

$$\sum_{p'\in P_{od}^{\mathrm{equi}}}\sum_{\varrho\in\hat{\mathcal{T}}}\sum_{t\in\mathcal{T}} h_{w,p',\varrho}^{\mathrm{EV}}(t) = \sum_{\varrho\in\hat{\mathcal{T}}} Q_{w,od,\varrho}^{\mathrm{EV}}, \forall w \in W, od \in OD \tag{12.88}$$

$$\theta_{w,p'}^{\mathrm{EV}}(t) = \sum_{t'\in\mathcal{T}_{p'}^{\mathrm{ch}}(t)} \gamma_{w,\upsilon(p')}^{\mathrm{ch}}(t')P_{\mathrm{ch}}\Delta t, \forall w \in W, od \in OD, p' \in P_{od}^{\mathrm{equi}}, t \in \mathcal{T} \tag{12.89}$$

By repeatedly solving the simplified bounded rational dynamic user equilibrium and dynamic network load assignment under representative scenarios, the charging load $P_{w,f}^{\mathrm{ch}}(t)$ for each scenario w can be obtained:

$$P_{w,f}^{\mathrm{ch}}(t) = \sum_{t_s\in\left[\max(t-\tau^{\mathrm{ch}},0),t\right)} Q_{w,f}^{\mathrm{begin}}(t_s)P_{w,f}^{\mathrm{ch}}(t,t_s), \forall w \in W, t \in \mathcal{T} \tag{12.90}$$

Taking into account the uncertainty in the upper limits of photovoltaic generation $P_{w,f}^{\mathrm{PV}}$ and wind power generation $P_{w,f}^{\mathrm{wind}}$, (12.70) and (12.71) regarding the upper limits of photovoltaic and wind generation are revised as follows:

$$0 \le P_f^{\mathrm{PV}}(t) \le \overline{P}_{w,f}^{\mathrm{PV}}(t), \forall w \in W, f \in F, t \in \mathcal{T} \tag{12.91}$$

$$0 \leq P_f^{\text{wind}}(t) \leq \overline{P}_{w,f}^{\text{wind}}(t), \forall w \in W, f \in F, t \in \mathcal{T} \tag{12.92}$$

The stochastic demand averaging modifications for the energy adjustment scheduling constraints of fast charging stations in (12.53)–(12.77) are similar and thus omitted.

The revenue of the fast charging aggregator in scenario w is defined as the total charging fees paid by all electric vehicle users to the fast charging stations:

$$R_w = \sum_{od \in OD} \sum_{p' \in P_{od}^{\text{equi}}} \sum_{\varrho \in \hat{\mathcal{T}}} \sum_{t \in \mathcal{T}} h_{w,p',\varrho}^{\text{EV}}(t) \theta_{w,p'}^{\text{EV}}(t), \forall w \in W \tag{12.93}$$

During the day-ahead phase, the operation and maintenance of renewable energy generation facilities and energy storage equipment incur certain costs, which are affected by photovoltaic and wind power outputs. The corresponding calculation method is as follows:

$$C_w^{\text{RT,RES}} = \sum_{f \in F} \sum_{t \in T} \left[\gamma^{\text{PV}} P_{w,f}^{\text{PV}}(t) \Delta t + \gamma^{\text{wind}} P_{w,f}^{\text{wind}}(t) \Delta t \right], \forall w \in W \tag{12.94}$$

$$C_w^{\text{RT,ESS}} = \sum_{f \in F} \sum_{t \in \mathcal{T}} \gamma^{\text{ESS}} \left| P_{w,f}^{\text{ESS}}(t) \right| \Delta t, \forall w \in W \tag{12.95}$$

In the above equations, $C_w^{\text{RT,RES}}$ and $C_w^{\text{RT,ESS}}$ represent the operating costs of renewable and storage resources for fast charging aggregator under scenario w, respectively. γ^{PV}, γ^{wind}, and γ^{ESS} denote the per-unit operating cost coefficients for photovoltaic, wind, and energy storage systems.

The total cost C_w of the fast charging aggregator under scenario w is calculated as:

$$C_w = C^{\text{DA,EM}} + C_w^{\text{RT,EM}} + C_w^{\text{RT,RES}} + C_w^{\text{RT,ESS}}, \forall w \in W \tag{12.96}$$

The first term in the above equation refers to the day-ahead cost, which does not vary with the scenario, while the latter three terms are real-time costs and are associated with scenario-dependent adjustments.

At this point, the optimization of the bidding and charging scheduling strategies for the fast charging aggregator can be formulated as the following stochastic optimization problem:

$$\max O = R - C = \sum_{w \in W} \rho_w R_w - \sum_{w \in W} \rho_w C_w \tag{12.97}$$

In this equation, O denotes the expected total profit across all scenarios for the fast charging aggregator; R_w is the expected total revenue under scenario w; C_w is the expected total cost under scenario w; and ρ_w is the probability of scenario w. Given a specific temporal distribution of EV charging demand, the stochastic optimization model can be expressed as a Mixed-Integer Linear Programming (MILP) problem.

12.4 Case Study and Simulation Analysis

Simulation Setup. To verify the effectiveness of the proposed simplified bounded rationality dynamic user equilibrium model based on the strategies of path merging and flow separation, a traffic network test system, denoted as Test System 3 is adopted and illustrated in Fig. 12.4. The performance of the model is evaluated in terms of computational time and approximation error in capturing the dynamic traffic flow through fast charging stations, comparing results before and after simplification.

The original traffic network of Test System 3 is a 14-node configuration, consisting of 2 fast charging station nodes, 4 origin–destination (OD) nodes, 8 intermediate nodes, and 19 road segments. The free-flow travel time on each segment is uniformly set to 0.15 h. After applying path merging and flow separation, the simplified model reduces the network to 6 nodes, retaining only the 2 fast charging station nodes and 4 OD nodes, with 8 constructed equivalent paths. Among them, the free-flow travel time for equivalent paths 1 through 4 is 0.45 h, while for paths 5 through 8 it is 0.3 h.

Details of the travel demand for Test System 3 are provided in Table 12.1. The travel path configuration of the original network is listed in Table 12.2, where both electric vehicles (EVs) and internal combustion engine vehicles (ICEVs) share the same path set. The key distinction lies in the fact that EVs temporarily stop at fast charging station nodes for charging, whereas ICEVs do not stop during travel. The travel path configuration of the simplified model is presented in Table 12.3, and the remaining key simulation parameters are summarized in Table 12.4.

To further validate the effectiveness of the proposed energy optimization and regulation strategy for fast charging station aggregators, this section introduces a

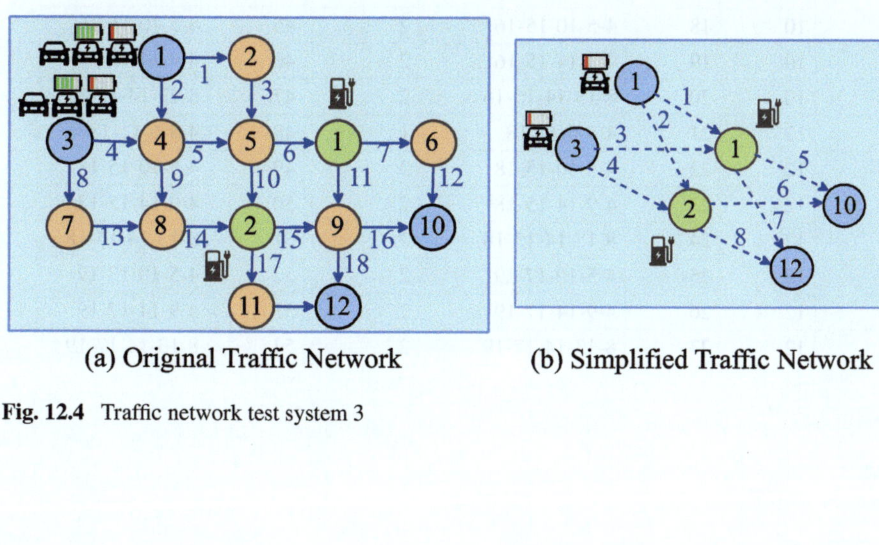

Fig. 12.4 Traffic network test system 3

Table 12.1 Traffic demand in traffic network test system 3

o	d	Q_{od}^{EV} (veh)	Q_{od}^{GV}
1	10	750	5250
1	12	750	5250
3	10	750	5250
3	12	750	5250

Table 12.2 Path in the original traffic network of Test System 3

o	d	EV			GV	
		#	P_{od}^{EV}	$\upsilon(p)$	#	P_{od}^{GV}
1	10	1	1-3-6-7-12	1	28	1-3-6-7-12
1	10	2	1-3-6-11-16	1	29	1-3-6-11-16
1	10	3	2-5-6-7-12	1	30	2-5-6-7-12
1	10	4	2-5-6-11-16	1	31	2-5-6-11-16
1	10	5	1-3-10-15-16	2	32	1-3-10-15-16
1	10	6	2-5-10-15-16	2	33	2-5-10-15-16
1	10	7	2-9-14-15-16	2	34	2-9-14-15-16
1	12	8	1-3-6-11-18	1	35	1-3-6-11-18
1	12	9	2-5-6-11-18	1	36	2-5-6-11-18
1	12	10	1-3-10-15-18	2	37	1-3-10-15-18
1	12	11	2-5-10-15-18	2	38	2-5-10-15-18
1	12	12	2-9-14-15-18	2	39	2-9-14-15-18
1	12	13	1-3-10-17-19	2	40	1-3-10-17-19
1	12	14	2-5-10-17-19	2	41	2-5-10-17-19
1	12	15	2-9-14-17-19	2	42	2-9-14-17-19
3	10	16	4-5-6-7-12	1	43	4-5-6-7-12
3	10	17	4-5-6-11-16	1	44	4-5-6-11-16
3	10	18	4-5-10-15-16	2	45	4-5-10-15-16
3	10	19	4-9-14-15-16	2	46	4-9-14-15-16
3	10	20	8-13-14-15-16	2	47	8-13-14-15-16
3	12	21	4-5-6-11-18	1	48	4-5-6-11-18
3	12	22	4-5-10-15-18	2	49	4-5-10-15-18
3	12	23	4-9-14-15-18	2	50	4-9-14-15-18
3	12	24	8-13-14-15-18	2	51	8-13-14-15-18
3	12	25	4-5-10-17-19	2	52	4-5-10-17-19
3	12	26	4-9-14-17-19	2	53	4-9-14-17-19
3	12	27	8-13-14-17-19	2	54	8-13-14-17-19

Table 12.3 Path in the simplified traffic network of Test System 3

o	d	EV		
		#	P_{od}^{equi}	$\upsilon(p)$
1	10	1	1–5	1
1	10	2	2–6	2
1	12	3	1–7	1
1	12	4	2–8	2
3	10	5	3–5	1
3	10	6	4–6	2
3	12	7	2–6	1
3	12	8	4–8	2

Table 12.4 Parameter settings in traffic network test system 3

Parameter	Value	Parameter	Value	Parameter	Value (veh/h)
Δt	0.05/0.25 h	N_f	50个	C_l	648

simplified traffic network model shown in Fig. 12.5. The model consists of a bidirectional traffic system with 9 nodes, including 4 fast charging station nodes and 5 origin–destination (OD) nodes, and constructs a total of 24 equivalent paths. Among them, the equivalent paths numbered 1–4, 11–14, and 21–24 have a free-flow travel time of 0.4 h, while the remaining paths have a travel time of 0.2 h.

Figure 12.6 illustrates the congestion factor curve reflecting the time-varying congestion characteristics of the traffic network, which indicates pronounced congestion during morning and evening peak periods. The corresponding travel demand and path configuration are provided in Tables 12.5 and 12.6, respectively, while other key simulation parameters are listed in Table 12.7.

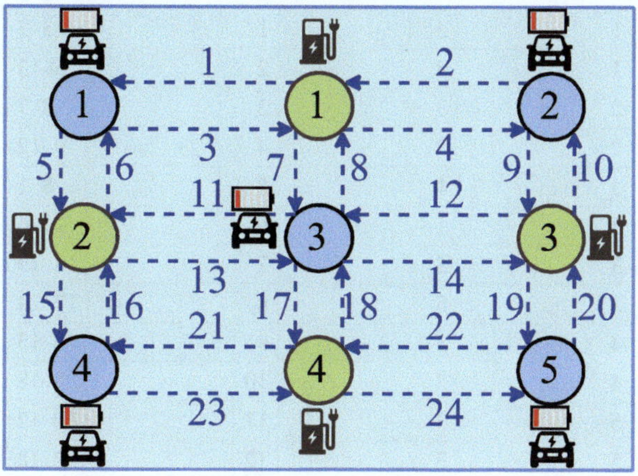

Fig. 12.5 Simplified traffic network of Test System 4

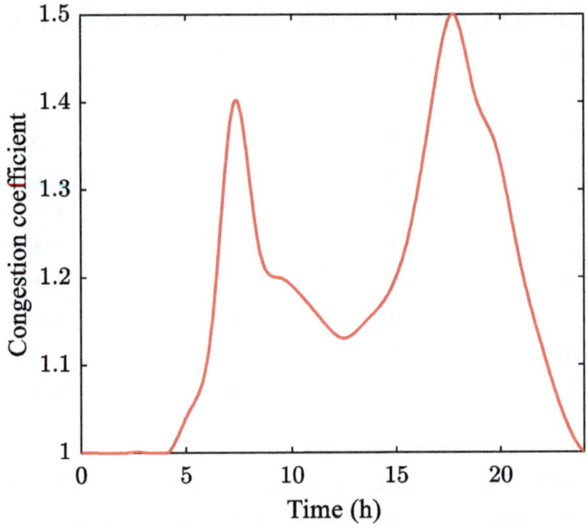

Fig. 12.6 Time-varying congestion coefficient

Table 12.5 Traffic demand in traffic network Test System 4

o	*d*	Q_{od}^{EV} (veh)
1	3	1500
2	3	1500
3	1	750
3	5	750
4	3	1500
5	3	1500

Table 12.6 Path in traffic network Test System 3

o	*d*	EV		
		#	P_{od}^{equi}	$\upsilon(p)$
1	3	1	3–7	1
1	3	2	5–13	2
2	3	3	2–7	1
2	3	4	9–12	3
3	1	5	8–1	1
3	1	6	11–6	2
3	5	7	14–19	3
3	5	8	17–24	4
4	3	9	6–13	2
4	3	10	23–18	4
5	3	11	20–12	3
5	3	12	22–18	4

Table 12.7 Parameter settings in traffic network Test System 4

Parameter	Value	Parameter	Parameter	Parameter	Parameter
Δt	0.05/0.25 h	τ^{ch}	0.5 h	N_f	90 ↑
P_{ch}	50 kW	$\overline{P}^{\text{ch}}$	100 kW	E^{ch}	25 kWh
E_f^{ESS}	500 kWh	η^+/η^-	95%	$\underline{\text{SoC}}_f^{\text{ESS}}$	10%
$\overline{\text{SoC}}_f^{\text{ESS}}$	90%	$\text{SoC}_f^{\text{ESS,init}}$	50%	K	3
α^{P}	1.5	$\overline{E}^{\text{DA}}/\underline{E}^{\text{DA}}$	50/300 MWh	$\omega^+/\,\omega^-$	1.05/0.95
γ^{PV}	0.06 \$/kWh	γ^{wind}	0.04 \$/kWh	γ^{ESS}	0.02 \$/kWh

The time step for the traffic network model is set to 0.05 h, whereas the time step for power bidding and charging scheduling by the fast charging station aggregator is set to 0.25 h. Uncertainty factors include travel demand, real-time electricity prices, photovoltaic (PV) generation power upper bounds, and wind power generation upper bounds. The time-varying profiles of these uncertainty scenarios are shown in Fig. 12.7. The electricity price data is derived from actual market measurements in a province in China.

Each type of uncertainty factor is clustered into 3 representative scenarios using the clustering method. A total of 81 combined operating scenarios are thus generated. The optimal electricity bidding and charging scheduling problem for the fast charging station aggregator is modeled and solved using the Gurobi optimization solver.

Performance Analysis of the Simplified Bounded Rational Dynamic User Equilibrium Model. In Traffic Network Test System 3, the iterative performance of the dynamic user equilibrium process is compared between the original traffic network model and the simplified traffic network model. The results are shown in Fig. 12.8. Using Step 2 of Algorithm 10.1 for computation, the original model converges after 32 iterations with a computation time of 69.2 s, while the simplified model satisfies the convergence condition after only 10 iterations, requiring merely 3.5 s. This indicates that the simplification reduces the number of iterations by 68.8% and the computational time by 94.9%, significantly improving computational efficiency.

Furthermore, a comparison of the charging traffic dynamics at fast-charging stations under both models is presented in Fig. 12.9.

It can be observed from the figure that the simplified model produces results closely matching those of the original model in terms of charging traffic flow (i.e., charging load), with a relative error of only 6.3%. This demonstrates that the proposed simplified traffic network model achieves a high level of accuracy while significantly improving modeling and computational efficiency.

Revenue and Cost Analysis of the Fast-Charging Station Aggregator. To assess the impact of the charging scheduling strategy and the integration of on-site renewable energy on the economic performance of the fast-charging station aggregator, this section compares four operational modes listed in Table 12.8, and evaluates their operational performance and economic indicators. The computation times for the

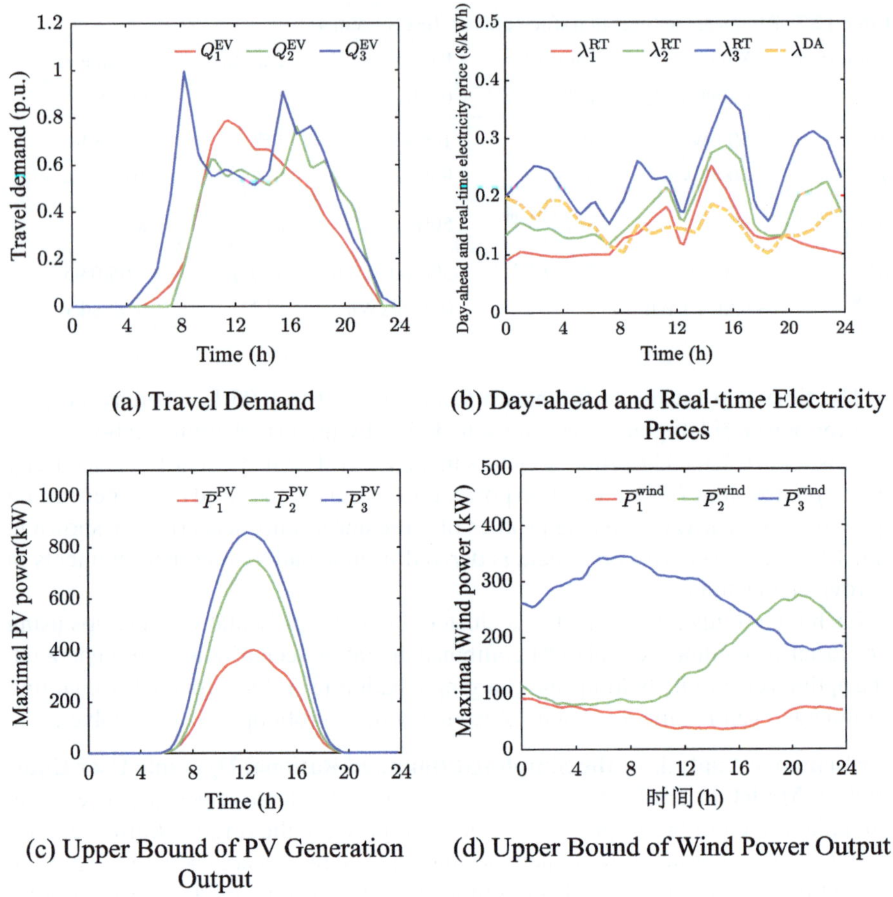

(a) Travel Demand

(b) Day-ahead and Real-time Electricity Prices

(c) Upper Bound of PV Generation Output

(d) Upper Bound of Wind Power Output

Fig. 12.7 Representative scenarios of multi-dimensional operational uncertainties

four modes are 197.3, 198.3, 197.9, and 198.6 s, respectively, all of which meet the computational time requirement for day-ahead scheduling. Figure 12.10 presents the aggregator's profit, total revenue, average cost, day-ahead cost, and real-time cost for each mode (where the day-ahead cost is a fixed value, and all other indicators are expected values across all scenarios). Figure 12.11 further illustrates the expected values of intra-day electricity purchase cost, renewable energy cost, storage cost, and renewable energy curtailment (i.e., wind and solar curtailment) under different modes.

Taking Operational Mode 1 as the baseline, the adoption of the charging scheduling strategy (Operational Mode 2) and the integration of on-site renewable energy (Operational Mode 3) can improve the expected profit of the fast-charging station aggregator by 3.5 and 11.2%, respectively. When both strategies are applied simultaneously (Operational Mode 4), the expected profit increases by 14.7%.

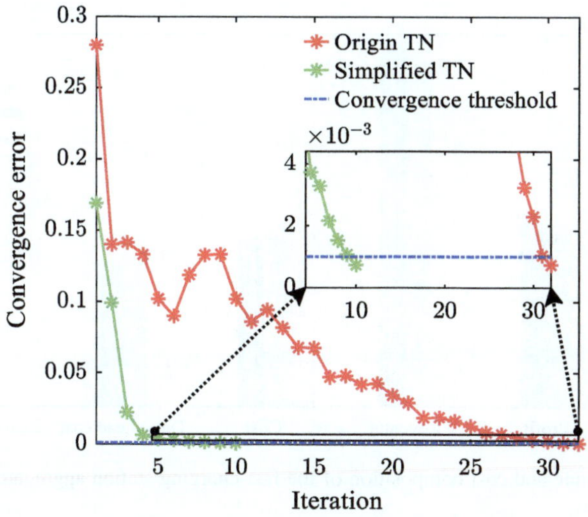

Fig. 12.8 Comparison of iterative convergence between original and simplified models in traffic network Test System 3

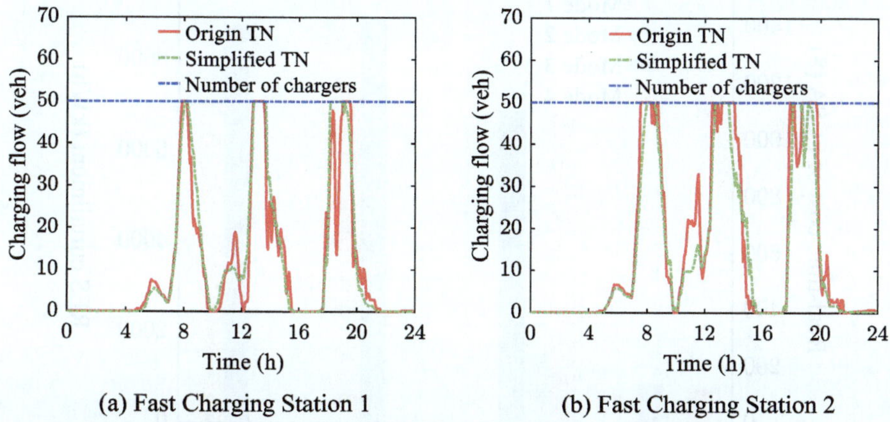

Fig. 12.9 Comparison of charging traffic dynamics between original and simplified models in traffic network Test System 3

Table 12.8 Four operational modes of the fast-charging station aggregator

Operational mode	Charging scheduling	RES
1	×	×
2	√	×
3	×	√
4	√	√

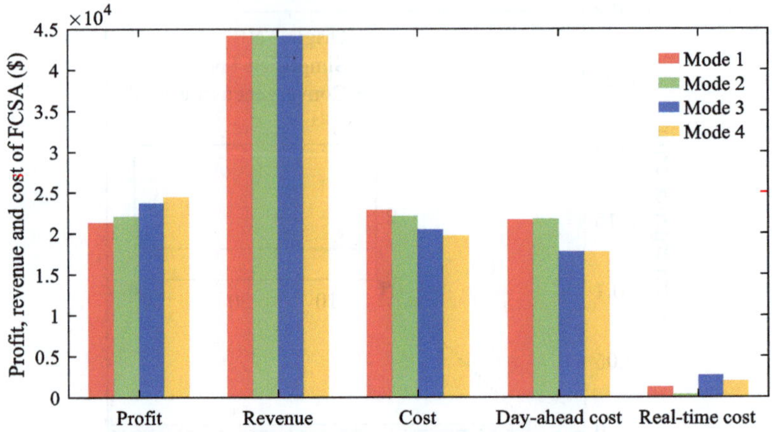

Fig. 12.10 Revenue and cost composition of the fast-charging station aggregator under different operational modes

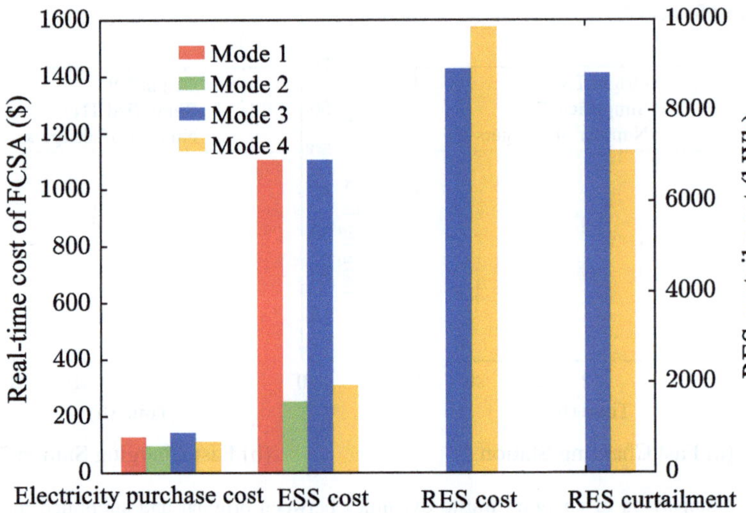

Fig. 12.11 Intra-day cost composition of the fast-charging station aggregator under different operational modes

These results indicate that both the charging scheduling and renewable integration significantly enhance economic performance, and their benefits are additive.

It is noteworthy that the total revenue (i.e., the total charging fees paid by EV users) remains unchanged across all four modes. This is because EV users are unaware of the implemented charging scheduling strategy, and the changes in operational mode do not affect their actual payments, which aligns with the conclusion. As illustrated in Fig. 12.10, the differences in profit across the modes primarily stem from variations in cost structure. Charging scheduling has a minimal effect on day-ahead costs but significantly reduces intra-day costs. On the other hand, renewable energy integration primarily reduces the day-ahead electricity bidding quantity, thereby lowering the corresponding cost. However, due to the uncertainty of renewable output, it leads to a slight increase in intra-day costs.

Further analysis of Fig. 12.11 reveals that charging scheduling effectively reduces intra-day electricity purchase costs. This is achieved by leveraging the flexibility of EV charging loads to make the real-time power consumption curve better match the day-ahead bidding curve, thus minimizing deviation-related penalty costs and improving the aggregator's ability to cope with operational uncertainties. In addition, charging scheduling serves as a partial substitute for energy storage. By controlling EV charging power and shifting loads in time, it facilitates the absorption of surplus renewable energy, reduces reliance on storage systems, cuts down storage costs, and significantly mitigates renewable energy curtailment (e.g., wind and solar curtailment).

On the other hand, the cost of renewable generation constitutes a major component of intra-day costs. Although renewable integration significantly reduces day-ahead bidding volumes, the uncertainty in renewable output leads to a slight increase in intra-day electricity costs. Therefore, a trade-off between renewable energy utilization and cost control is essential to achieve a more optimal economic control strategy.

Day-Ahead Electricity Bidding and Intra-Day Charging Scheduling Strategy of Fast-Charging Aggregators. The electricity bidding strategies of fast-charging station aggregators under different operational modes in the day-ahead and real-time electricity markets, along with the corresponding electricity prices, are illustrated in Fig. 12.12.

In the absence of charging scheduling (i.e., Operational Mode 1), the electricity bids of the fast-charging aggregator are primarily concentrated in three time intervals: 08:00, 13:00, and 19:00. These time slots coincide with both peak EV charging demand and relatively lower day-ahead electricity prices. Under this scenario, EVs charge at a constant power rate, and the resulting charging load profile aligns closely with the distribution of traffic flow. Consequently, the real-time electricity consumption curve of the aggregator exhibits a relatively smooth trend.

Upon introducing the charging scheduling strategy, the original charging load curve is reconstructed. A substantial portion of the charging demand is shifted to off-peak periods with lower electricity prices, resulting in a more fluctuating and spiked day-ahead bidding curve. This strategy takes advantage of the temporal flexibility of EV charging loads to redistribute charging behavior and smooth the load profile, thereby facilitating the orderly integration of large-scale charging loads into the grid. Meanwhile, the implementation of charging scheduling significantly reduces the average real-time deviation in electricity consumption, indicating that the actual consumption curve aligns more closely with the day-ahead bidding curve. This leads to a reduction in intra-day procurement costs arising from deviations and is consistent

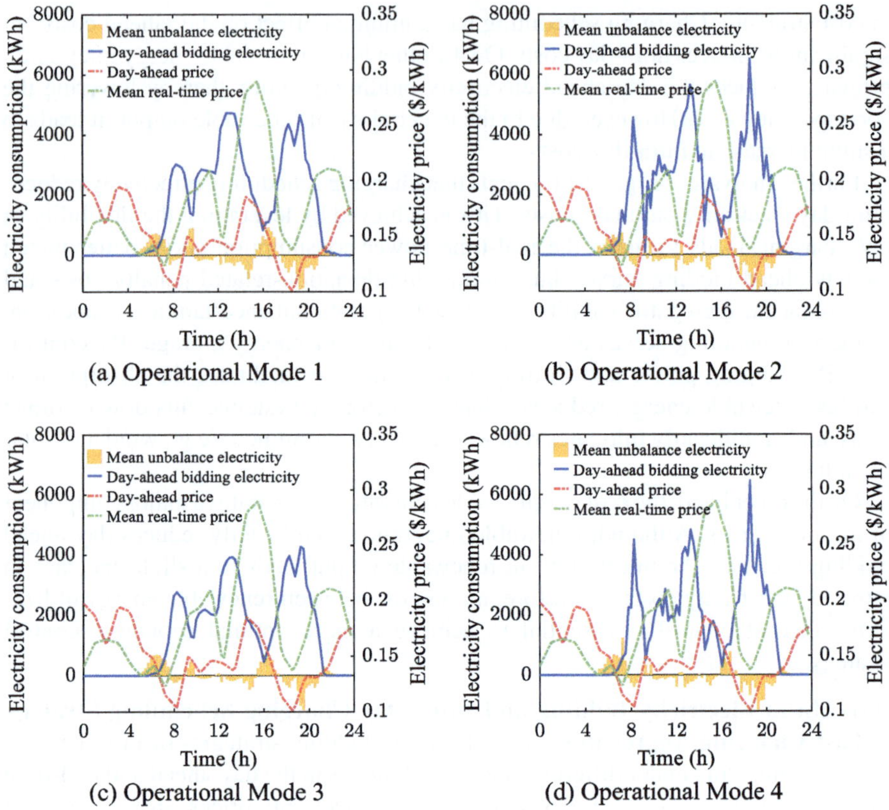

Fig. 12.12 Electricity market trading strategies and corresponding prices of fast-charging aggregators under different operational modes

with the analysis, further confirming the effectiveness of charging scheduling in enhancing market participation efficiency.

The integration of on-site renewable energy provides the aggregator with additional energy supply during the day-ahead stage, especially during periods of high solar output. This allows the aggregator to reduce external electricity procurement and day-ahead bidding quantities, ultimately lowering electricity costs. Notably, since partial curtailment of renewables (wind/solar) is permitted, the integration of renewables does not lead to a significant increase in real-time deviation, ensuring overall operational flexibility.

Figure 12.12 also shows that all operational modes exhibit significant positive deviations in real-time electricity consumption around 07:15, i.e., real-time consumption exceeds the corresponding day-ahead bidding quantities (which are nearly zero during this time). This indicates that the aggregator deliberately refrains from bidding in the day-ahead market for this period, preserving operational flexibility for real-time scenarios. Although positive deviations may occur in the real-time stage, the real-time and day-ahead prices during this time are relatively similar. As indicated by (12.86), the penalty costs induced by the deviation are negligible. Hence, the aggregator is not constrained by day-ahead commitments at this time, and greater flexibility in scheduling is allowed, contributing to overall cost optimization.

Additionally, all operational modes show significant negative deviations around 19:15, where the actual consumption is lower than the day-ahead bidding quantity. During this period, real-time prices generally exceed day-ahead prices. According to the derivation in (12.83), such negative deviations can yield positive returns through price arbitrage, thus creating additional profit opportunities for the aggregator.

The charging scheduling strategies of Fast-Charging Station 2 under Operational Scenario 45 across different modes are presented in Fig. 12.13. This scenario corresponds to representative uncertainty scenarios (Q_3^{EV}, λ_3^{RT}, $\overline{P}_2^{\mathrm{PV}}$, $\overline{P}_2^{\mathrm{wind}}$) in Fig. 12.7, simplified as scenario (3, 3, 2, 2), representing the highest travel demand, the highest real-time electricity price, medium photovoltaic generation level, and medium wind generation level. Under this scenario, the charging power at Fast-Charging Station 2 is jointly supplied by grid electricity, photovoltaic power, wind power, and energy storage systems.

Consistent with the results presented in Fig. 12.12, under the average power mode (Operational Modes 1 and 3), the charging load profile of the fast-charging station is relatively smooth, which reflects the characteristics of a constant charging power strategy. When the charging scheduling strategy is introduced (Operational Modes 2 and 4), a substantial portion of the charging load is shifted to time periods with lower electricity prices, forming multiple pronounced peaks and valleys, thereby demonstrating a strong responsiveness to dynamic price signals.

Meanwhile, the integration of on-site renewable energy partially substitutes electricity procured from the grid. During nighttime low-load periods, the energy storage system absorbs excess wind power generation and releases energy during high-load daytime periods to meet users' charging demands, effectively reducing wind curtailment events. Further comparing Figs. 12.13c and d, it can be observed that when charging scheduling is allowed, the number of charge–discharge cycles of the energy storage system is significantly reduced. This indicates that the charging scheduling strategy can partially replace the role of the storage system by dynamically balancing the load through optimized charging power allocation, thereby reducing reliance on storage equipment.

Under Operational Mode 4, the spatiotemporal distribution of internal traffic flow for each fast-charging station in Operational Scenario 45 is shown in Fig. 12.14.

Due to the symmetric topology of Test System 4, Fast-Charging Stations 1 and 4 are symmetrically located and exhibit highly similar traffic flow dynamics. Similarly, Stations 2 and 3 are symmetrically distributed and are closer to the origin points of traffic flows, thereby showing comparable traffic characteristics. During the three main peak hours (08:00, 13:00, and 19:00), all fast-charging stations experience peak charging periods. The corresponding retail charging prices during these hours are relatively low, attracting a large number of electric vehicles to charge simultaneously. At these times, all stations experience full-load operation and queuing phenomena.

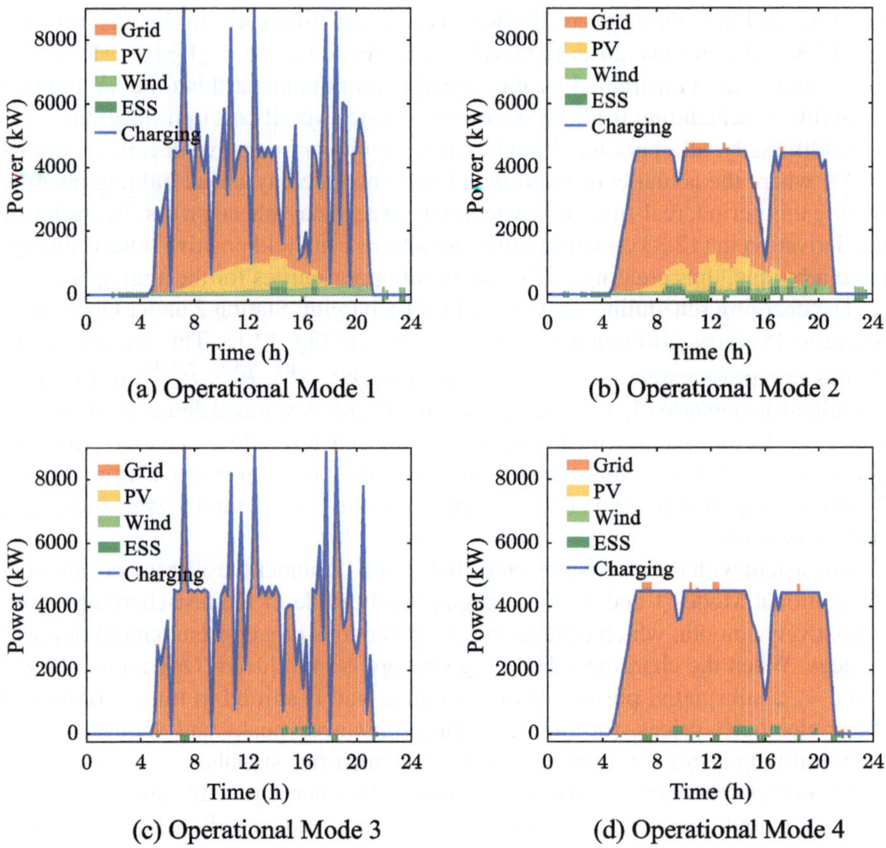

Fig. 12.13 Charging scheduling strategy of Fast-Charging Station 2 under operational scenario (3, 3, 2, 2) across different operational modes

In comparison, Stations 2 and 3 serve a significantly larger number of charging and queuing vehicles than Stations 1 and 4, due to their proximity to traffic origins and their greater availability of charging piles. This reflects the influence of geographic location and resource allocation on the spatiotemporal distribution of traffic flow and charging load.

Sensitivity Analysis of Bounded Rationality Coefficient. To evaluate the impact of electric vehicle (EV) users' bounded rationality coefficient on the economic performance of the fast-charging station aggregator, this section calculates the expected profit under all scenarios for different user tolerance levels (ranging from \$0 to \$2, with a step size of \$0.5). The results are shown in Fig. 12.15.

The results indicate that the aggregator's expected profit increases with the rise in user tolerance. Taking the case of fully rational users (i.e., user tolerance of \$0) as the baseline, when the tolerance increases to \$2, the expected profit under all scenarios

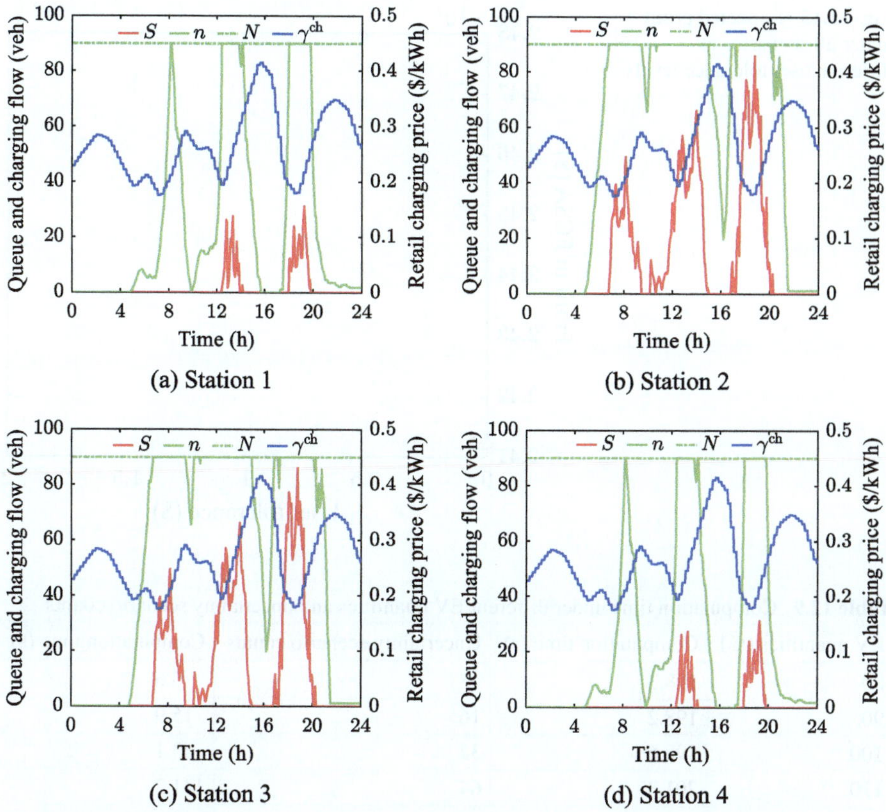

Fig. 12.14 Spatiotemporal distribution of internal traffic flow at each fast-charging station under operational scenario (3, 3, 2, 2) in Operational Mode 4

improves by approximately 2.6%. This phenomenon suggests that higher user tolerance provides the aggregator with more flexibility in pricing strategies, which subsequently leads to higher economic returns. Therefore, to expand profit margins, aggregators should consider enhancing user price tolerance through well-designed operational strategies (e.g., reward points, service discounts, behavioral nudges), thereby improving users' acceptance of price volatility and increasing customer stickiness. This conclusion is consistent with the findings regarding the pricing model analysis, further confirming the significant influence of user rationality on aggregator profitability.

Computational Performance Analysis. To assess the applicability of the proposed models and algorithms in scenarios involving large-scale EV integration, and to further analyze the influence of the number of operational scenarios on computational performance, this section evaluates the optimization solving time under different combinations of EV access scale and uncertainty scenario quantity. The results are summarized in Table 12.9.

Fig. 12.15 Expected profit under all scenarios at different user tolerance levels

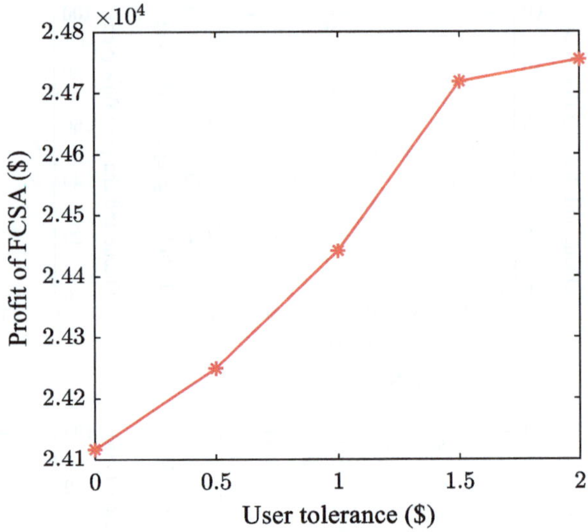

Table 12.9 Computation time under different EV quantities and uncertainty scenario counts

EV quantities (%)	Computation time (s)	Uncertainty scenario counts	Computation time (s)
80	198.3	8	6.6
90	198.2	16	15.9
100	198.6	32	43.1
110	201.3	64	141.2
120	212.6	81	198.6

12.5 Summary

In the context of considering both bounded rationality behaviors of users and multi-dimensional operational uncertainties, this chapter proposes a stochastic optimization-based energy regulation method for fast-charging station aggregators, aiming to achieve orderly grid integration and flexible scheduling of fast-charging station clusters in the competitive electricity spot market. First, based on a bounded rationality dynamic user equilibrium model, a path-merging and flow-splitting strategy is introduced to construct a simplified dynamic user equilibrium model with bounded rationality. By simplifying the transportation network topology and focusing on traffic flows with charging demands, the model efficiently predicts the spatiotemporal distribution of charging loads for large-scale fast-charging station clusters. Second, a grouped charging scheduling strategy is designed, where electric vehicles with similar charging needs are aggregated for joint scheduling, which significantly reduces the dimensionality of decision variables and improves computational efficiency. Finally, an aggregator model that integrates local renewable energy resources is developed. A scenario-based stochastic optimization method is employed to model and solve for travel demand, real-time electricity prices, and renewable generation uncertainties. Based on this, an energy regulation strategy is proposed for fast-charging station aggregators participating in the two-stage competitive electricity spot market, determining the optimal day-ahead bidding and real-time charging scheduling strategies.

The proposed modeling methods and algorithms are validated through numerical simulations based on empirical data from a provincial electricity market in China. The main conclusions are summarized as follows: (1) The simplified bounded rationality dynamic user equilibrium model, based on path-merging and flow-splitting strategies, significantly reduces the number of simulation iterations and computation time, while maintaining high modeling accuracy for charging traffic flow and load within fast-charging stations; (2) The Charging scheduling and local renewable energy integration are key approaches to improving aggregator profitability. Charging scheduling directly reshapes charging loads to flatten peaks and fill valleys, partially replacing the role of energy storage and thereby reducing intra-day costs; renewable energy integration decreases dependency on external electricity procurement, effectively reducing day-ahead costs; (3) The proposed stochastic optimization-based energy regulation method effectively addresses multi-dimensional uncertainties, including travel demand, real-time electricity prices, and renewable generation outputs. It enables fast-charging station aggregators to maintain small real-time load deviations across all scenarios, ensuring orderly response to market mechanisms; (4) The bounded rationality coefficient of EV users is closely related to aggregator profitability. To expand profit margins, aggregators should adopt operational strategies (e.g., pricing policy optimization, behavioral guidance mechanisms) to enhance user tolerance to price fluctuations; and finally (5) The proposed stochastic optimization scheduling method demonstrates excellent computational performance and scalability, making it applicable to large-scale EV integration scenarios. It provides theoretical and technical support for the large-scale operation of fast-charging stations and their participation in electricity market transactions.

Chapter 13
Risk-Averse Energy Dispatch for Integrated Energy Service Stations

Abstract This chapter proposes a risk-averse energy optimization and control method for integrated energy service station (IESS) aggregators participating in electricity markets, considering heterogeneous vehicle types and extreme scenario uncertainties. First, a boundedly rational heterogeneous user equilibrium (BRHUE) model is developed to characterize the spatiotemporal energy demand of electric vehicles, hydrogen fuel cell vehicles, natural gas vehicles, and conventional vehicles. Then, a multi-energy coordination model for IESSs is constructed, integrating electricity, hydrogen, and natural gas systems with on-site renewables to provide one-stop refueling services. To manage operational risks under uncertainties in travel demand, renewable generation, and electricity prices, a conditional value-at-risk (CVaR)-based risk-averse optimization approach is introduced. This enables the IESS aggregator to jointly optimize day-ahead bidding and real-time multi-energy dispatch while balancing expected profit and tail-end losses. Case studies validate the effectiveness of the proposed method, showing that the risk-averse strategy reduces extreme-scenario profit loss by 21.8% with only a 0.3% decrease in overall expected profit. The results also highlight the economic potential of synthetic natural gas production and carbon capture, as well as the impact of user bounded rationality on load distribution and system complexity. The proposed framework offers theoretical and practical support for low-carbon, multi-energy-coordinated transportation systems.

13.1 Introduction

Chapter 12 focused on the energy optimization and control strategies for fast charging station (FCS) clusters within the context of transportation networks where electric vehicles (EVs) and internal combustion engine vehicles (ICEVs) coexist. However, in real-world urban transportation systems, in addition to EVs and ICEVs, hydrogen fuel cell vehicles (HFCVs) and compressed natural gas vehicles (CNGVs) are gradually emerging as important modes of transportation. These vehicles are playing an increasingly prominent role in the development of a green, low-carbon, and

Q. Yang et al., *Integration of Massive Electric Vehicles to Smart Grid*, Smart Energy Systems, https://doi.org/10.1007/978-981-92-0557-8_13

sustainable integrated energy-transportation system. HFCVs are particularly suited for heavy-duty transportation due to their advantages of zero carbon emissions and long driving range, while CNGVs offer a cost-effective pathway to low-carbon transition through fuel switching. Together with EVs, they form the key pillars of future integrated transportation energy systems.

In line with this trend of deep integration between energy and transportation, fast charging stations, hydrogen refueling stations, and gas refueling stations are expected to jointly serve as key hubs coupling the power and transportation systems. These hubs will become critical infrastructures for coordinating the operation of electricity–hydrogen–gas multi-energy systems. To adapt to this development, it is necessary to expand upon the work of Chapter 12 and extend the research scope to address the complexities introduced by the coexistence of electric, hydrogen, and gas-powered vehicles in a unified energy-transportation network.

Currently, the number of HFCVs and CNGVs remains relatively small. To improve land use efficiency and energy service quality, newly constructed hydrogen and gas refueling stations are often co-located with fast charging stations, and are supported by on-site renewable generation and energy storage facilities, effectively forming integrated energy service stations (IESSs). These stations provide one-stop energy services for EVs, HFCVs, and CNGVs. Their operation and scheduling must coordinate electricity, hydrogen, and gas resources, and the customer base expands from a single vehicle type to three heterogeneous user groups, significantly increasing system complexity. In such a multi-vehicle-type environment, transportation network modeling must consider the refueling behaviors and bounded rational decision-making features of each vehicle type. This greatly increases the computational burden of user equilibrium modeling and numerical solving.

In addition, since electricity remains the dominant energy source within IESSs, the total power demand of such stations is typically higher than that of standalone FCSs, making them eligible participants in electricity markets. However, due to the complexity of multi-energy coordination and the uncertainties in user demand and renewable generation, IESS operators face greater exposure to electricity market volatility and operational risk.

To address these challenges, this chapter proposes a risk-averse energy optimization and control method for IESS aggregators participating in electricity markets. Specifically, a boundedly rational heterogeneous user equilibrium model is first developed to characterize the dynamic behavior and spatiotemporal energy demand of various vehicle types, including EVs, HFCVs, CNGVs, and ICEVs. Then, an IESS model with multi-energy coordination capabilities is constructed to provide integrated charging, hydrogen refueling, and gas refueling services for heterogeneous users based on available renewable energy resources. Furthermore, a conditional value-at-risk (CVaR)-based risk-averse optimization approach is proposed to quantify tail-end profit losses under extreme scenarios driven by uncertainties in travel demand, electricity prices, and renewable output. Based on this, coordinated day-ahead and real-time electricity market participation strategies are designed for the IESS aggregator to achieve optimal and risk-controlled operation. Through case studies, the effectiveness of the proposed methods is validated. This chapter aims to provide theoretical support and methodological reference for developing risk-averse participation mechanisms for IESSs in electricity markets, as well as practical guidance for building a low-carbon, multi-energy-coordinated transportation system.

The remainder of this chapter is organized as follows. Section 13.2, based on the modeling framework proposed in Sect. 10.3, develops a boundedly rational heterogeneous user equilibrium model that accounts for diversified energy demands. Section 13.3 presents the multi-energy coordination model of IESSs, enabling one-stop energy services powered by on-site renewables. Section 13.4 introduces a CVaR-based risk-averse energy control optimization method and designs two-stage electricity market participation strategies for IESS aggregators. Section 13.5 validates the proposed modeling and optimization methods through case studies. Finally, Sect. 13.6 summarizes the main findings and contributions of this chapter.

13.2 Model for Transportation Network Model Based on Boundedly Rational Heterogeneous User Equilibrium

The transportation network models developed in the previous chapters primarily focused on the charging demand of electric vehicles (EVs). However, in practical applications, heterogeneous refueling demands, e.g., charging, hydrogen refueling, and gas refueling, often coexist, resulting in greater structural and analytical complexity in transportation network modeling. The core objective of such modeling is to estimate the spatial and temporal distribution patterns of various types of refueling demands. Both multi-period static models and dynamic models can provide valuable analytical insights.

While the previous sections have systematically established and applied a boundedly rational dynamic user equilibrium (BRDUE) model, this class of models typically relies on iterative numerical solvers. When facing heterogeneous traffic flows with diverse energy demands, the computational cost of such models may significantly increase due to excessive iteration steps. Although techniques such as path merging and network structure simplification can partially improve computational performance, the incorporation of heterogeneous traffic flows still substantially increases modeling complexity and computational burden.

In contrast, multi-period static user equilibrium models with bounded rationality allow for parallel solving across time periods, offering superior computational efficiency when analyzing the spatial–temporal distribution of heterogeneous traffic flows. Therefore, this chapter adopts a multi-period static modeling approach to characterize the spatial–temporal evolution of multiple types of energy demand within the transportation network. The resulting distribution profiles also serve as essential input for the subsequent coordination of electricity–hydrogen–gas (E–H-G) multi-energy scheduling and electricity market participation strategies for integrated energy service station (IESS) clusters.

Heterogeneous Traffic Flow Modeling. The multi-period static transportation network model proposed in Sect. 10.3 considered only the charging demands of EV traffic flows. Building upon that foundation, this section proposes a heterogeneous traffic network model by incorporating hydrogen fuel cell vehicles (HFCVs) and natural gas vehicles (NGVs), thereby constructing a bounded rational heterogeneous user equilibrium (BRHUE) model that supports the coexistence of multiple user types.

This model not only captures the behavior and route selection mechanisms of EVs, HFCVs, and CNGVs in the transportation network but also accounts for scenario-based uncertainties present in real-world operating environments. As such, it provides efficient and robust input data for modeling the spatial–temporal distribution of energy demands and optimizing the energy coordination strategies of IESS clusters.

Here, the vehicles are classified into electric vehicles, hydrogen fuel cell vehicles, natural gas vehicles, and conventional internal combustion engine vehicles as follows:

$$V = \{v|\text{EV}, \text{HFCV}, \text{NGV}, \text{GV}\} \tag{13.1}$$

The proportions of different vehicle types across various OD travel demand pairs sum up to 1. The proportion of vehicle type v is denoted as α^v, satisfying:

$$\sum_{v \in V} \alpha^v = 1 \tag{13.2}$$

In addition, following Sect. 10.3.1, it is assumed that electric vehicles (EVs), hydrogen fuel cell vehicles (HFCVs), and natural gas vehicles (NGVs) with energy replenishment needs will choose routes that pass through a designated integrated energy service station. In contrast, EVs, HFCVs, and NGVs without energy replenishment needs are uniformly regarded as internal combustion engine vehicles, and their routes do not pass through any integrated energy service station.

For each vehicle type, route choice is modeled as path flow, with the corresponding travel demand constraint expressed as:

$$\sum_{p \in P^v_{od}} f^v_{w,p}(t) = \alpha^v Q_{w,od}(t), \quad \forall v \in V, w \in W, od \in OD, t \in \mathcal{T} \tag{13.3}$$

where $f^v_{w,p}(t)$ is the path flow of vehicle type v on path p under scenario w at time t, P^v_{od} is the set of paths available to vehicle type v between OD pair od, and $Q_{w,od}(t)$ is the total traffic demand between od under scenario w at time t.

The traffic flow on each road link is composed of the flows of different vehicle types, given by:

$$x_{w,l}(t) = \sum_{v \in V} x^v_{w,l}(t), \quad \forall w \in W, l \in L^{\text{TN}}, t \in \mathcal{T} \tag{13.4}$$

where $x_{w,l}(t)$ denotes the traffic flow on link l under scenario w at time t, and $x_{w,l}^{v}(t)$ represents the traffic flow component of vehicle type v within $x_{w,l}(t)$.

The traffic flow of vehicle type v on link l equals the sum of all path flows of that type which pass through link l, i.e.,

$$x_{w,l}^{v}(t) = \sum_{p \in P} \Lambda_{p,l}^{v} f_{w,p}^{v}(t), \quad \forall v \in V, w \in W, l \in L^{\mathrm{TN}}, t \in \mathcal{T} \tag{13.5}$$

where $\Lambda_{p,l}^{v}$ is a 0–1 indicator denoting whether path p in the path set of vehicle type v passes through link l, as follows:

$$\Lambda_{p,l}^{v} = \begin{cases} 0, \ l \notin p \\ 1, \ l \in p \end{cases}, \forall v \in V, p \in P^{v}, l \in L^{\mathrm{TN}} \tag{13.6}$$

In heterogeneous traffic networks, there are regular road links L^{road} and integrated energy service station (IESS) links L^{IESS}. The travel time on these links is computed using the BPR model and the M/M/c/K queueing theory model, respectively, as follows:

$$\varsigma_{w,l}^{v}(t) = \begin{cases} \tau_l^{\mathrm{free}} + 0.15\tau_l^{\mathrm{free}}\left[\frac{x_{w,l}(t)}{C_l}\right]^4, & l \in L^{\mathrm{road}} \\ \tau_l^{\mathrm{free}} + \tau_l^{q,v}\left[\frac{x_{w,l}^{v}(t)}{N_l^{v}}\right]^3 + \frac{D^{v}}{R^{v}}, & l \in L^{\mathrm{IESS}} \end{cases}, \forall v \in V, w \in W, t \in \mathcal{T} \tag{13.7}$$

Here, $\varsigma_{w,l}^{v}(t)$ represents the travel time for vehicle type v on link l under scenario w at time t; τ_l^{free} is the free-flow travel time without congestion; C_l is the capacity of link l; $\tau_l^{q,v}$ is the queuing delay coefficient for vehicle type v; N_l^{v} is the number of refueling/charging devices available for vehicle type v at the IESS; D^{v} denotes the energy demand of vehicle type v; R^{v} is the charging/refueling rate of vehicle type v; and D^{v}/R^{v} gives the energy service time for vehicle type v.

For regular links L^{road}, the travel time includes both movement and queuing times, and it is shared across all vehicle types. For IESS links L^{IESS}, the travel time consists of movement time, queuing time, and energy service time. Due to differences in energy requirements and equipment among electric vehicles, hydrogen fuel cell vehicles, and natural gas vehicles (e.g., chargers, hydrogen dispensers, gas pumps), the queuing time within the energy service station is related to vehicle types.

To model the differentiated service requirements of heterogeneous vehicle flows, this section, based on the modeling framework in Sect. 10.3.1, proposes a service-region structure for IESS that supports simultaneous service for multiple energy types. As illustrated in Fig. 13.1, the IESS includes three parallel service lanes dedicated to charging, hydrogen refueling, and gas refueling, respectively. Paths of electric vehicles, hydrogen fuel cell vehicles, and natural gas vehicles all pass through the IESS link; once entering the IESS, they separately enter their corresponding service lanes to complete their respective refueling/charging operations. After the service is completed, the vehicles merge at the IESS exit and continue along regular road links.

Fig. 13.1 Traffic network model with IESSs

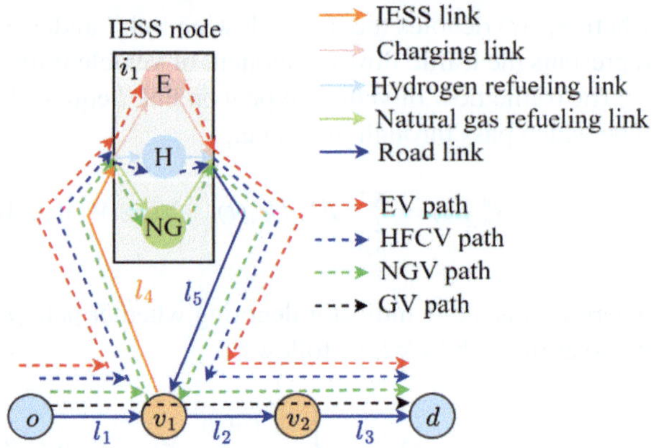

To ensure that the traffic flow leaving the integrated energy service station stays within a reasonable upper limit, this section considers the capacity constraint of the integrated energy service station, expressed as:

$$s_l^v = \frac{N_l^v R^v \Delta t}{D^v}, \ \forall v \in V, l \in L^{\text{ESS}} \tag{13.8}$$

where s_l^v denotes the maximum number of type-v users that can be served by the integrated energy service station on link l during a time interval Δt.

Further, the capacity constraint of the integrated energy service station can be formulated as:

$$x_{w,l}^v(t) \le s_l^v, \ \forall v \in V, w \in W, l \in L^{\text{ESS}}, t \in \mathcal{T} \tag{13.9}$$

This constraint ensures that the charging, hydrogen refueling, and gas refueling load on the integrated energy service station link does not exceed the corresponding upper limit.

The delay time of the path traffic flow for type-v vehicles at time t can be expressed as:

$$\xi_{w,p}^v(t) = t + \sum_{l \in p} \varsigma_{w,l}^v(t), \ \forall v \in V, w \in W, od \in OD, p \in P_{od}^v, t \in \mathcal{T} \tag{13.10}$$

where $\xi_{w,p}^v(t)$ is the delay time of type-v path traffic flow $f_{w,p}^v(t)$ along path p during time t.

Similarly, the delay time requirement must not exceed the path duration, i.e.,

$$\sum_{l \in p} \varsigma_{w,l}^v(t) \le \Delta t, \ \forall v \in V, w \in W, od \in OD, p \in P_{od}^v, t \in \mathcal{T} \tag{13.11}$$

The path travel cost of type-v vehicle traffic flow under scenario w includes travel time cost and refueling cost. For fuel vehicles, the total cost consists of both parts, expressed as:

$$\phi_{w,p}^{v}(t) = \begin{cases} \gamma^{\text{time}}\left[\xi_{w,p}^{v}(t) - t\right] + \gamma_{w,\upsilon(p)}^{\text{refuel},v}(t)D^{v}, & v \in V/\{\text{GV}\} \\ \gamma^{\text{time}}\left[\xi_{w,p}^{v}(t) - t\right], & v = \text{GV} \end{cases},$$
$$\forall w \in W, od \in OD, p \in P_{od}^{v}, t \in \mathcal{T} \quad (13.12)$$

where $\gamma_{w,u(p)}^{\text{refuel},v}(t)$ denotes the type-v refueling price at the integrated energy service station located on path p at time t under scenario w.

Multi-Period Boundedly Rational User Equilibrium Model. Considering the competitive behavior among multiple types of vehicle traffic flows under bounded rationality, the following multi-period boundedly rational user equilibrium (BRUE) conditions can be established:

$$\forall v \in V, w \in W, od \in OD, p \in P_{od}^{v}, t \in \mathcal{T}$$
$$\Rightarrow \begin{cases} f_{w,p}^{v}(t) > 0 \Rightarrow \phi_{w,p}^{v}(t) \in \left[\mu_{od}^{v}(t), \mu_{od}^{v}(t) + \varepsilon_{od}^{v}\right] \\ \phi_{w,p}^{v}(t) > \mu_{od}^{v}(t) + \varepsilon_{od}^{v} \Rightarrow f_{w,p}^{v}(t) = 0 \end{cases} \quad (13.13)$$

where $\varepsilon_{od}^{v}(t)$ denotes the user tolerance parameter representing the rationality level of type-v users; $\mu_{w,od}^{v}(t)$ is the minimum travel cost of type-v traffic flow between OD pair od under scenario w at time t.

The minimum travel cost for type-v traffic flow under scenario w at time t is calculated as:

$$\mu_{w,od}^{v}(t) = \min_{p \in P_{od}^{v}} \phi_{w,p}^{v}(t), \ \forall v \in V, w \in W, od \in OD, t \in \mathcal{T} \quad (13.14)$$

According to Sect. 10.3.4, the above BRUE conditions can be reformulated into a set of mutually equivalent linearized constraints for easier computation, as follows:

$$0 \le f_{w,p}^{v}(t) \le M\left[1 - z_{w,p}^{v}(t)\right], \ \forall v \in V, w \in W, od \in OD, p \in P_{od}^{v}, t \in \mathcal{T} \quad (13.15)$$

$$0 \le \phi_{w,p}^{v}(t) - \pi_{w,od}^{v}(t) + \sigma_{w,p}^{v}(t) \le M z_{w,p}^{v}(t),$$
$$\forall v \in V, w \in W, od \in OD, p \in P_{od}^{v}, t \in \mathcal{T} \quad (13.16)$$

$$0 \le \sigma_{w,p}^{v}(t) \le \varepsilon_{od}^{v}(t), \ \forall v \in V, w \in W, od \in OD, p \in P_{od}^{v} \quad (13.17)$$

where $\pi^{v}_{w,od}(t)$ and $\sigma^{v}_{w,p}(t)$ represent the auxiliary continuous variables for travel cost and cost deviation, respectively, and $z^{v}_{w,p}(t)$ is a 0–1 binary variable indicating whether the bounded rationality condition is satisfied.

Since different types of traffic flows may share common regular links and compete for limited link capacity, this will cause queuing and congestion effects, which in turn affect the travel time of other traffic flows. Therefore, the BRUE model for each type of traffic flow should be constructed independently, rather than jointly modeled with interaction.

13.3 Integrated Energy Service Station (IESS) Multi-Energy Coordinated Dispatching Model for Electricity-Hydrogen-Gas Complementarity

In response to the previously obtained spatiotemporal distribution of heterogeneous energy demands, the integrated energy service station (IESS) must perform coordinated dispatching of multiple energy types to meet the differentiated refueling requirements of electric vehicles (EVs), hydrogen fuel cell vehicles (HFCVs), and natural gas vehicles (NGVs). This section presents the electricity–hydrogen–gas (E–H-G) complementary multi-energy coordinated dispatching model.

Basic Configuration of Integrated Energy Service Station. The proposed IESS model is illustrated in Fig. 13.2. The station is equipped with a variety of energy conversion and storage facilities, including photovoltaic and wind power generation units, electrolyzers, carbon capture systems, synthetic natural gas (SNG) generators, hydrogen fuel cells, hydrogen storage tanks, natural gas storage systems (NGSS), hydrogen compressors, and natural gas compressors.

Specifically, the electrolyzer utilizes renewable electricity to produce hydrogen. The SNG generator conducts the Sabatier reaction to synthesize natural gas from hydrogen and carbon dioxide, the latter of which is extracted from the atmosphere by the carbon capture unit. When renewable energy generation is insufficient, the IESS allows moderate curtailment of energy services such as EV charging, hydrogen refueling, or gas refueling. Such curtailment incurs a penalty cost, representing the economic loss due to decreased service quality.

In addition, the IESS maintains a connection to the main power grid. When on-site renewable power is insufficient to meet the total electricity demand, the grid provides supplementary power to ensure continuous service delivery.

Electricity Dispatch Model. The charging power of the integrated energy service station (IESS) is supplied by the grid, photovoltaic (PV) and wind power generation, and hydrogen fuel cell storage systems. In addition, the operation of the electrolyzer, hydrogen compressor, natural gas compressor, and carbon capture equipment also consumes electricity. The power supply–demand balance equation is given as follows:

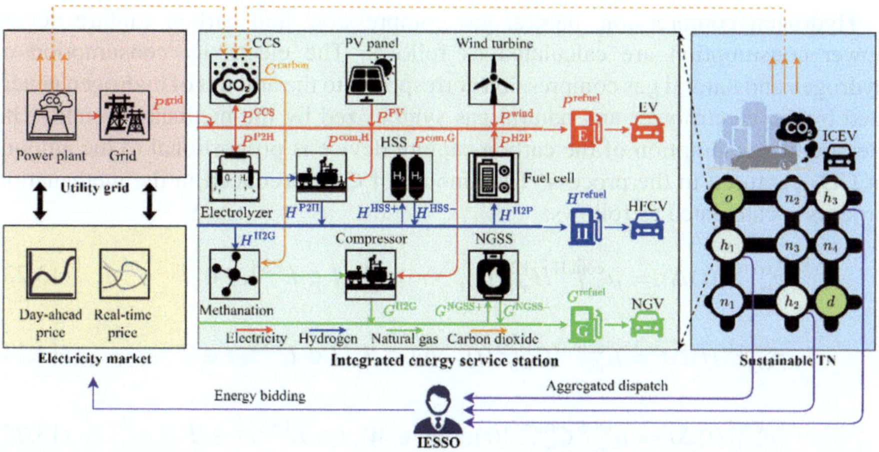

Fig. 13.2 IESS model

$$
\begin{aligned}
p_{w,l}^{\text{refuel}}(t) = {} & p_{w,l}^{\text{grid}}(t) + p_{w,l}^{\text{PV}}(t) + p_{w,l}^{\text{wind}}(t) + p_{w,l}^{\text{H2P}}(t) - p_{w,l}^{\text{P2H}}(t) - p_{w,l}^{\text{com,H}}(t) \\
& - p_{w,l}^{\text{com,G}}(t) - p_{w,l}^{\text{CCS}}(t), \quad \forall w \in W, l \in L^{\text{ESS}}, t \in \mathcal{T}
\end{aligned} \tag{13.18}
$$

where $p_{w,l}^{\text{refuel}}(t)$ is the total charging power provided to EV users on IESS link l at time t under scenario w; $p_{w,l}^{\text{grid}}(t)$ is the grid power; $p_{w,l}^{\text{PV}}(t)$ and $p_{w,l}^{\text{wind}}(t)$ denote photovoltaic and wind power generation; $p_{w,l}^{\text{H2P}}(t)$ is the hydrogen fuel cell power; $p_{w,l}^{\text{P2H}}(t)$ is the electrolyzer power; $p_{w,l}^{\text{com,H}}(t)$ and $p_{w,l}^{\text{com,G}}(t)$ are the hydrogen and natural gas compressor power; $p_{w,l}^{\text{CCS}}(t)$ is the power consumed by the carbon capture equipment.

The upper and lower bounds of grid power, photovoltaic power, and wind power are defined as follows:

$$
\omega D_{w,l}^{\text{EV}}(t) \le p_{w,l}^{\text{refuel}}(t)\Delta t \le D_{w,l}^{\text{EV}}(t) \le s_l^{\text{EV}}, \quad \forall w \in W, l \in L^{\text{ESS}}, t \in \mathcal{T} \tag{13.19}
$$

$$
0 \le p_{w,l}^{\text{grid}}(t) \le \overline{P}_l^{\text{grid}}, \quad \forall w \in W, l \in L^{\text{ESS}}, t \in \mathcal{T} \tag{13.20}
$$

$$
0 \le p_{w,l}^{\text{PV}}(t) \le \overline{P}_{w,l}^{\text{PV}}, \quad \forall w \in W, l \in L^{\text{ESS}}, t \in \mathcal{T} \tag{13.21}
$$

$$
0 \le p_{w,l}^{\text{wind}}(t) \le \overline{P}_{w,l}^{\text{wind}}, \quad \forall w \in W, l \in L^{\text{ESS}}, t \in \mathcal{T} \tag{13.22}
$$

Here, $D_{w,l}^{\text{EV}}(t)$ is the expected charging demand of EV users, $\omega \in [0, 1]$ is the adjustable curtailment factor, and s_l^{EV} is the maximum charging capacity on link l. The upper bounds $\overline{P}_{w,l}^{\text{PV}}$ and $\overline{P}_{w,l}^{\text{wind}}$ are affected by solar irradiance and wind speed under scenario w.

Hydrogen compression, natural gas compression, and carbon capture device power consumption are calculated as follows. The electricity consumption of hydrogen and natural gas compression corresponds to the amount of hydrogen generated by the electrolyzer and natural gas synthesized by the methanation unit. The electricity consumption of the carbon capture device is proportional to the amount of CO_2 captured in the process. The amount of CO_2 produced in the methanation process is calculated as follows:

$$p_{w,l}^{\mathrm{com.H}}(t)\Delta t = \mu_{w,l}^{\mathrm{com.H}} H_{w,l}^{\mathrm{P2H}}(t),\ \forall w \in W, l \in L^{\mathrm{ESS}}, t \in \mathcal{T} \tag{13.23}$$

$$p_{w,l}^{\mathrm{com.G}}(t)\Delta t = \mu_{w,l}^{\mathrm{com.G}} G_{w,l}^{\mathrm{H2G}}(t),\ \forall w \in W, l \in L^{\mathrm{ESS}}, t \in \mathcal{T} \tag{13.24}$$

$$p_{w,l}^{\mathrm{CCS}}(t)\Delta t = \mu_{w,l}^{\mathrm{CCS}} C_{w,l}^{\mathrm{carbon}}(t),\ \forall w \in W, l \in L^{\mathrm{ESS}}, t \in \mathcal{T} \tag{13.25}$$

$$C_{w,l}^{\mathrm{carbon}}(t) = r H_{w,l}^{\mathrm{H2G}}(t),\ \forall w \in W, l \in L^{\mathrm{ESS}}, t \in \mathcal{T} \tag{13.26}$$

where $H_{w,l}^{\mathrm{P2H}}(t)$ is the amount of hydrogen produced by electrolysis; $G_{w,l}^{\mathrm{H2G}}(t)$ is the amount of natural gas synthesized by the methanation unit; $C_{w,l}^{\mathrm{carbon}}(t)$ is the amount of CO_2 captured during methanation; $H_{w,l}^{\mathrm{H2G}}(t)$ is the amount of hydrogen consumed in the methanation reaction; $\mu_{w,l}^{\mathrm{com.H}}$ and $\mu_{w,l}^{\mathrm{com.G}}$ are the unit energy consumption coefficients for hydrogen and natural gas compression, respectively; $\mu_{w,l}^{\mathrm{CCS}}$ is the unit energy consumption coefficient for the carbon capture device; r is the stoichiometric ratio of CO_2 to hydrogen in the methanation process.

Hydrogen Regulation Model. The hydrogen supplied by the integrated energy service station for hydrogen fuel cell vehicles primarily comes from the electrolyzer. Additionally, some hydrogen is sourced from hydrogen storage tanks, either discharged directly or synthesized and then stored. The hydrogen balance is expressed as:

$$\begin{aligned} H_{w,l}^{\mathrm{refuel}}(t) = {} & H_{w,l}^{\mathrm{P2H}}(t) - H_{w,l}^{\mathrm{H2P}}(t) + H_{w,l}^{\mathrm{HSS-}}(t) \\ & - H_{w,l}^{\mathrm{HSS+}}(t),\ \forall w \in W, l \in L^{\mathrm{ESS}}, t \in \mathcal{T} \end{aligned} \tag{13.27}$$

where $H_{w,l}^{\mathrm{refuel}}(t)$ is the amount of hydrogen supplied by the integrated energy service station on link l at time t for hydrogen fuel cell vehicles under scenario w; $H_{w,l}^{\mathrm{H2P}}(t)$ is the amount of hydrogen consumed by hydrogen fuel cell vehicles; $H_{w,l}^{\mathrm{HSS-}}(t)$ and $H_{w,l}^{\mathrm{HSS+}}(t)$ are the hydrogen discharged from and charged into the storage tanks, respectively.

To ensure that the hydrogen supplied by the integrated energy service station meets the demand of hydrogen fuel cell vehicles (HFCVs), the following constraint must be satisfied:

$$\varpi D^{\mathrm{HFCV}} x_{w,l}^{\mathrm{HFCV}}(t) \leq H_{w,l}^{\mathrm{refuel}}(t) \leq D^{\mathrm{HFCV}} x_{w,l}^{\mathrm{HFCV}}(t) \leq D^{\mathrm{NGV}} s_{l}^{\mathrm{NGV}}, \\ \forall w \in W, l \in L^{\mathrm{IESS}}, t \in \mathcal{T} \quad (13.28)$$

The operation of electrolyzers and hydrogen fuel cells must satisfy constraints (13.29)–(13.32). Due to the conversion losses in the electrolysis and hydrogen fuel cell generation processes, these two devices are not allowed to operate simultaneously. Since hydrogen fuel cells serve as a backup power source, they are prohibited from running in parallel with electrolyzers.

$$P_{w,l}^{\mathrm{P2H}}(t)\Delta t \kappa^{\mathrm{P2H}} \eta^{\mathrm{P2H}} = H_{w,l}^{\mathrm{P2H}}(t), \forall w \in W, l \in L^{\mathrm{IESS}}, t \in \mathcal{T} \quad (13.29)$$

$$H_{w,l}^{\mathrm{H2P}}(t) \kappa^{\mathrm{H2P}} \eta^{\mathrm{H2P}} = P_{w,l}^{\mathrm{H2P}}(t)\Delta t, \forall w \in W, l \in L^{\mathrm{IESS}}, t \in \mathcal{T} \quad (13.30)$$

$$0 \leq H_{w,l}^{\mathrm{P2H}}(t) \leq \overline{H}_{l}^{\mathrm{P2H}} z_{w,l}^{\mathrm{P2H}}(t), \forall w \in W, l \in L^{\mathrm{IESS}}, t \in \mathcal{T} \quad (13.31)$$

$$0 \leq H_{w,l}^{\mathrm{H2P}}(t) \leq \overline{H}_{l}^{\mathrm{H2P}} \left[1 - z_{w,l}^{\mathrm{P2H}}(t)\right], \forall w \in W, l \in L^{\mathrm{IESS}}, t \in \mathcal{T} \quad (13.32)$$

In the above equations, κ^{P2H} is the conversion coefficient of electricity-to-hydrogen in electrolysis; η^{P2H} is the electrolysis efficiency; κ^{H2P} is the conversion coefficient of hydrogen-to-electricity in hydrogen fuel cells; η^{H2P} is the efficiency of hydrogen fuel cell generation; $z_{w,l}^{\mathrm{P2H}}(t)$ is a 0–1 auxiliary variable indicating whether electrolysis and hydrogen fuel cells are forbidden from simultaneous operation; $\overline{H}_{l}^{\mathrm{P2H}}$ is the maximum hydrogen production by the electrolyzer during time interval t; $\overline{H}_{l}^{\mathrm{P2H}}$ is the maximum hydrogen consumption by the hydrogen fuel cell during time interval t.

The hydrogen storage process must satisfy constraints (13.33)–(13.37). Similar to the battery storage device in Chapter 12, hydrogen storage is not allowed to charge and discharge simultaneously. Due to the sealing nature of the tank, a certain amount of loss occurs during the inflow and outflow of hydrogen. The hydrogen storage process must also satisfy the constraint that the hydrogen level at the beginning and end of the period matches the scheduled amount.

$$0 \leq H_{w,l}^{\mathrm{HSS+}}(t) \leq \overline{H}_{l}^{\mathrm{HSS+}} z_{w,l}^{\mathrm{HSS+}}(t), \forall w \in W, l \in L^{\mathrm{IESS}}, t \in \mathcal{T} \quad (13.33)$$

$$0 \leq H_{w,l}^{\mathrm{HSS-}}(t) \leq \overline{H}_{l}^{\mathrm{HSS-}} \left[1 - z_{w,l}^{\mathrm{HSS+}}(t)\right], \forall w \in W, l \in L^{\mathrm{IESS}}, t \in \mathcal{T} \quad (13.34)$$

$$\underline{S}_{l}^{\mathrm{HSS}} \leq S_{w,l}^{\mathrm{HSS}}(t) \leq \overline{S}_{l}^{\mathrm{HSS}}, \forall w \in W, l \in L^{\mathrm{IESS}}, t \in \mathcal{T} \quad (13.35)$$

$$S_{w,l}^{\mathrm{HSS}}(t + \Delta t) = S_{w,l}^{\mathrm{HSS}}(t) + H_{w,l}^{\mathrm{HSS+}}(t)\eta^{\mathrm{HSS+}} \\ - H_{w,l}^{\mathrm{HSS-}}(t)/\eta^{\mathrm{HSS-}}, \forall w \in W, l \in L^{\mathrm{IESS}}, t \in \mathcal{T} \quad (13.36)$$

$$S_{w,l}^{\mathrm{HSS}}(0) = S_{w,l}^{\mathrm{HSS}}(T) = S_{l}^{\mathrm{HSS,init}}, \forall w \in W, l \in L^{\mathrm{IESS}} \tag{13.37}$$

In the above equations, $H_{w,l}^{\mathrm{HSS}+}$ denotes the maximum amount of hydrogen injected into the hydrogen storage tank per time period; $H_{w,l}^{\mathrm{HSS}-}$ denotes the maximum amount of hydrogen released from the hydrogen storage tank per time period; $z_{w,l}^{\mathrm{HSS}}(t)$ is a 0–1 auxiliary variable indicating whether hydrogen charging and discharging are prohibited simultaneously; $\eta^{\mathrm{HSS}+}$ is the efficiency of hydrogen injection; $\eta^{\mathrm{HSS}-}$ is the efficiency of hydrogen release; $H_{w,l}^{\mathrm{HSS}}$ denotes the lower limit of hydrogen stored in the hydrogen tank; $H_{\mathrm{init}}^{\mathrm{HSS}}$ denotes the initial and final hydrogen level requirement in the hydrogen storage tank.

Natural Gas Regulation Model. The natural gas supplied by the integrated energy service station to NGVs comes from synthetic natural gas production equipment. In addition, due to the limitation of the synthesis rate, part of the natural gas is produced in advance and stored in the gas storage tank. The natural gas supply balance equation is formulated as:

$$G_{w,l}^{\mathrm{refuel}}(t) = G_{w,l}^{\mathrm{H2G}}(t) + G_{w,l}^{\mathrm{NGSS}-}(t) - G_{w,l}^{\mathrm{NGSS}+}(t), \forall w \in W, l \in L^{\mathrm{IESS}}, t \in \mathcal{T} \tag{13.38}$$

where $G_{w,l}^{\mathrm{refuel}}(t)$ denotes the total amount of natural gas supplied to NGVs on link l by the integrated energy service station under scenario w at time t; $G_{w,l}^{\mathrm{NGSS}-}(t)$ and $G_{w,l}^{\mathrm{NGSS}+}(t)$ denote the amount of natural gas released from and injected into the gas storage tank, respectively.

The amount of natural gas supplied by the integrated energy service station to NGVs must meet the following constraint:

$$\varpi D^{\mathrm{NGV}} x_{w,l}^{\mathrm{NGV}}(t) \le G_{w,l}^{\mathrm{refuel}}(t) \le D^{\mathrm{NGV}} x_{w,l}^{\mathrm{NGV}}(t) \le D^{\mathrm{NGV}} s_{l}^{\mathrm{NGV}},$$
$$\forall w \in W, l \in L^{\mathrm{IESS}}, t \in \mathcal{T} \tag{13.39}$$

The operation of the synthetic natural gas device must satisfy the constraints (13.40) and (13.41). Similarly, carbon dioxide and hydrogen consumption in the synthesis process are considered, leading to the following constraints:

$$H_{w,l}^{\mathrm{H2G}}(t)\kappa^{\mathrm{H2G}}\eta^{\mathrm{H2G}} = G_{w,l}^{\mathrm{H2G}}(t), \forall w \in W, l \in L^{\mathrm{IESS}}, t \in \mathcal{T} \tag{13.40}$$

$$0 \le G_{w,l}^{\mathrm{H2G}}(t) \le \overline{G}_{l}^{\mathrm{H2G}}, \forall w \in W, l \in L^{\mathrm{IESS}}, t \in \mathcal{T} \tag{13.41}$$

In the above equations, κ^{H2G} denotes the molar conversion coefficient of hydrogen to synthetic natural gas; η^{H2G} is the efficiency of the synthetic natural gas device; $G_{w,l}^{\mathrm{H2G}}(t)$ is the amount of synthetic natural gas produced by the device per unit period.

The hydrogen storage tank operation must satisfy the constraints (13.42)–(13.46), similar to hydrogen storage, and is omitted here for brevity.

$$0 \le G_{w,l}^{\text{NGSS}+}(t) \le \overline{G}_l^{\text{NGSS}+} z_{w,l}^{\text{NGSS}+}(t), \forall w \in W, l \in L^{\text{IESS}}, t \in \mathcal{T} \tag{13.42}$$

$$0 \le G_{w,l}^{\text{NGSS}-}(t) \le \overline{G}_l^{\text{NGSS}}\left[1 - z_{w,l}^{\text{NGSS}+}(t)\right], \forall w \in W, l \in L^{\text{IESS}}, t \in \mathcal{T} \tag{13.43}$$

$$\underline{S}_l^{\text{NGSS}} \le S_{w,l}^{\text{NGSS}}(t) \le \overline{S}_l^{\text{NGSS}}, \forall w \in W, l \in L^{\text{IESS}}, t \in \mathcal{T} \tag{13.44}$$

$$S_{w,l}^{\text{NGSS}}(t + \Delta t) = S_{w,l}^{\text{NGSS}}(t) + G_{w,l}^{\text{NGSS}+}(t)\eta^{\text{NGSS}+} - G_{w,l}^{\text{NGSS}-}(t)/\eta^{\text{NGSS}-}, \\ \forall w \in W, l \in L^{\text{IESS}}, t \in \mathcal{T} \tag{13.45}$$

$$S_{w,l}^{\text{NGSS}}(0) = S_{w,l}^{\text{NGSS}}(T) = S_l^{\text{NGSS,init}}, \forall w \in W, l \in L^{\text{IESS}} \tag{13.46}$$

In the above equations, $G_l^{\text{NGSS}+}$ denotes the maximum amount of natural gas that can be injected into the storage tank per unit time period; $G_l^{\text{NGSS}-}$ denotes the maximum amount of natural gas that can be released from the storage tank per unit time period; $z_{w,l}^{\text{NGSS}+}(t)$ is a 0–1 auxiliary variable indicating whether simultaneous injection and release of natural gas are prohibited; $S_{w,l}^{\text{NGSS}}(t)$ denotes the current amount of natural gas stored in the tank under scenario w; S_l^{NGSS} and $\overline{S}l^{\text{NGSS}}$ represent the lower and upper bounds of the allowable natural gas storage; $\eta^{\text{NGSS}+}$ is the efficiency of natural gas injection into the storage tank; $\eta^{\text{NGSS}-}$ is the efficiency of natural gas release from the storage tank; $Sl^{\text{NGSS,init}}$ denotes the initial and final required natural gas volume in the storage tank.

13.4 Risk-Averse Energy Optimization Strategy for Integrated Energy Service Station Aggregators Participating in the Two-Stage Electricity Market

Under a centralized dispatching model, clusters of integrated energy service stations (IESSs) can also participate in electricity market transactions through a unified aggregator. However, due to the coordinated operation of multiple energy carriers, including electricity, hydrogen, and natural gas, the IESS aggregator must manage greater uncertainty and operational risks when engaging in market participation. Therefore, this section investigates a risk-averse energy optimization strategy for IESS aggregators in the context of two-stage electricity market trading, aiming to ensure profitability while mitigating exposure to risks under extreme scenarios.

Model of the Integrated Energy Service Station Aggregator. The model of the IESS aggregator is illustrated in Fig. 13.2. All IESSs within the transportation network are centrally managed and dispatched by a single aggregator. By designing a global energy scheduling strategy, the aggregator coordinates its managed IESS cluster to participate in the two-stage competitive electricity spot market, including the day-ahead and real-time markets. As an intermediary, the aggregator must balance two key objectives: on one hand, it must rationally allocate multiple energy resources, e.g., electricity, hydrogen and natural gas, according to heterogeneous refueling demands from users; on the other hand, it must address multiple sources of uncertainty, such as electricity price fluctuations, intermittent renewable generation, and load forecasting errors. The ultimate goal is to strike a balance between economic efficiency and operational robustness while ensuring the quality of energy refueling services.

Assuming the retail pricing schemes for charging, hydrogen refueling, and gas refueling at the integrated energy service station are consistent with those in Chapter 12, we have:

$$\gamma_{w,l}^{\mathrm{refuel,EV}}(t) = \alpha^{\mathrm{s}}\lambda_w^{\mathrm{RT}}(t), \forall w \in W, l \in L^{\mathrm{IESS}}, t \in \mathcal{T} \tag{13.47}$$

$$\gamma_{w,l}^{\mathrm{refuel,HCEV}}(t) = \alpha^{\mathrm{s}}\lambda^{\mathrm{H}}, \forall w \in W, l \in L^{\mathrm{IESS}}, t \in \mathcal{T} \tag{13.48}$$

$$\gamma_{w,l}^{\mathrm{refuel,NGV}}(t) = \alpha^{\mathrm{s}}\lambda^{\mathrm{NG}}, \forall w \in W, l \in L^{\mathrm{IESS}}, t \in \mathcal{T} \tag{13.49}$$

where $\gamma_{w,l}^{\mathrm{refuel,EV}}(t)$, $\gamma_{w,l}^{\mathrm{refuel,HFCV}}(t)$, and $\gamma_{w,l}^{\mathrm{refuel,NGV}}(t)$ denote the retail prices for charging, hydrogen refueling, and gas refueling, respectively, at the integrated energy service station; α^s is the retail markup coefficient; λ^H and λ^{NG} are the market wholesale prices of hydrogen and natural gas, respectively.

The revenue from charging, hydrogen refueling, and gas refueling services provided by the cluster of integrated energy service stations is calculated as:

$$R_w^{\mathrm{E}} = \sum_{l \in L^{\mathrm{IESS}}} \sum_{t \in \mathcal{T}} \gamma_{w,l}^{\mathrm{refuel,EV}}(t) P_{w,l}^{\mathrm{refuel}} \Delta t, \forall w \in W \tag{13.50}$$

$$R_w^{\mathrm{H}} = \sum_{l \in L^{\mathrm{IESS}}} \sum_{t \in \mathcal{T}} \gamma_{w,l}^{\mathrm{refuel,HFCV}}(t) H_{w,l}^{\mathrm{refuel}}(t), \forall w \in W \tag{13.51}$$

$$R_w^{\mathrm{G}} = \sum_{l \in L^{\mathrm{IESS}}} \sum_{t \in \mathcal{T}} \left[\gamma_{w,l}^{\mathrm{refuel,NGV}}(t) G_{w,l}^{\mathrm{refuel}}(t) + \gamma^{\mathrm{carbon}} C_{w,l}^{\mathrm{carbon}}(t)\right], \forall w \in W \tag{13.52}$$

where R_w^E is the charging service revenue of the aggregator; R_w^H is the hydrogen refueling service revenue; R_w^G is the revenue from gas refueling and carbon capture; γ^{carbon} is the incentive price per unit of carbon dioxide captured.

The operating cost of the aggregator mainly includes electricity, hydrogen, and natural gas-related components, as shown in formulas:

$$\begin{aligned} C_w^{\mathrm{E}} = &\sum_{t \in \mathcal{T}} \left[\gamma^{\mathrm{DA}}(t) E^{\mathrm{DA}}(t) + \gamma_w^{\mathrm{RT}}(t) \Delta E_w^{\mathrm{RT}}(t) + C_w^{\mathrm{penalty}}(t)\right] \\ &+ \sum_{l \in L^{\mathrm{IESS}}} \sum_{t \in \mathcal{T}} \left[\gamma^{\mathrm{PV}} P_{w,l}^{\mathrm{PV}}(t) \Delta t + \gamma^{\mathrm{wind}} P_{w,l}^{\mathrm{wind}}(t) \Delta t + \gamma^{\mathrm{H2P}} P_{w,l}^{\mathrm{H2P}}(t) \Delta t\right] \end{aligned}$$

$$+\sum_{l\in L^{\text{IESS}}}\sum_{t\in\mathcal{T}}\gamma^{\text{penalty,E}}\left[D^{\text{EV}}x_{w,l}^{\text{EV}}(t)-P_{w,l}^{\text{refuel}}(t)\Delta t\right],\forall w\in W \tag{13.53}$$

$$C_w^{\text{H}}=\sum_{l\in L^{\text{IESS}}}\sum_{t\in\mathcal{T}}(\gamma^{\text{P2H}}+\gamma^{\text{com,H}})H_{w,l}^{\text{P2H}}(t)$$
$$+\sum_{l\in L^{\text{IESS}}}\sum_{t\in\mathcal{T}}\gamma^{\text{penalty,H}}\left[D^{\text{HFCV}}x_{w,l}^{\text{HFCV}}(t)-H_{w,l}^{\text{refuel}}(t)\right],\forall w\in W \tag{13.54}$$

$$C_w^{\text{G}}=\sum_{l\in L^{\text{IESS}}}\sum_{t\in\mathcal{T}}\left[(\gamma^{\text{H2G}}+\gamma^{\text{com,G}})G_{w,i,t}^{\text{H2G}}(t)+\gamma^{\text{CCS}}C_{w,l}^{\text{carbon}}(t)\right]$$
$$+\sum_{l\in L^{\text{IESS}}}\sum_{t\in\mathcal{T}}\gamma^{\text{penalty,G}}\left[D^{\text{NGV}}x_{w,l}^{\text{NGV}}(t)-G_{w,l}^{\text{refuel}}(t)\right],\forall w\in W \tag{13.55}$$

where C_w^E denotes the electricity cost; $\gamma^{\text{penalty},E}$ is the penalty cost coefficient for unmet electricity demand; $\gamma^{\text{penalty},H}$ is the penalty cost coefficient for unmet hydrogen demand; $\gamma^{\text{penalty},G}$ is the penalty cost coefficient for unmet gas demand; γ^{P2H}, $\gamma^{\text{com,H}}$, γ^{H2P}, γ^{H2G}, $\gamma^{\text{com,G}}$, γ^{CCS} denote the unit operating cost coefficients of electrolysis, hydrogen compressor, fuel cell, gas synthesis device, natural gas compressor, and carbon capture equipment, respectively.

The mathematical model for real-time electricity market constraints used in this section is consistent with Sect. 12.4.2 and is given by:

$$0\le E^{\text{DA}}(t),\forall t\in\mathcal{T} \tag{13.56}$$

$$\underline{E}^{\text{DA}}\le\sum_{t\in\mathcal{T}}E^{\text{DA}}(t)\le\overline{E}^{\text{DA}} \tag{13.57}$$

$$E_w^{\text{RT}}(t)=\sum_{l\in L^{\text{IESS}}}P_{w,l}^{\text{grid}}(t)\Delta t,\forall w\in W,t\in\mathcal{T} \tag{13.58}$$

$$\Delta E_w^{\text{RT}}(t)=E_w^{\text{RT}}(t)-E^{\text{DA}}(t),\forall w\in W,t\in\mathcal{T} \tag{13.59}$$

$$C_w^{\text{penalty}}(t)=\begin{cases}\max\{0,[E^{\text{DA}}(t)-\omega^{+}E_w^{\text{RT}}(t)]\Delta\lambda_w(t)\}, & \Delta\lambda_w(t)\ge 0\\ \max\{0,[E^{\text{DA}}(t)-\omega^{-}E_w^{\text{RT}}(t)]\Delta\lambda_w(t)\}, & \Delta\lambda_w(t)<0\end{cases},$$
$$\forall w\in W,t\in\mathcal{T} \tag{13.60}$$

$$\Delta\lambda_w(t)=\gamma_w^{\text{RT}}(r)-\gamma^{\text{DA}}(t),\forall w\in W,t\in\mathcal{T} \tag{13.61}$$

In summary, the profit O_w of the aggregator under scenario w is defined as:

$$O_w=R_w^{\text{E}}+R_w^{\text{H}}+R_w^{\text{G}}-C_w^{\text{E}}-C_w^{\text{H}}-C_w^{\text{G}},\forall w\in W \tag{13.62}$$

CVaR-Based Risk-Averse Optimization Strategy for Day-Ahead Bidding and Intra-Day Multi-Energy Coordination. Traditional scenario-based stochastic planning methods often aim to maximize the expected profit of the integrated energy service station aggregator under multiple scenarios. However, this method lacks sufficient risk aversion capability and may lead to significant profit loss under extreme conditions. To enhance the robustness of the aggregator in the face of uncertainties, this section proposes a CVaR-based risk-averse optimization approach. This approach is designed to help the aggregator jointly optimize its bidding strategy in the day-ahead market and coordinate the electricity, hydrogen and gas resources in the intra-day operation. The method guarantees a satisfactory expected return while controlling losses in extreme scenarios, thereby improving the aggregator's resilience in managing uncertainty and its capability for robust operation.

Given the multi-dimensional uncertainties, we define the risk of profit loss for the aggregator by taking the negative profit O_w under scenario w as the loss value X_w:

$$X_w = -O_w, \forall w \in W \tag{13.63}$$

To measure the risk of uncertain profit loss, a quantitative risk metric must first be determined. Let ρ_w be the probability distribution of scenario-based profit loss. A common strategy is to compute the expected value of loss under all scenarios:

$$\text{Average} = \sum_{w \in W} \rho_w X_w \tag{13.64}$$

This expectation-based metric fails to capture losses in extreme scenarios. Therefore, in the field of economics, Value-at-Risk (VaR) and Conditional Value-at-Risk (CVaR) are commonly used to evaluate and control risks under tail scenarios. The VaR value estimates the maximum tolerable loss at a given confidence level α, and is defined as:

$$P(x) = \Pr\{X_w \leq x | w \in W\} = \sum_{X_w \leq x, w \in W} \rho_w, \forall x \in R \tag{13.65}$$

$$\text{VaR}_\alpha = \min x : P(x) \geq \alpha, \forall x \in R, \alpha \in [0, 1] \tag{13.66}$$

where $P(x)$ denotes the cumulative probability distribution of loss less than or equal to x, and R is the discrete scenario set.

The above procedure sorts all scenario losses in ascending order, then sums their probabilities until the cumulative value reaches α, and finally takes the corresponding loss value of the last scenario as the VaR value. Hence, VaR satisfies the following property:

$$P(\text{VaR}_\alpha) = \alpha, \forall \alpha \in [0, 1] \tag{13.67}$$

That is, the cumulative probability of loss not exceeding VaR_α equals α, and the probability of loss exceeding VaR_α equals $1-\alpha$.

The risk value estimates the potential loss induced by uncertainty factors. Although it provides the maximum allowable loss under a given confidence level, it cannot measure risks within the tail region of the risk distribution. Moreover, the risk value is neither coherent nor convex, which limits its application in distributed computation and mathematical optimization. To address these shortcomings, the conditional value-at-risk (CVaR) has been proposed for tail risk assessment. CVaR possesses the desirable properties of subadditivity, coherence, and convexity, and exhibits favorable mathematical characteristics such as continuity and differentiability, making it a consistent risk quantification metric. CVaR is defined as the expected loss exceeding the risk value VaR_α:

$$\text{CVaR}_\alpha = \frac{1}{1-\alpha} \sum_{X_w > \text{VaR}_\alpha, w \in W} \rho_w X_w, \forall \alpha \in [0, 1] \tag{13.68}$$

For ease of computation, the following lemma is introduced [179].

Lemma 5.1 Given the following auxiliary function:

$$F_\alpha(x) = x + \frac{1}{1-\alpha} \sum_{w \in W} \rho_w (X_w - x)_+, \forall x \in R, \alpha \in [0, 1] \tag{13.69}$$

where $(x)+ = \max(0, x)$, the risk value VaRα and conditional value-at-risk CVaR_α can be computed as follows:

$$\text{VaR}_\alpha = \arg\min_{x \in R} F_\alpha(x), \forall \alpha \in [0, 1] \tag{13.70}$$

$$\text{CVaR}_\alpha = \min_{x \in R} F_\alpha(x), \forall \alpha \in [0, 1] \tag{13.71}$$

For the profit maximization problem, VaRα and CVaRα can be used to evaluate the expected profit under extreme scenarios. The higher the VaRα and CVaRα, the lower the potential loss risk faced by the aggregator of the integrated energy service station. Thus, the CVaR can be reformulated into the following maximization form:

$$\text{CVaR}_\alpha = \max(Z - \frac{1}{1-\alpha} \sum_{w \in W} \rho_w \psi_w), \forall \alpha \in [0, 1] \tag{13.72}$$

$$\psi_w \geq 0, \forall w \in W \tag{13.73}$$

$$\psi_w \geq Z - O_w, \forall w \in W \tag{13.74}$$

where Z is a continuous decision variable representing the overall profit; ψ_w is a continuous auxiliary variable denoting the portion of profit O_w that is below the risk value VaR_α under scenario w, and it is constrained to be non-negative.

The expected profit of the aggregator over all scenarios is defined as O^{net}, that is:

$$O^{\text{net}} = \sum_{w \in W} \rho_w O_w \tag{13.75}$$

The risk-averse weighting coefficient β is defined to apply a weighted combination of the expected profit O^{net} and the tail risk CVaR_α of the aggregator of integrated energy service stations, resulting in the risk-averse objective function O^{risk}. Accordingly, the risk-averse formulation of the integrated energy service station aggregator's day-ahead bidding and multi-energy coordinated dispatch problem is given as follows:

$$\max O^{\text{risk}} = (1 - \beta)O^{\text{net}} + \beta \text{CVaR}_\alpha, \forall \alpha \in [0, 1], \beta \in [0, 1] \tag{13.76}$$

The aggregator of integrated energy service stations can obtain different levels of risk aversion by adjusting the value of β. When $\beta = 0$, the above model degenerates into a stochastic optimization model (risk-neutral), which only optimizes the expected profit under all scenarios without considering the tail risk of extreme scenarios, and thus lacks robustness under risk. When $\beta > 0$, the model becomes a risk-averse optimization model (risk-averse), which simultaneously optimizes the expected profit and reduces the potential losses under extreme scenarios, thereby possessing a certain degree of risk resistance. When $\beta = 1$, the model solely focuses on optimizing the tail profit and ignores the overall expected profit, making it close to a worst-case optimization.

13.5 Case Study

Simulation Setup. This section uses a traffic network simulation test system 5, as shown in Fig. 13.3. The system is built based on the Nguyen traffic network, including 17 nodes in total: 4 integrated energy service station nodes, 4 traffic origin nodes, 9 intermediate nodes, as well as 44 integrated energy service station links and 23 regular links. The travel demand settings are shown in Table 13.1, and the remaining main simulation parameters are listed in Table 13.2. Parameters that are the same as those in Sect. 10.5 will not be repeated here. The simulation time step of the traffic flow model is set to 0.05 h, while the bidding and multi-energy coordinated dispatch time step of the integrated energy service stations is set to 1 h.

Considering uncertainty factors such as demand prediction errors and renewable output fluctuations during operation, this section introduces four types of uncertain parameters: travel demand d_w [121], PV output upper bound, wind power output upper bound [177], and real-time electricity prices. The uncertainty intervals for

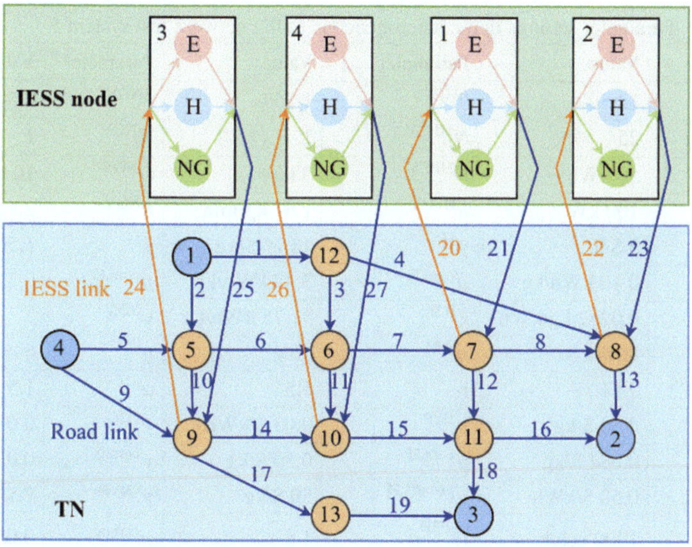

Fig. 13.3 Heterogeneous traffic network test system 5

Table 13.1 Traffic demand in the heterogeneous traffic network test system 5

o	d	Q_{od}^{EV}(veh)	Q_{od}^{HFCV}(veh)	Q_{od}^{NGV}(veh)	Q_{od}^{GV}(veh)
1	2	360	36	72	2400
1	3	360	36	72	2400
4	2	360	36	72	2400
4	3	360	36	72	2400

these parameters are listed in Fig. 13.4, where the electricity price data is based on the actual market data released by China Southern Power Grid. To assess the impact of different uncertainty levels, 4 scenario clusters are extracted according to the classification method in Sect. 12.4.1, each containing 64 representative operating scenarios. Combining these scenarios yields a total of 256 representative operating scenarios. The stochastic optimization problem is solved using Gurobi under these background scenarios to determine the optimal bidding strategy and multi-energy coordinated dispatch plan of the integrated energy service station aggregator.

Revenue and Cost Analysis of Integrated Energy Service Station Aggregator. To evaluate the effectiveness of the risk-averse optimization method proposed in this chapter, we compare the revenue and cost structure of the integrated energy service station aggregator under two models: risk-neutral ($\beta = 0$, i.e., random optimization) and risk-averse ($\beta = 0.5$), as shown in Table 13.3. Except for the expected overall profit, other indicators represent the average values over 256 operating scenarios.

Table 13.2 Parameter settings in the heterogeneous traffic network test system 5

Parameter	Value	Parameter	Value	Parameter	Value
θ^{EV}	12.50%	θ^{HFCV}	1.25%	θ^{NGV}	2.50%
N_l^{EV}	25	N_l^{HFCV}	1	N_l^{NGV}	1
D^{EV}	50 kWh	D^{HFCV}	5 kg	D^{NGV}	10 kg
R^{EV}	100 kW	R^{HFCV}	1.67 kg/min	R^{NGV}	2 kg/min
γ^{time}	5 $/h	γ^{H}	18.70 $/kg	γ^{G}	1.50 $/kg
μ^{CCS}	0.40 kWh/kg	$\mu^{com,H}$	3.47 kWh/kg	$\mu^{com,G}$	0.22 kWh/kg
κ^{P2H}	0.0252 kg/kWh	κ^{H2P}	31.75 kWh/kg	κ^{H2G}	2
η^{P2H}	80%	η^{H2P}	60%	η^{H2G}	64%
η^{HSS}/η^{NGSS}	98%	r	5.5	α^{P}	1.5
γ^{P2H}	0.50 $/kg	γ^{H2P}	0.03 $/kWh	γ^{H2G}	0.05 $/kg
γ^{CCS}	0.084 $/kg	$\gamma^{com,H}$	0.30 $/kg	$\gamma^{com,G}$	0.02 $/kg
$\gamma^{penalty,E}$	0.50 $/kWh	$\gamma^{penalty,H}$	50 $/kg	γ^{carbon}	7 $/kg
$\overline{P}^{grid}$	1000 veh/h	$\overline{H}_i^{P2H}$	1 $	γ^{carbon}	0.05 $/kg
$\overline{P}^{grid}$	2 MW	$\overline{H}_i^{P2H}$	20 kg/h	$\overline{G}_i^{H2G}$	75 kg/h
$\overline{H}_i^{HSS}$	100 kg/h	$\underline{S}_i^{HSS}/\overline{S}_i^{HSS}$	20/180 kg	$S_i^{HSS,init}$	100 kg
$\overline{G}_i^{NGSS}$	100 kg/h	$\underline{S}_i^{NGSS}/\overline{S}_i^{NGSS}$	40/360 kg	$S_i^{NGSS,init}$	200 kg
ω^{+}/ω^{-}	1.02/0.98	$\underline{E}^{DA}$	100 MWh	$\overline{E}^{DA}$	150 MWh
α	0.95	β	0.5		

In terms of revenue structure, as electric vehicle users represent the majority of the integrated energy service station user base, the charging service accounts for 53.5% of the aggregator's total revenue. Although hydrogen fuel cell vehicles are relatively few, the higher price of hydrogen leads to hydrogen refueling services contributing 34.1% of the revenue, which is much higher than the 12.4% contribution from gas refueling. This indicates that even with a small scale of usage, the high added value of hydrogen brings considerable revenue to the station.

From the perspective of cost structure, electricity consumption is the main contributor, accounting for 73.9% of the total operating cost. This includes electricity for EV charging, electrolysis, and synthetic natural gas production. The operation costs of hydrogen and natural gas are 8.7% and 17.5% respectively. Overall, the integrated energy service station's operating cost structure is centered around electricity, while electricity also serves as the main carrier supporting multi-energy collaborative services, highlighting the importance of energy-saving optimization.

Further analysis of the two models reveals notable differences in their economic robustness. While the charging and hydrogen services achieve full user demand satisfaction in both models, the gas refueling service that is subject to higher marginal costs, especially under insufficient renewable generation, may result in unmet demand for 79.5% of gas refueling cases in the risk-neutral model. This

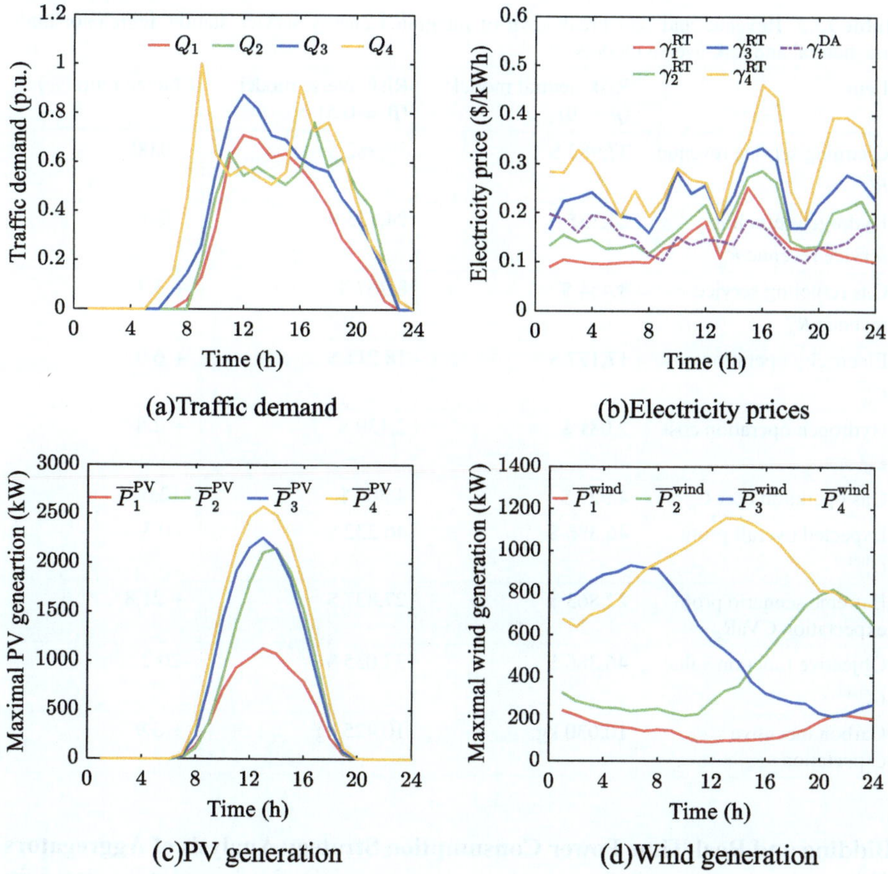

Fig. 13.4 Representative scenarios of multi-dimensional operational uncertainties

leads to significant penalties in extreme scenarios. Under such conditions, risk-averse optimization reduces the uncertainty associated with natural gas demand response through two-stage energy control strategies.

On the revenue side, EV users contributed 53.5% of total revenue, followed by hydrogen refueling (34.1%) and natural gas refueling (12.4%). On the cost side, electricity consumption accounted for 73.9%, hydrogen 8.7%, and natural gas 17.5%, reflecting the cost structure dominated by electricity.

Under the risk-averse model, extreme profit loss (CVaR) is reduced by 21.8%, with only a 0.3% drop in expected profit, demonstrating improved risk resilience. The cost of natural gas operation decreases by 12.7%, while carbon capture rewards increase by 3.9%, indicating enhanced carbon synergy.

Further analysis shows that the electricity and hydrogen services are fully satisfied under both models, while the natural gas service shows unmet demand in 79.5% of scenarios, indicating that gas demand-side flexibility is more affected by risk.

Table 13.3 Revenue and cost breakdown of integrated energy service station aggregator under risk-neutral and risk-averse models

Item	Risk-neutral model ($\beta = 0$)	Risk-averse model ($\beta = 0.5$)	Change ratio (%)
Charging service revenue R_w^E	37,982 $	37,982 $	0.00
Hydrogen refueling service revenue R_w^H	24,135 $	24,135 $	0.00
Gas refueling service revenue R_w^G	8,454 $	8,767 $	+ 3.7
Electricity operation cost C_w^E	17,177 $	18,211 $	+ 6.0
Hydrogen operation cost C_w^H	2,081 $	2,139 $	+ 2.8
Gas operation cost C_w^G	4,927 $	4,306 $	–12.6
Expected overall profit O_w^{net}	46,386 $	46,232 $	–0.3
Extreme scenario profit expectation CVaR_α	22,863 $	27,837 $	+ 21.8
Objective function value O_w^{risk}	46,386 $	37,035 $	–20.2
Carbon incentive expectation	10,030 kg	10,425 kg	+ 3.9

Bidding and Real-Time Power Consumption Strategy Analysis of Aggregators. This section compares the impact of aggregator participation in two-stage electricity markets under different demand and price scenarios. Fig. 13.5 illustrates bidding and real-time usage under different demand scenarios.

From Fig. 13.5a and b, under low-demand scenarios, the aggregator shows real-time power usage less than day-ahead bidding in early hours (before 8:00), indicating a conservative strategy. In high-demand scenarios (Fig. 13.5c and d, deviations occur, but overall bias is small.

Further, Fig. 13.6 shows bidding and real-time usage under different price scenarios.

To reduce overall electricity cost, the aggregator tends to generate actual electricity consumption deviations in periods with relatively low real-time electricity prices. For example, in Scenario (2,2,2,1) in Fig. 13.6a and Scenario (2,2,2,2) in Fig. 13.6b, the real-time electricity is close to the day-ahead level before 8:00, leading to lower penalty costs according to (13.60). In contrast, for high-price deviation scenarios such as (2,3,2,2) in Fig. 13.6c and (2,4,2,2) in Fig. 13.6d, the actual electricity consumption remains consistent with the day-ahead bidding results, thereby avoiding the penalties due to high deviations.

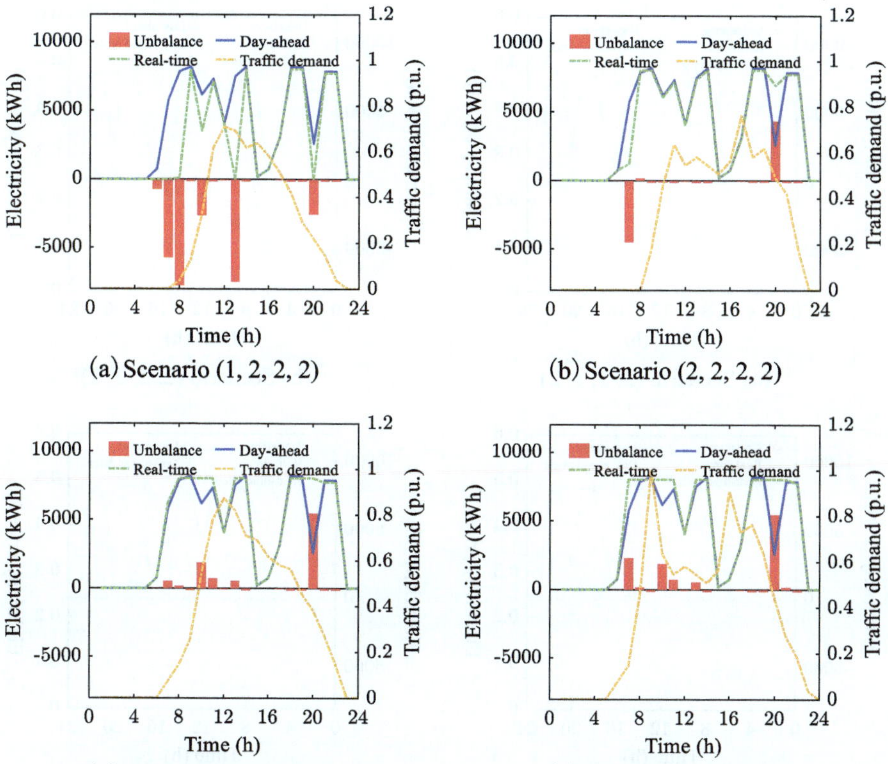

(a) Scenario (1, 2, 2, 2)

(b) Scenario (2, 2, 2, 2)

(c) Scenario (3, 2, 2, 2)

(d) Scenario (4, 2, 2, 2)

Fig. 13.5 Bidding and real-time power consumption strategies of the integrated energy service station aggregator under different travel demand scenarios

Finally, Fig. 13.7 compares the aggregator's bidding and real-time electricity usage strategies under different renewable energy generation scenarios. The scenario indices are (2,2,1,1), (2,2,2,2), (2,2,3,3), and (2,2,4,4), corresponding to Fig. 13.4c and d. Scenarios 1 through 4 represent increasing solar and wind power availability, while other parameters are set to medium levels.

The results show a significant negative correlation between the variation in renewable energy resources and the electricity consumption deviation. In scenarios with relatively scarce renewable resources (e.g., Fig. 13.7a and b), the aggregator relies more on external electricity purchases, resulting in positive deviations. In contrast, in scenarios with abundant renewable energy (e.g., Fig. 13.7c and d), the aggregator prioritizes the use of renewable resources over grid electricity, showing a tendency for negative deviations. This strategy helps improve the utilization of renewable energy while reducing electricity procurement costs.

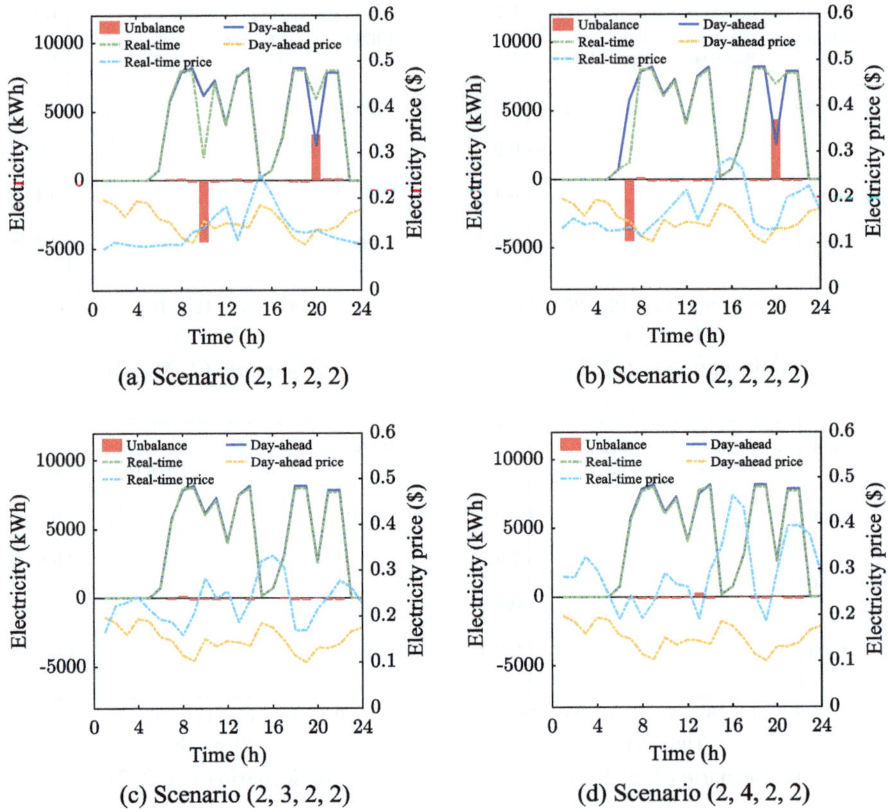

(a) Scenario (2, 1, 2, 2)

(b) Scenario (2, 2, 2, 2)

(c) Scenario (2, 3, 2, 2)

(d) Scenario (2, 4, 2, 2)

Fig. 13.6 Bidding and real-time power consumption strategies of the aggregator under different real-time electricity price scenarios

In summary, the aggregator dynamically adjusts its bidding and electricity usage strategy according to different types of uncertainties, fully demonstrating the advantage of the proposed stochastic optimization model in terms of adaptability and flexibility.

Coordinated Electricity-Hydrogen-Gas Dispatching Analysis for Integrated Energy Service Station Aggregators. Considering that the costs of electrolytic hydrogen production and synthetic natural gas are significantly affected by renewable energy availability, this section evaluates the impact of different renewable power scenarios on the coordinated electricity-hydrogen-gas dispatch strategy of integrated energy service stations.

Figure 13.8 illustrates the electricity dispatch strategy of the integrated energy service station No. 2 under four different renewable energy scenarios. The selected scenarios are consistent with those in Fig. 13.7 and are therefore not repeated here.

The electricity supply for the integrated energy service station primarily relies on grid power and renewable energy, with priority given to renewable sources. In

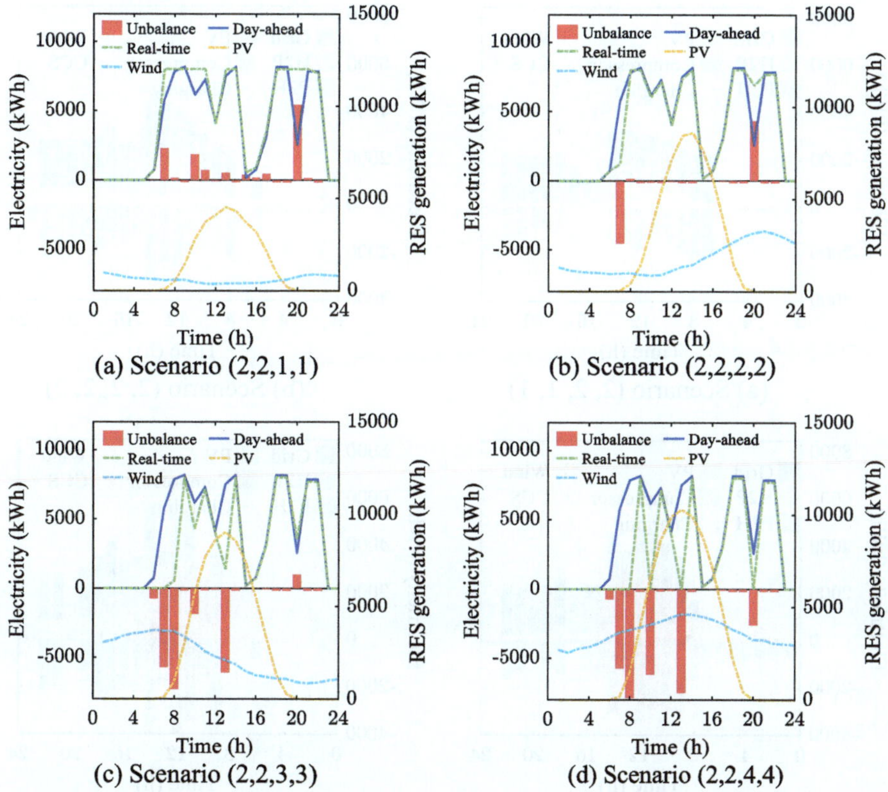

Fig. 13.7 Comparison of the aggregator's day-ahead bidding and real-time electricity usage strategies under different renewable energy generation scenarios

renewable-abundant scenarios (e.g., Fig. 13.8d), the system utilizes midday renewable energy for electrolytic hydrogen production and stores the hydrogen, which is then used in fuel cells (H2P) around 20:00 to supply power to the station. On the demand side, electricity consumption mainly consists of charging services and electrolytic hydrogen production (P2H), while compressors and carbon capture units account for a smaller share. Since the demand for charging is eventually fully satisfied, the power required for this service remains relatively stable and is not influenced by fluctuations in renewable energy. In contrast, electricity consumption for hydrogen production is significantly affected.

In renewable-scarce scenarios (e.g., Fig. 13.8a), the electricity consumed by P2H and charging services remains roughly equal. In renewable-rich scenarios (e.g., Fig. 13.8d), the electricity consumed by P2H increases to twice that of charging. Notably, during the period from 0:00 to 8:00 in Fig. 13.8c, electricity from photovoltaic and wind sources is available but not used for charging due to insufficient user demand, prompting the system to prioritize storing excess electricity for hydrogen production, thereby enabling peak shaving and demand-side flexibility.

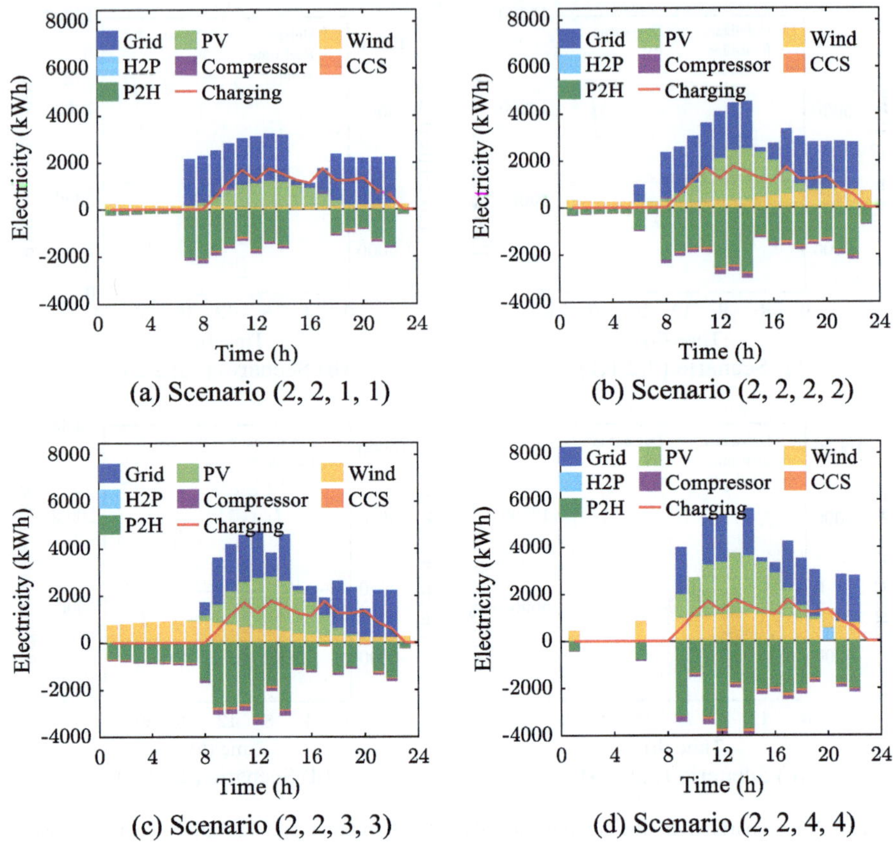

Fig. 13.8 Electricity dispatch strategies of integrated energy service station No. 2 under different renewable energy scenarios

Figure 13.9 shows the hydrogen dispatch strategies of IESS#2 under different renewable energy scenarios.

The hydrogen supply relies entirely on electrolytic hydrogen production, and its consumption mainly stems from hydrogen refueling services and synthetic natural gas (H2G), with the latter occupying a relatively higher share. Similar to charging services, the hydrogen demand for refueling is eventually fully satisfied, making its consumption relatively stable and unaffected by fluctuations in renewable energy availability. In contrast, hydrogen used for synthetic natural gas production varies more noticeably with renewable power conditions.

As shown in Fig. 13.9c, during periods of abundant nighttime wind power, the system prioritizes electrolytic hydrogen production and storage, which is subsequently used in H2G reactions during the day. Given the limited rate of electrolysis, hydrogen storage becomes critical during the time window from 22:00 to 9:00 the next day, playing a key role in balancing hydrogen supply and demand.

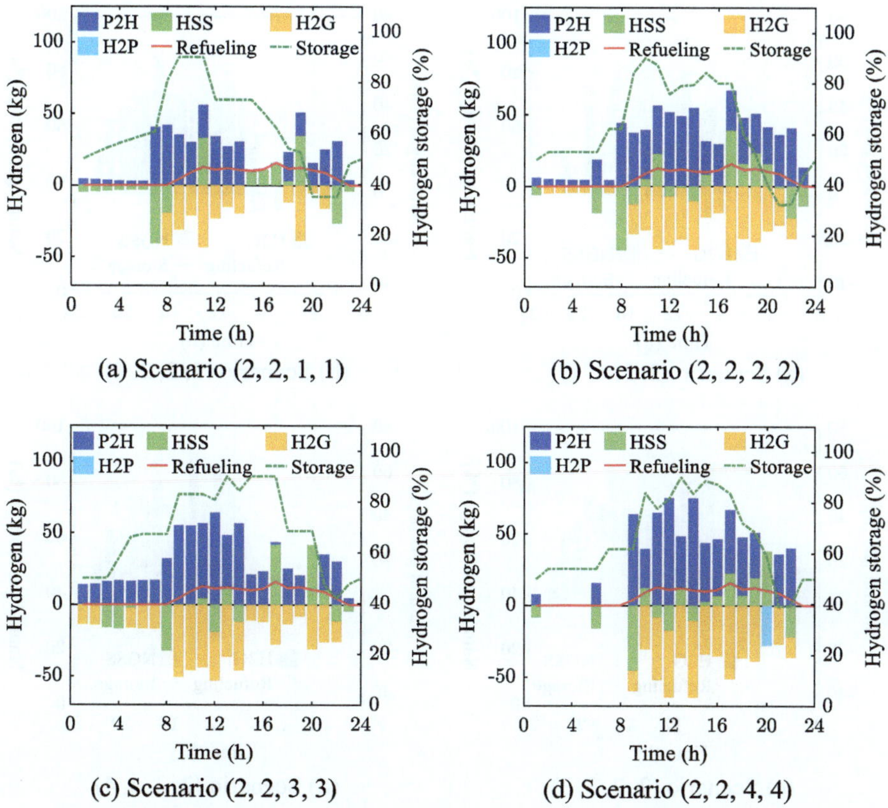

Fig. 13.9 Hydrogen dispatch strategies of IESS#2 under different renewable energy scenarios

As shown in Fig. 13.10, under the same renewable energy scenarios, IESS#2 adopts different natural gas dispatch strategies.

Natural gas supply is entirely generated from synthetic natural gas (SNG) systems and used to satisfy gas refueling service demand. Renewable power availability directly affects gas refueling: in scenarios with limited resources, part of the gas demand is curtailed to reduce cost. For example, in Fig. 13.10c, the system stores excess renewable power at night as synthetic natural gas to meet peak gas demand during the day. In contrast, Fig. 13.10d shows that under high renewable output and gas demand conditions, the system achieves "just-in-time production," reducing reliance on gas storage.

Figure 13.11 further presents the overall energy flow of electricity–hydrogen–gas in the integrated energy service station under scenarios (2,2,1,1) and (2,2,4,4). The percentage of each energy type is normalized to the total electricity supply under scenario (2,2,4,4) (set as 100%).

The energy flow results in Fig. 13.11 are consistent with the analysis results from Figs. 13.8, 13.9, 13.10. Taking Fig. 13.11b as an example, under scenarios with

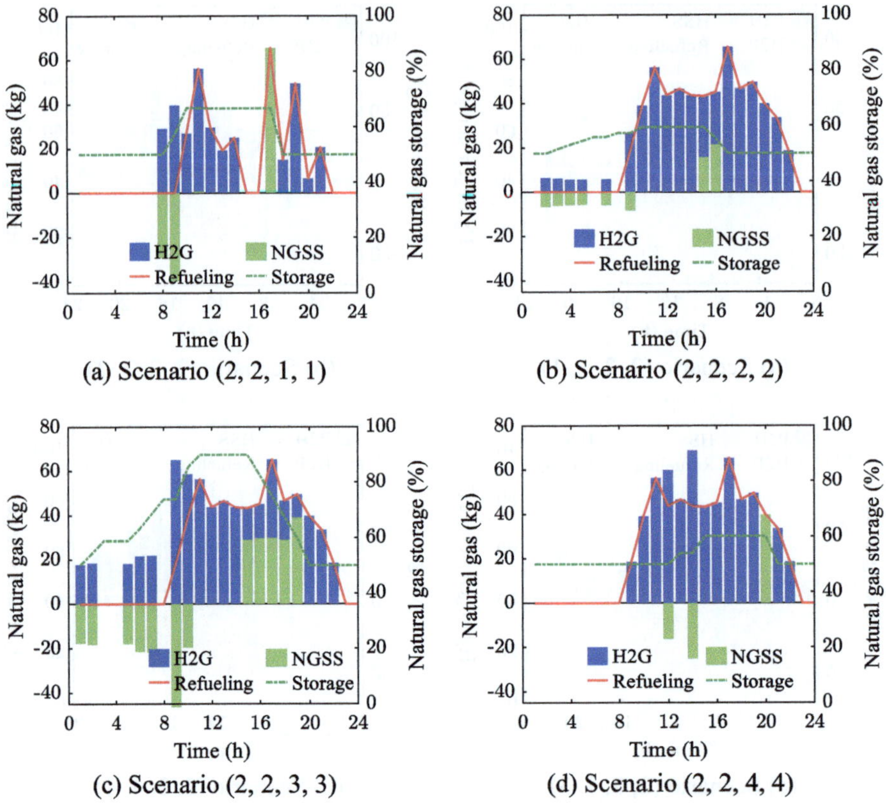

Fig. 13.10 Natural gas dispatch strategies of IESS#2 under the same set of renewable energy scenarios

relatively abundant renewable resources, the total energy consumption of electricity, hydrogen, and natural gas decreases by 25.5%, 18.3%, and 10.4%, respectively. Among them, the energy share for charging and hydrogen refueling services accounts for 33.0% and 11.0%, respectively, remaining essentially unaffected by renewable energy constraints. In contrast, the energy share for gas refueling increases to 22.3% in resource-rich scenarios. Additionally, energy losses are incurred during processes such as electrolysis (P2H), hydrogen fuel cell discharge (H2P), natural gas synthesis (H2G), and hydrogen/natural gas storage and conversion. Taking Fig. 13.11b as an example, the energy loss ratios of each process in total electricity supply are ranked as follows: electrolysis 12.2%, hydrogen fuel cell 1.1%, natural gas 12.6%, hydrogen storage 0.3%, and natural gas storage 0.1%. Among them, the high loss ratio of natural gas synthesis is mainly due to multiple energy conversions in the synthesis pathway.

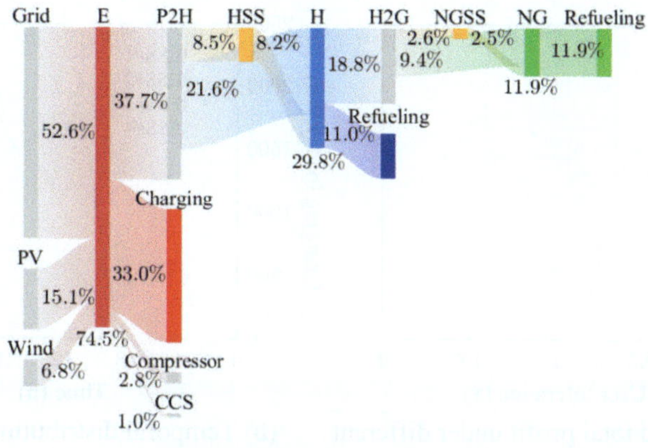

(a) Scenarios (2,2,1,1)

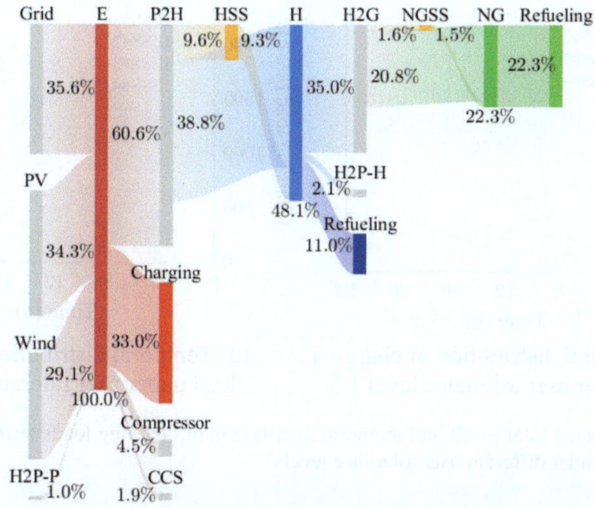

(b) Scenarios (2,2,4,4)

Fig. 13.11 Overall energy flow of electricity–hydrogen–gas in the integrated energy service station

Sensitivity Analysis of Bounded Rationality Parameter and Risk Aversion Coefficient. To evaluate the impact of user bounded rationality parameters on the operational performance of the integrated energy service station aggregator, this section calculates the expected total profit of the aggregator under different user tolerance levels (from 0 $ to 2 $, in steps of 0.5 $), and analyzes the temporal distribution of charging loads at each IESS under scenario (2, 2, 2, 2). The results are shown in Fig. 13.12.

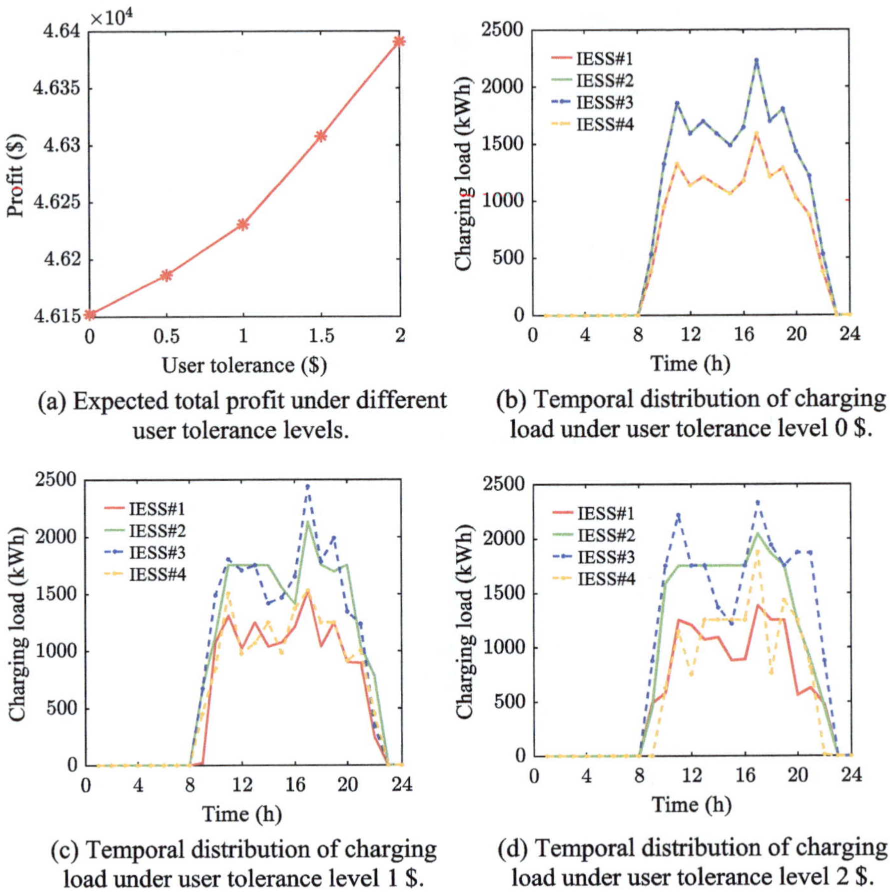

Fig. 13.12 Expected total profit and temporal distribution of charging loads at integrated energy service stations under different user tolerance levels

It is shown that the expected total profit of the aggregator increases significantly with the growth of user tolerance. Taking perfectly rational users (tolerance level of 0 $) as the baseline, when the tolerance increases to 2 $, the expected profit increases by approximately 0.5%.

At the same time, with the increase in user tolerance, the spatial and temporal distribution of electric vehicle charging loads becomes more dispersed and uneven, revealing a more significant degree of imbalance. This trend is also reflected in the distribution of hydrogen refueling and gas refueling loads, indicating that bounded rationality not only affects the behavioral patterns of traffic flow but also introduces challenges for the optimal dispatch of refueling infrastructure.

To further analyze the impact of the risk aversion coefficient, this section investigates the expected overall profit and optimization objective value of all scenarios under different risk aversion coefficients (ranging from 0 to 1 with a step size of 0.1),

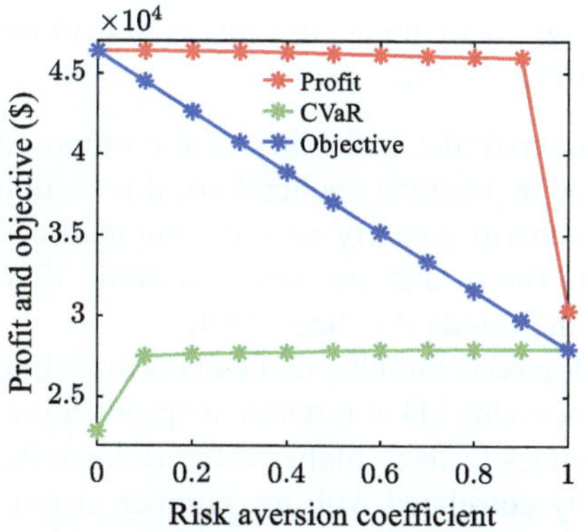

(a) Expected profit and objective value under different risk aversion coefficients.

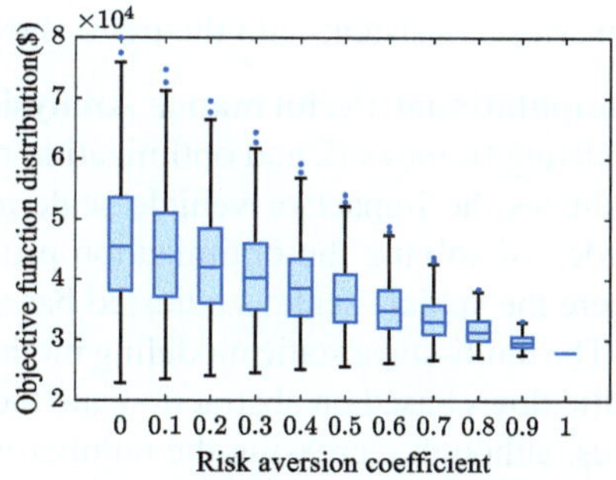

(b) Distribution of objective values across scenarios under different risk aversion coefficients.

Fig. 13.13 Expected profit and objective value under different risk aversion coefficients, and scenario distribution of objective value

and plots the distribution of the objective function across all scenarios, as shown in Fig. 13.13.

When the risk aversion coefficient is 0, the model degenerates into a risk-neutral stochastic optimization model, and the objective function is equal to the expected profit across all scenarios. The aggregator in this case adopts an aggressive strategy focused on maximizing the overall expected benefit. However, this leads to inadequate consideration of losses in extreme scenarios, resulting in lower expected profits under those cases.

When the risk aversion coefficient increases to 0.1, i.e., the model starts incorporating mild risk aversion, the expected profit in extreme scenarios rises significantly. Even though only 5% of all scenarios are extreme, the aggregator only needs to sacrifice 0.02% of total expected profit to raise expected profit in extreme cases by 2.04%, indicating enhanced capability to hedge against risks.

As the coefficient continues to increase from 0.1 to 0.9, the expected profit in extreme scenarios steadily increases by 1.6%, while overall profit decreases only slightly by 0.8%, indicating the system has entered a phase of effective risk-return balancing.

However, once the risk aversion coefficient further increases to 1.0 (fully risk-averse), the objective function becomes the worst-case expected profit. At this point, while the aggregator's extreme-scenario profits improve slightly (only 0.01%), the overall expected profit across all scenarios drops by 34.1%, reflecting a sharp decline in efficiency.

In Fig. 13.13b, the distribution of objective values across all scenarios demonstrates typical trade-off behavior. As the risk aversion coefficient increases from 0 to 1, the minimum value of the objective (i.e., the profit in the worst scenario) increases significantly, and both the highest and average values decrease. The distribution narrows, showing that the system is increasingly hedging against losses in extreme scenarios, consistent with the trend shown in Fig. 13.13a.

Computational Performance Analysis. To verify the scalability of the proposed modeling framework and optimization method in practical applications, this section evaluates the impact of vehicle scale and scenario quantity on the computational burden of solving the optimization problem. The results are shown in Table 13.4, where the vehicle scale is adjusted based on the standard in Sect. 13.4.

The multi-stage static modeling method adopted in this chapter involves modeling traffic flows based on abstraction and then processing via non-iterative optimization. Thus, although increasing the number of vehicles leads to higher network load, the increase in model complexity is only linearly correlated with the number of road segments, resulting in relatively mild computational growth and good scalability. This feature demonstrates that the proposed method is suitable for large-scale new energy vehicle scenarios, especially under complex and variable traffic environments.

However, compared to the application scale of electric vehicles, the number of scenarios has a greater impact on computational burden. As the number of scenarios increases, the model must handle more uncertainty sets during the solution process, which significantly increases the complexity of the decomposition-based optimization process and, consequently, the computational time.

In practical deployment, the integrated energy service station aggregator must complete energy scheduling and electricity market bidding timely. To ensure that optimization accuracy is met while maintaining acceptable solution time, it is necessary to control the scenario quantity. This can be done by combining real historical data and scenario generation techniques, such as K-means or clustering-based scenario reduction, to extract a small number of representative scenarios from the original set. This effectively reduces the optimization computational burden while meeting practical scheduling requirements.

Table 13.4 Computational burden under different vehicle scales and scenario quantities

Vehicle size (%)	Computational time (s)	Scenario size	Computational time (s)
80	391.0	16	7.7
90	396.6	32	18.0
100	409.6	64	38.8
110	413.5	128	121.9
120	418.8	256	409.6

13.6 Summary

Considering user bounded rationality and the risk of profit loss under extreme scenarios, this chapter proposes a CVaR-based risk-averse energy optimization and regulation method for IESS aggregators, aiming to achieve orderly grid participation and coordinated electricity-hydrogen-gas scheduling of IESS clusters in competitive electricity spot markets. First, a mixed user equilibrium model incorporating heterogeneous refueling demands and bounded rationality is constructed to characterize the dynamic behavior of traffic flow and the spatiotemporal distribution of energy demand under the coexistence of electric vehicles (EVs), hydrogen fuel cell vehicles (HFCVs), natural gas vehicles (NGVs), and fuel-powered vehicles. Second, a coordinated scheduling model integrating electricity, hydrogen, and gas energy systems is designed. By leveraging on-site photovoltaic (PV) and wind power resources, the model enables one-stop energy services such as EV charging, hydrogen refueling, and gas fueling for various types of traffic users. Finally, a CVaR-based risk-averse optimization method is proposed to quantify the tail-end profit risks caused by multidimensional uncertainties such as travel demand, real-time electricity prices, and renewable generation. Accordingly, a risk-averse energy regulation strategy for IESS cluster participation in the two-stage electricity spot market is formulated.

The proposed model and method are evaluated and validated through simulation cases based on electricity market data from a province in China. The main conclusions are as follows: (1) IESSs are capable of multi-energy coordination involving electricity, hydrogen, and gas, and can provide integrated energy services for EVs, HFCVs, and NGVs. Among these, the economic viability of synthetic natural gas is highly dependent on on-site renewable resources. Its production process enables carbon dioxide recycling, showing potential for achieving net-zero carbon emissions; (2) Through centralized energy regulation by the aggregator, IESS clusters can flexibly participate in the two-stage competitive electricity spot market, achieving orderly grid connection of power loads, reducing overall electricity costs, and improving market transaction efficiency; (3) The CVaR-based risk-averse optimization method significantly enhances the expected profit of aggregators under extreme scenarios (with an increase of 21.8%), while having minimal impact on the overall expected profit across all scenarios (only a 0.3% decrease), indicating that the method effectively improves system risk resilience while maintaining high overall profitability; (4) A reduction in user rationality (i.e., increased tolerance level) leads to an increase in the overall expected profit of the IESS aggregator, while the spatiotemporal behavior of heterogeneous traffic users becomes more complex and diverse, increasing the scheduling complexity of energy supply; and finally (5) The proposed risk-averse optimization method demonstrates high computational efficiency and is capable of supporting large-scale access of heterogeneous refueling vehicles, thus providing essential technical support and methodological foundation for the clustered, intelligent, and low-carbon operation of IESSs.

Chapter 14
Conclusions and Perspectives

Abstract This book presents a comprehensive study on the collaborative optimization planning and scheduling methods for electric vehicles (EVs) and active distribution networks, addressing the challenges posed by large-scale EV grid integration. The research systematically investigates three core areas: EV charging demand forecasting, charging facility planning, and charging/discharging scheduling strategies. Key contributions include the development of Markov-based models for predicting EV charging behavior and battery state dynamics, and a hierarchical optimization framework for the coordinated siting and sizing of charging stations and distributed generation under coupled transportation and distribution network constraints. A spatiotemporal scheduling strategy is proposed to achieve peak shaving and valley filling, alongside a master–slave game model to balance the interests of EV users and grid operators. The study further explores the operational mechanisms of EV aggregators in electricity market environments, proposing stochastic optimization methods for day-ahead bidding and real-time regulation, incorporating risk-averse strategies based on Conditional Value at Risk to manage market uncertainties. Finally, the book extends the analysis to the dynamic coupling of power and transportation systems, introducing bounded rationality into user equilibrium models and developing Stackelberg game frameworks for fast charging station pricing and aggregation. The findings provide theoretical and technical support for enhancing grid stability, economic efficiency, and the low-carbon transformation of integrated energy systems.

14.1 Conclusions

In recent years, the increasingly severe problems of global fossil energy depletion and environmental pollution have attracted widespread attention to energy-saving and environmentally friendly electric vehicles. With the increasingly severe issues of climate change, environmental constraints, and energy security, promoting the use of electric vehicles to replace traditional fuel vehicles is not only a key measure for China to build a green and low-carbon transportation system, but also a strategic choice for China to achieve sustainable transportation development. However, the

Q. Yang et al., *Integration of Massive Electric Vehicles to Smart Grid*, Smart Energy Systems, https://doi.org/10.1007/978-981-92-0557-8_14

large-scale grid-connected electric vehicles in the future will bring new challenges to the planning and scheduling of active distribution networks. In this context, this book studied some of the problems faced by electric vehicles after widespread integration into the distribution network from three aspects: electric vehicle charging demand forecasting, electric vehicle charging facility planning, and electric vehicle charging and discharging scheduling strategies.

Regarding the research on collaborative optimization planning and scheduling methods for electric vehicles and distribution networks, the following conclusions have been drawn through the study:

(1) A quantitative analysis method for the travel characteristics and charging demand of electric vehicles based on time electricity quantity, as well as a Markov model for electric vehicle state transition and an invisible Markov model for electric vehicles considering battery electricity quantity are proposed. Based on historical data and Monte Carlo simulation methods, the inherent laws of daily travel and charging status changes of electric vehicles were analyzed and summarized in depth, thus accurately predicting the charging load of large-scale electric vehicles and the distribution and variation of battery power, and evaluating the negative impact of disorderly charging of large-scale electric vehicles on the power grid.
(2) A method for optimal location and capacity determination of economical electric vehicle fast charging stations has been proposed. The various constraints of the transportation network and distribution network were converted into the economic cost of the fast charging system, and an optimization model was established with the objective function of maximizing the total profit of the system. The optimal number and location of fast charging stations, as well as the number of charging piles and waiting positions, were solved to reasonably balance the service quality and economic profit of fast charging stations, balance the density of the fast charging station network, and meet the fast charging needs of as many electric vehicles as possible.
(3) A hierarchical optimization planning method for multiple charging facilities and distributed power sources of electric vehicles under the coupling of transportation and distribution networks has been proposed. A collaborative optimization planning model for conventional charging parking lots, fast charging stations, and distributed power sources of electric vehicles with multiple layers and optimization objectives has been established and solved layer by layer. While ensuring the convenience and economy of electric vehicle charging facilities, as well as the safety and stability of the distribution network and other constraints, the complexity of the problem has been reduced, and the scalability of the algorithm has been improved.
(4) A scheduling strategy for time–space segmentation has been proposed. Based on real-time information in the intelligent charging scheduling system, while ensuring the satisfaction of electric vehicle charging needs, electric vehicles are scheduled to perform regular charging in the optimal time interval on a time scale to achieve peak shaving and valley filling effects. At the spatial scale,

a driving path planning method considering fast charging for electric vehicle users has been proposed, which can schedule electric vehicles to the optimal charging station for fast charging while ensuring their travel needs.

(5) A master–slave game based optimal charging and discharging strategy for electric vehicle users and grid operators in active distribution networks has been proposed. Firstly, the costs of disorderly charging and discharging, non-cooperative game, and cooperative game among electric vehicles under the influence of dynamic electricity prices were analyzed and compared. Secondly, considering the development trend of energy storage application in future grid frequency regulation and peak shaving, a master–slave game model between electric vehicle users and grid operators was proposed, and the equilibrium point between the charging and discharging costs of electric vehicle users and the cost of energy storage equipment used for peak shaving by grid operators was determined. The proposed planning method and scheduling strategy are expected to provide certain technical support for the grid connection of large-scale electric vehicles and address the problems caused by the widespread grid connection of electric vehicles.

A large number of electric vehicles entering the grid for charging will greatly increase the load on the power system, thereby putting pressure on all aspects of power generation, transmission, distribution, transformation, and utilization. The integration and interaction between electric vehicles and the power grid can not only alleviate the impact of large-scale electric vehicle charging on the power grid but also improve the stability and economy of power grid operation. In recent years, China has promoted the reform of the electricity market through multi-regional pilot projects, focusing on building a modern electricity market system with spot trading as the core. The trial operation of the electricity spot market provides a practical and feasible application scenario for the large-scale implementation of vehicle network interaction. Under the framework of a competitive electricity market, electric vehicle aggregators can be introduced as third-party organizations to guide or control the orderly charging of electric vehicles, achieving collaborative interaction between large-scale electric vehicles and the power grid. However, electric vehicle aggregators face significant challenges in the process of market participation, including complex spot trading rules and various uncertain factors such as market electricity prices, electric vehicle charging behavior, and user willingness. Therefore, this book conducts in-depth research on the operation mechanism and charging optimization scheduling methods of electric vehicle aggregators in the electricity market environment from the perspectives of recent bidding, real-time regulation, and user interaction.

Based on the research on the operation mechanism and charging optimization scheduling methods of electric vehicle aggregators in the competitive electricity market environment, the following conclusions have been drawn through the study:

(1) A stochastic optimization-based method for an electric vehicle aggregator's day-ahead bidding and electric vehicle cluster charging regulation is proposed to address the uncertainties faced by the spot market system, such as multiple

factors and matching between day-ahead bidding and electricity consumption. In order to capture the flexibility of electric vehicle charging, two proxy models between electric vehicle aggregators and electric vehicle users have been proposed, namely the centralized management model and the decentralized incentive model. Based on these two proxy models, a day-ahead bidding model and a real-time charging regulation model for electric vehicle aggregators were established. Among them, the recent bidding model integrates factors such as the randomness of electric vehicle charging, the uncertainty of real-time clearing electricity prices, and user response. The real-time charging scheduling model adopts a rolling optimization method to dynamically adjust the charging power of electric vehicles based on their charging demand and real-time electricity price signals, in order to maximize the economic benefits of electric vehicle aggregators.

(2) A risk-averse day-ahead bidding strategy based on conditional value at risk is proposed to address the issue of uncontrollable risks in bidding strategies aimed at minimizing expected costs, to avoid the huge losses caused by extreme real-time clearing results. In order to address the issue of real-time electricity price volatility, the risk measurement criterion of Conditional Value at Risk (CVR) was introduced and incorporated into the pre-day bidding model of electric vehicle aggregators. Based on the probability distribution of real-time clearance scenarios, the expression form of CVR for electric vehicle aggregation bidding was mathematically derived, and a stochastic optimization model to minimize CVR was constructed. Compared with the risk-neutral day-ahead bidding strategy, this paper elaborates on the superiority of the risk-averse day-ahead bidding strategy in risk management.

(3) A pricing strategy for electric vehicle aggregators based on master–slave game theory and a bidding and charging coordination scheduling strategy for electric vehicle aggregators considering market risk are proposed to address the interaction mechanism between aggregators and users, as well as the coordination problem of electric vehicle charging under different proxy modes. Under the proposed operational framework, electric vehicle aggregators need to reserve a portion of the bidding electricity for allocation to electric vehicles with decentralized incentive modes. The charging demand of decentralized incentive mode electric vehicles will change with the fluctuation of electricity prices. To evaluate the charging demand of electric vehicles in this mode, a one master, many slave Stackelberg game model is proposed to analyze the bidirectional game relationship between the pricing strategy of electric vehicle aggregators and the charging demand of users. By utilizing the KKT optimality condition, the optimal user response is incorporated into the electric vehicle aggregator decision model, thereby transforming the two-layer optimization problem into a single-layer mathematical optimization problem to obtain a game equilibrium solution. In addition, a day-ahead bidding model that balances the risk cost, and return was constructed by combining conditional value at risk, quantifying the

potential risks of electricity market uncertainty on day-ahead bidding. By coordinating the charging demand of different modes of electric vehicles, the optimal day-ahead bidding strategy for electric vehicle aggregators was formulated.

(4) A multi-agent Stackelberg game model is proposed to address the balance of competition among multiple agents in electric vehicles, to explore the energy trading mechanisms and market operation strategies between multiple electric vehicle aggregators and users. In the leader model, a mathematical model was established to maximize profits for electric vehicle aggregators. This model considers the randomness of electric vehicle charging and the uncertainty of real-time clearance results, in order to develop a day-ahead bidding strategy for electric vehicle aggregators and a pricing strategy for potential real-time clearance scenarios. In the follower model, a logarithmic function was used to characterize the satisfaction cost of electric vehicle users under certain charging demands, and a mathematical model was proposed to minimize the comprehensive charging cost to formulate the optimal charging plan for electric vehicles. For the equilibrium problem of multi master and multi slave Stackelberg games, this master–slave game problem is transformed into a non-cooperative game model among electric vehicle aggregators by introducing the optimal response of electric vehicle users into the decision model of electric vehicle aggregators. Based on the potential game theory, the Nash equilibrium of the electric vehicle aggregator game was verified, and the existence and uniqueness of the equilibrium solution of the Stackelberg game were proved. The Stackelberg equilibrium solution of the game model was obtained to maximize the potential function.

The power and transportation systems are the main sources of direct carbon emissions in China, and their low-carbon collaborative transformation is the key path to achieving the "dual carbon" strategic goals. With the rapid increase in the penetration rate of electric vehicles, urban power and transportation networks are deeply coupled through fast charging stations, forming a dynamic, interactive power transportation coupling network. The charging behavior of electric vehicles drives the spatiotemporal evolution of power load, and their travel path selection also affects the distribution of traffic flow. At the same time, users' decision-making behavior is influenced by their level of rationality, further exacerbating the operational complexity of the coupled network. In the instant charging mode, fast charging stations have become a key hub for the interaction and transformation of power flow and traffic flow, as well as the core facility for regulating the coordinated operation of coupled networks. Therefore, in the context of considering the coupling relationship between electric power and transportation and rational user decision-making, exploiting the optimization scheduling of fast charging stations has important theoretical value and practical significance. To this end, this book firstly constructed an equilibrium modeling framework for the electric power transportation coupling network from the perspective of user bounded rationality. Based on this, in-depth research is conducted on the pricing, power regulation, and aggregation mechanisms of fast charging stations to participate in electricity market transactions. Furthermore, the research object is expanded to comprehensive energy service stations that integrate charging, hydrogen injection, and gas filling functions. With the optimization and scheduling of coupling hubs, e.g., fast charging stations and integrated energy systems, the dynamic collaborative operation of the electric power transportation coupling network can be promoted.

Based on the research on optimizing the scheduling of fast charging stations considering the coupling of power transportation and user bounded rationality, the following conclusions have been drawn through the study:

(1) A modeling method for power transportation coupling networks considering user bounded rationality is proposed to address the issues of incomplete user information perception and non-absolute sensitivity to travel costs in power transportation coupling network modeling. By introducing tolerance parameters to quantify the rationality level of users, a multiple time-scale bounded rationality static user equilibrium model and a bounded rationality dynamic user equilibrium model were established, revealing the suboptimal decision-making characteristics of bounded rationality users' travel and charging behavior in scenarios with incomplete information. Design an improved point queue model based on a differential equation system to track the dynamic propagation characteristics of traffic flow on the road network, such as movement, queuing, and charging, accurately depicting the conversion process of traffic flow to charging load in fast charging stations. Using the optimal power flow model to model the power flow distribution of the distribution network, where the marginal electricity price of nodes serves as the charging price for fast charging stations, which increases with the increase of charging load. Based on the fixed point theory, a dynamic coupling equilibrium model of traffic flow and power flow is established to analyze the coupling mechanism between the two networks under user-bounded rationality. A multi-stage static and dynamic modeling method for the power transportation coupling network applicable to different demand scenarios is proposed, providing key model support for subsequent research.

(2) A dynamic pricing method for fast charging stations based on a three-layer Stackelberg game framework of the vehicle station network is proposed to address the issues of low flexibility in static pricing methods and insufficient market supervision effectiveness in price setting. Constructing a tripartite interaction mechanism among distribution network operators, fast charging station operators, and users through a Stackelberg game model, where the distribution network operator determines the node marginal electricity price based on the optimal power flow model, the fast charging station operator sets the retail charging price with the goal of maximizing profits, and users adjust their charging behavior based on a bounded rationality dynamic user equilibrium model. The game equilibrium of the three parties is solved through the Gauss–Seidel iterative algorithm. Introducing market regulatory constraints to balance the interests of multiple parties, while ensuring the profitability of fast charging stations, using price signals to guide charging load peak shaving and valley filling, alleviating traffic congestion, and achieving collaborative and efficient operation of the electric power transportation coupling network.

(3) A simplified dynamic transportation network model and a stochastic energy optimization control method for fast charging station aggregation merchants are proposed to address the problems of low efficiency in large-scale electric vehicle scheduling and multidimensional uncertainty in operating scenarios when participating in electricity market transactions with fast charging station clusters. Designed path merging and flow separation strategies to achieve lightweight application of bounded rationality dynamic user equilibrium models in complex transportation networks. Based on the grouping charging strategy, electric vehicles with similar charging needs are aggregated into groups for unified scheduling, achieving decision variable dimensionality reduction. Using representative scenarios to depict the uncertainty from travel demand, real-time electricity prices, and renewable energy generation, combined with the stochastic optimization methods to solve the optimal pre-day electricity bidding strategy and intraday charging scheduling strategy for fast charging station aggregators, to achieve orderly online transactions of fast charging clusters and promote on-site consumption of new energy within the station.

(4) A risk avoidance energy optimization and control method based on conditional value at risk is proposed for the aggregation of electricity, hydrogen gas, and integrated energy service station clusters to address the issues of significant differences in heterogeneous user energy replenishment characteristics and high operational risks when participating in electricity market transactions. A bounded rational heterogeneous user equilibrium model was constructed, which includes hybrid energy vehicles such as electric vehicles, hydrogen fuel cell vehicles, natural gas vehicles, and fuel vehicles, to uniformly characterize the spatiotemporal distribution characteristics of various energy replenishment demands. This book has designed a comprehensive energy service station model that can provide charging, hydrogen injection, and refueling services simultaneously. We have introduced technologies such as electrolytic hydrogen production, carbon capture, and synthetic natural gas to achieve multi-energy synergy and complementarity among electricity, hydrogen, and gas. In addition, we utilized conditional value at risk to assess the profit loss risk of integrated energy service station aggregators in extreme scenarios, and designed a risk-averse strategy for day-ahead power bidding and day-ahead multi-energy collaborative scheduling.

14.2 Perspectives

Firstly, this book exploited the optimization planning and scheduling methods of vehicle network coordination for electric vehicles. There are still some outstanding problems caused by large-scale electric vehicle grid connection, but there are also some shortcomings and areas that need improvement.

(1) In terms of predicting the demand for electric vehicle charging, the method proposed in this book can be further developed into an electric vehicle charging behavior prediction model that considers spatiotemporal distribution. By combining parameters such as the number of electric vehicles and traffic flow in the transportation network, it can be refined to predict the charging load for a specific parking lot or charging station, laying the foundation for more accurate planning and scheduling of electric vehicle charging stations and charging behavior.
(2) In terms of planning electric vehicle charging facilities, the collaborative planning method proposed in this book can further consider various types of distributed power sources, such as photovoltaics, energy storage, combined cooling, heating, and power supply. At the same time, synchronous planning and expansion of parameters such as substations and lines in the distribution network can be considered to establish a complete theoretical method system for active distribution network planning.
(3) In terms of electric vehicle charging and discharging scheduling strategies, the proposed method can further consider electric vehicles providing various types of auxiliary services through V2G mode. In addition to peak shaving and frequency regulation, it can also serve as a power source for black start, thereby alleviating the scale and impact of power system failures. The master–slave game model between electric vehicle clusters and grid operators can also be extended to more complex situations, while considering the usage strategies of multiple power equipment. In addition, studies have shown that frequent charging and discharging of batteries can affect their service life. How to control the orderly charging of electric vehicles to minimize the impact on battery life is also a key research issue at this stage.

Secondly, this book focused on the operational strategies and charging optimization control methods of electric vehicle aggregators in the electricity market environment, and has achieved some results in theoretical modeling and strategy optimization. However, this study still has certain limitations and further in-depth research is needed. This book has only focused on the charging scheduling strategy of electric vehicles and considers only the load attributes of electric vehicles. Electric vehicles are not only adjustable loads, but also serve as a mobile energy storage device to supply power to the grid. Therefore, in the electricity market environment, electric vehicle aggregators can use V2G technology to participate in both the electric energy and ancillary service markets simultaneously. However, V2G may cause certain attenuation to the power batteries of electric vehicles, which will pose greater challenges to the charging and discharging scheduling of electric vehicles. In summary, future research can focus on the following aspects:

(1) With the opening of the future ancillary services market, electric vehicle aggregators can further enhance their economic benefits in market participation by utilizing the flexibility of electric vehicle charging and discharging. In the electric energy and ancillary services market, the market operation strategies of electric vehicle aggregators will become increasingly complex. How to allocate

the charging power of electric vehicles and develop a collaborative strategy for participating in the electric energy and ancillary services market is one of the directions for future research.

(2) Electric vehicle V2G may cause a decrease in its battery capacity, which can lead to a lack of willingness among users to participate in V2G. Electric vehicle aggregators need to optimize the management of battery losses and establish incentive mechanisms to obtain flexibility in charging and discharging electric vehicle batteries. In addition, electric vehicle aggregators also need to consider how to allocate the profits obtained from the electricity market to electric vehicle users. In addition, while participating in the electric energy and ancillary services market, how to reduce the negative impact of V2G on batteries is also one of the issues that needs to be considered.

(3) The application of vehicle network interaction in the electricity market environment involves multiple stakeholders, including electric vehicle users, charging service providers, electric vehicle aggregators, and grid operators. How to balance the interests of all parties, how to attract electric vehicle users, and how to innovate business models to promote the market-oriented implementation of vehicle network interaction technology are also key directions for future research.

Finally, within the research framework of dynamic modeling of power transportation coupling networks and analysis of user bounded rational behavior, this paper conducted systematic research on the optimization scheduling problem of fast charging stations and comprehensive energy service stations, and achieved phased results. However, with the continuous advancement of energy transformation and the construction of new power systems, the operation scenarios of coupled networks will become more complex, and further expansion is still needed in the following directions in the future:

(1) Dynamic pricing mechanism in a multi-operator competitive environment: Current research assumes a single operator and does not involve a market structure with multiple operators coexisting. In the future, based on the bounded rationality of users, we can study the price game and market mechanism design among multiple operators, explore their impact on the spatiotemporal distribution of charging demand and network operation efficiency, and analyze the dynamic evolution laws under multi-agent games.

(2) Integration mechanism of vehicle network interaction and auxiliary service market: With the promotion of V2G technology, electric vehicles have the potential to participate in the auxiliary service market. Further research is needed on the feasibility of fast charging station aggregators participating in frequency regulation, backup services, and other services by regulating electric vehicle charging and discharging. An optimization model considering user behavior, battery life, and revenue feedback should be constructed to achieve coordinated scheduling between the electricity market and auxiliary service market.

(3) Design of User Incentive and Benefit Distribution Mechanism: In practical operation, a reasonable profit distribution and user incentive mechanism are crucial

to ensuring the enthusiasm of all parties involved. In the future, based on cooperative game theory, a fair profit distribution model between aggregators and various charging stations can be constructed, dynamic incentive strategies for users can be designed, and the impact of incentive feedback on user behavior patterns and bounded rationality coefficients can be studied to enhance the practical feasibility of the strategy.

(4) Collaborative optimization of electric commercial vehicles and urban logistics: Current research mainly focuses on passenger cars, which can be expanded to commercial vehicles such as electric trucks and logistics vehicles in the future. Combining urban distribution routes and charging demand characteristics, an integrated model of is constructed to study the layout, dynamic pricing, and energy regulation coordination mechanism of dedicated charging stations, and promote the development of green urban logistics networks.

(5) Research on digital twin and multi-market collaboration mechanism: With the help of digital twin technology, a real-time simulation platform for coupled networks can be constructed to achieve state perception and strategy adaptive optimization. At the same time, the coupling mechanism between carbon trading, green certificate trading, and the electricity market can be studied to explore the joint clearing strategy of charging station aggregators in multiple markets, quantify the feedback of carbon emission reduction on economic efficiency and policy implementation, and help achieve the goal of carbon neutrality.

MIX
Papier aus verantwortungsvollen Quellen
Paper from responsible sources
FSC® C105338

If you have any concerns about our products,
you can contact us on
ProductSafety@springernature.com

In case Publisher is established outside the EU,
the EU authorized representative is:
Springer Nature Customer Service Center GmbH
Europaplatz 3, 69115 Heidelberg, Germany

Printed by Libri Plureos GmbH
in Hamburg, Germany